VI International Conference "Modern Nanotechnologies and Nanophotonics for Science and Industry"

Journal of Physics: Conference Series Volume 951

Suzdal, Russia
9 – 13 November 2017

ISBN: 978-1-5108-5753-7
ISSN: 1742-6588

Printed from e-media with permission by:

Curran Associates, Inc.
57 Morehouse Lane
Red Hook, NY 12571

Some format issues inherent in the e-media version may also appear in this print version.

Printed by Curran Associates, Inc. (2018)

For permission requests, please contact the Institute of Physics at the address below.

Institute of Physics
Dirac House, Temple Back
Bristol BS1 6BE UK

Phone: 44 1 17 929 7481
Fax: 44 1 17 920 0979

techtracking@iop.org

Additional copies of this publication are available from:

Curran Associates, Inc.
57 Morehouse Lane
Red Hook, NY 12571 USA
Phone: 845-758-0400
Fax: 845-758-2633
Email: curran@proceedings.com
Web: www.proceedings.com

Table of contents

Volume 951

VI International conference "Modern nanotechnologies and nanophotonics for science and industry"

9–13 November 2017, Suzdal, Russian Federation

Accepted papers received: 27 December 2017
Published online: 30 January 2018

Preface

Papers

IOP Conf. Series: Journal of Physics: Conf. Series **951** (2017) 011001 doi:10.1088/1742-6596/951/1/011001

Preface

The results of the 6th International Conference "Modern nanotechnology and nanophotonics for science and industry" conducted by the Vladimir State University named after A.G. and N.G. Stoletov during the period from 9 to 13 November 2017. Presented 34 articles of conference participants.

IOP Conf. Series: Journal of Physics: Conf. Series **951** (2017) 011002 doi:10.1088/1742-6596/951/1/011002

Peer review statement

All papers published in this volume of *Journal of Physics: Conference Series* have been peer reviewed through processes administered by the proceedings Editors. Reviews were conducted by expert referees to the professional and scientific standards expected of a proceedings journal published by IOP Publishing.

Conductivity switching of labyrinth metal films at the percolation threshold

M G Gushchin[1], I A Gladskikh[1] and T A Vartanyan[1]

[1]ITMO University, Saint-Petersburg, Russia

magusch@gmail.com

Abstract. Electrical properties of silver, gold and copper films at the percolation threshold were investigated experimentally. A convenient method to obtain films at the percolation threshold that consists of two phases: deposition of metal in vacuum on a dielectric substrate and subsequent thermal annealing has been developed. The metallic films produced in this way exist in two different states: a low-conductivity state and a high-conductivity state. The films can be switched between these states by the applied voltage with hysteresis of current-voltage characteristic curves. The conductivity difference between the states reaches seven orders of magnitude. The switching threshold voltage depends on the annealing time. The switching times differ considerably for different metals. They were 200 ns for silver, 2 μs for gold, and 60 μs for copper. A plausible explanation of the switching mechanism based on the voltage induced fine mechanical deformations is suggested and discussed.

1. Introduction

The thin metallic labyrinth films at the percolation threshold attract interest by virtue of their unique characteristic. The labyrinth film at the percolation threshold consists of separate nanoparticles that are at the edge to merge, i.e. to form an unbroken structure. These films conductivity depends on the applied voltage and the previous states in the recent past, i.e. the current-voltage (I-V) characteristic has hysteresis and the structure exists in two different states: a low-conductivity state and a high-conductivity state. The reason for this unusual behavior is an extremely small distance between nanoparticles in the film [1,2]. The threshold voltage depends on the distance between nanoparticles. The results of this study can be used in the future development of novel memory devices [3,4].

The conductivity switching behavior of silver films at the percolation threshold was observed and explained by the electric field induced film structure changes in [5]. According to this explanation at the voltages that exceed the threshold value the bridges between nanoparticles are formed. Thus a continuous metallic path between electrodes establishes and conductivity increases.

Because it is hard to control the structure of the film during the physical vapor deposition in high vacuum and it is almost impossible to catch the percolation transition while monitoring the film conductivity [6,7] another technique for obtaining the films at the percolation threshold has to be employed. Previously it was demonstrated that thermal annealing of conductive films is a convenient way to obtain silver nanostructure at the percolation threshold [8]. Thermal annealing foster the formation of island film with separated spheroidal nanoparticles [9-11]. In this process the height of nanoparticles and distance between them are rising and that leads to the formation of the film at the percolation threshold. It was also shown that the process of thermal annealing is well controlled.

2. Preparation

The preparation of the structure at the percolation threshold occurs in two phases. The first phase is the physical vapor deposition of metals on an isolating support. It was performed in the vacuum chamber PVD-75 (Kurt J. Lesker) at the residual pressure of ~10^{-7} Torr. The substrate was kept at the room temperature. Three types of metal (silver, gold and copper) were deposited on the sapphire substrates with pre-applied metallic electrodes. The size of electrodes was 5 mm, the distance between electrodes was 3 mm. The deposition rate was monitored by using a quartz microbalance. The average deposition rate was 0.5 Å/s for all metals. The films growth occurs via the of Volmer-Weber growth mode. The deposition process starts with the formation of separate nanoparticles. Discontinuous films have a very high resistance. Its conductivity is due to tunneling of electrons between particles or through electronic levels of the substrate (traps, impurity levels, etc.) depending on the distance between the particles [5]. With the material accumulation the particles start to coalescence (merge) and finally create a continuous highly conductive structure.

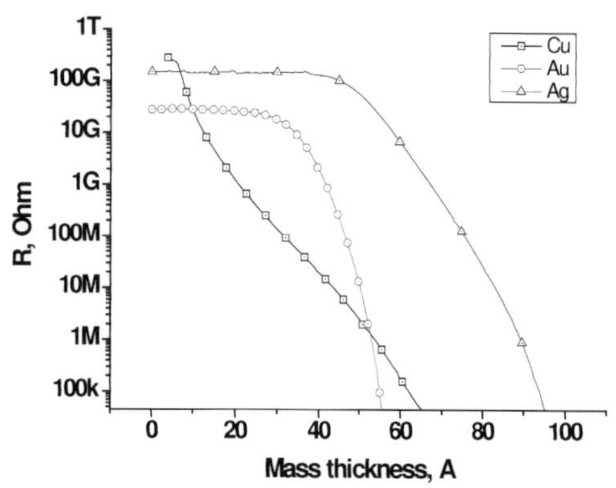

Figure 1. Evolution of the metal film resistances in the course of the deposition process as measured in situ by quartz crystal microbalance. The amount of the deposited material is expressed in terms of the equivalent thickness of a homogeneous film.

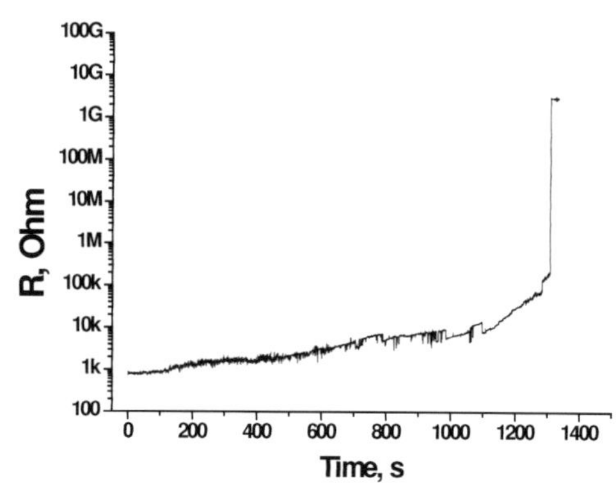

Figure 2. Evolution of the silver film resistance in the course of the thermal annealing. Annealing was immediately terminated as soon as the percolation threshold manifested by the steep rise of the resistivity was reached.

We controlled the deposited film conductivity by the picoammeter Keithley 6487. The deposition was stopped at the moment when the film resistance drops to 100 kΩ. The film resistance continued decreasing for some time after the flap closing.

The dependence of the film resistance on the film mass thickness is presented in figure 1. With the increasing amount of the deposited material expressed in the figure 1 in terms of the films mass thickness resistance is reducing. At the moment when the deposition process was stopped the films of different materials reached different mass thicknesses: 10 nm for silver and 6 nm for gold and copper. With these parameters, we obtained conductive films with the labyrinth structure.

Further, to break down the continuous metal paths that lead from one electrode to another and to obtain separate nanoparticles with small distances between them thermal annealing was used. The silver and gold films are noble metals, so the heating is carried out both in the air and in the vacuum, but for the copper films it could be done only in vacuum, because copper has strong oxidation in the

IOP Conf. Series: Journal of Physics: Conf. Series **951** (2017) 012001 doi:10.1088/1742-6596/951/1/012001

air. For silver the annealing temperature is 90-120 °C, while for gold it is 160-200 °C and for copper it is 180-220 °C. The process of thermal annealing was controlled by picoammeter Keithley 6487. The measured resistance of the silver film is presented in figure 2. The thermal annealing process was stopped when the conductivity dropped sharply, and the film resistance exceeded 100 GΩ. The time variation of the conductivity during the annealing process was similar for all studied metals. Firstly, the conductivity reduces slowly. The reason for that is further narrowing of the narrowest parts of the labyrinth film. Then, conductivity decreases sharply because the narrowest parts were finally broken. The threshold voltage (the voltage of switching) depends on the time of annealing after this sharp increase of the film resistance.

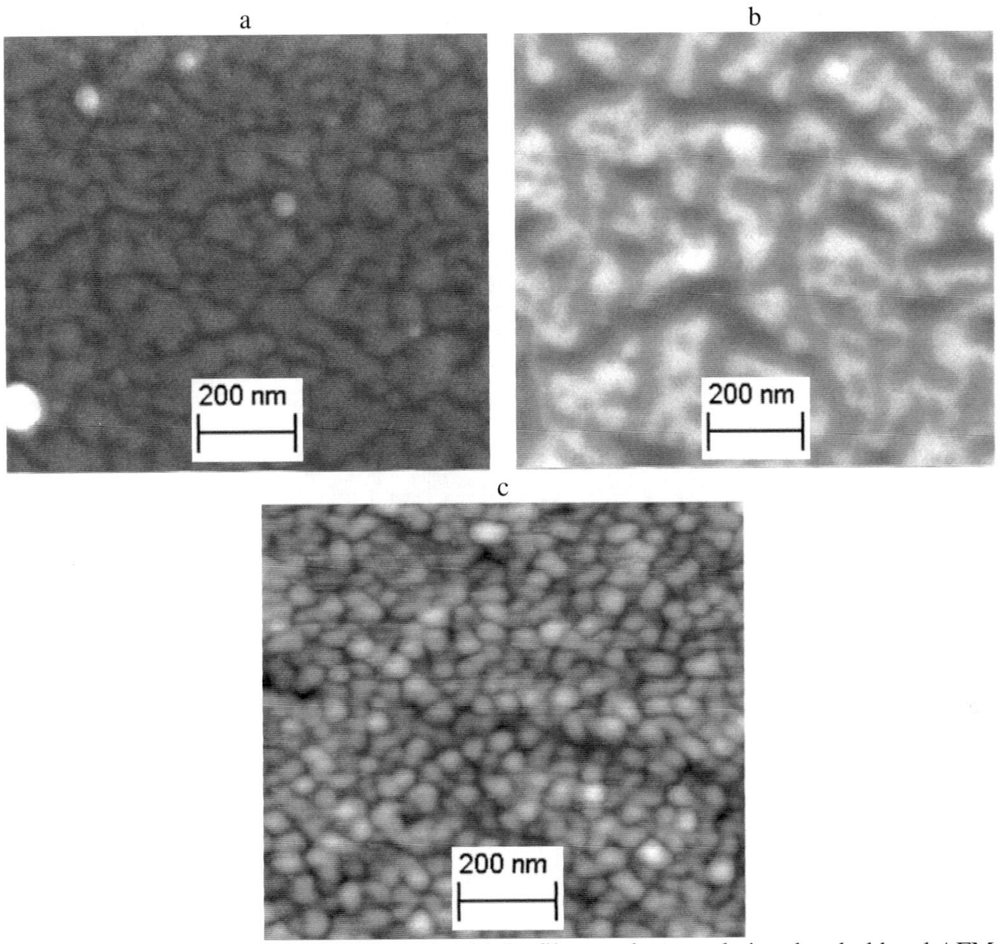

Figure 3. SEM images of silver (a) and gold (b) films at the percolation threshold and AFM image of copper film (c).

Figure 3 shows SEM and AFM images of the metal films at the percolation threshold. The structures of films look like labyrinths. In the gold film separated nanoparticles are more elongated, and in the silver and the copper film nanoparticles have more rounded shape.

3. Results

The films at the percolation threshold exist in two different states. The first state is the low-conductivity state which is obtained after the two-step preparation procedure described above. The film is preserved in this state provided the applied voltage is less than a definite threshold value. When

IOP Conf. Series: Journal of Physics: Conf. Series **951** (2017) 012001 doi:10.1088/1742-6596/951/1/012001

the applied voltage exceeds the threshold value the films pass to a high-conductivity state. After the switching the film stays in the high-conductivity state at lower voltages as well. After the voltage removal the films pass to the low-conductivity state. Hence, the hysteresis of conductivity is observed. Figure 4 plots the corresponding I-V characteristics.

The silver film differs from gold and copper films because switching of silver films is very sharp (figure 4a). The threshold voltage for silver films equals to 8 V. Bellow this voltage the film remains in the low-conductivity state (R~10^{12} Ω). At the voltage of 8 V the film conductivity rises sharply and the film passes to a high-conductivity state (R~10^{5} Ω). The difference of conductivities in two states reaches seven orders of magnitude. The film remains in the high-conductivity state even at the voltages that are less than the threshold voltage and returns to the low-conductivity state only when the voltage drops almost to zero. The voltage cycling leads to the slight reduction of the threshold voltage.

Similar behavior was observed in copper and gold films although the transition between the states in these metals is less sharp. The corresponding I-V characteristics are presented in figure 4 (b) and (c).

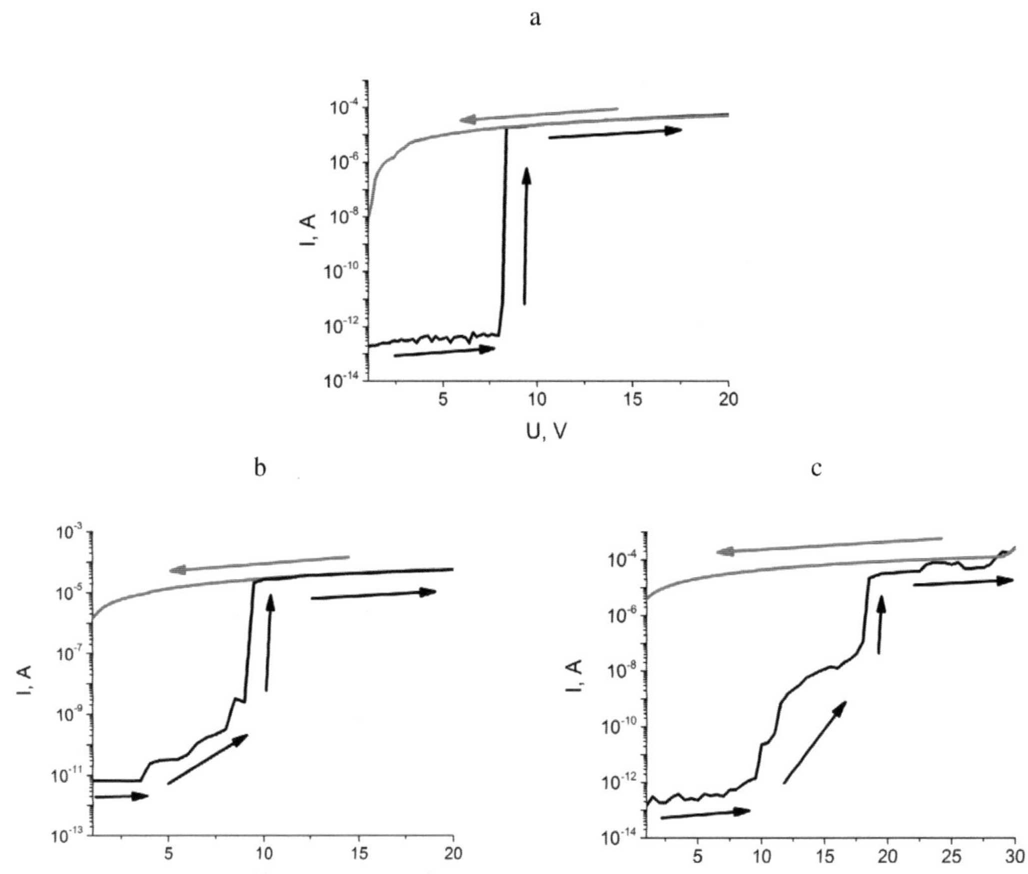

Figure 4. Current-Voltage characteristics of silver(a), gold(b) and copper(c) films at the percolation threshold.

The switching times were measured by the oscilloscope Rigol DS4052 that possesses the bandwidth of 500 MHz. Figure 5 shows how fast the films pass to a new state after a voltage in excess of the threshold voltage is applied. The measurements were performed with the series-connected resistor and the voltage source. The characteristic switching times for silver is close to 200 ns, for gold it is about 2 μs, and for copper it is about 60 μs. Figure 5 shows the kinetics of both the voltage and the current for films made of each metal. On the time scale presented in this figure it is evident that in the case of silver switching between the states proceeds in two steps due to the build up of two continuous paths formed at different voltages.

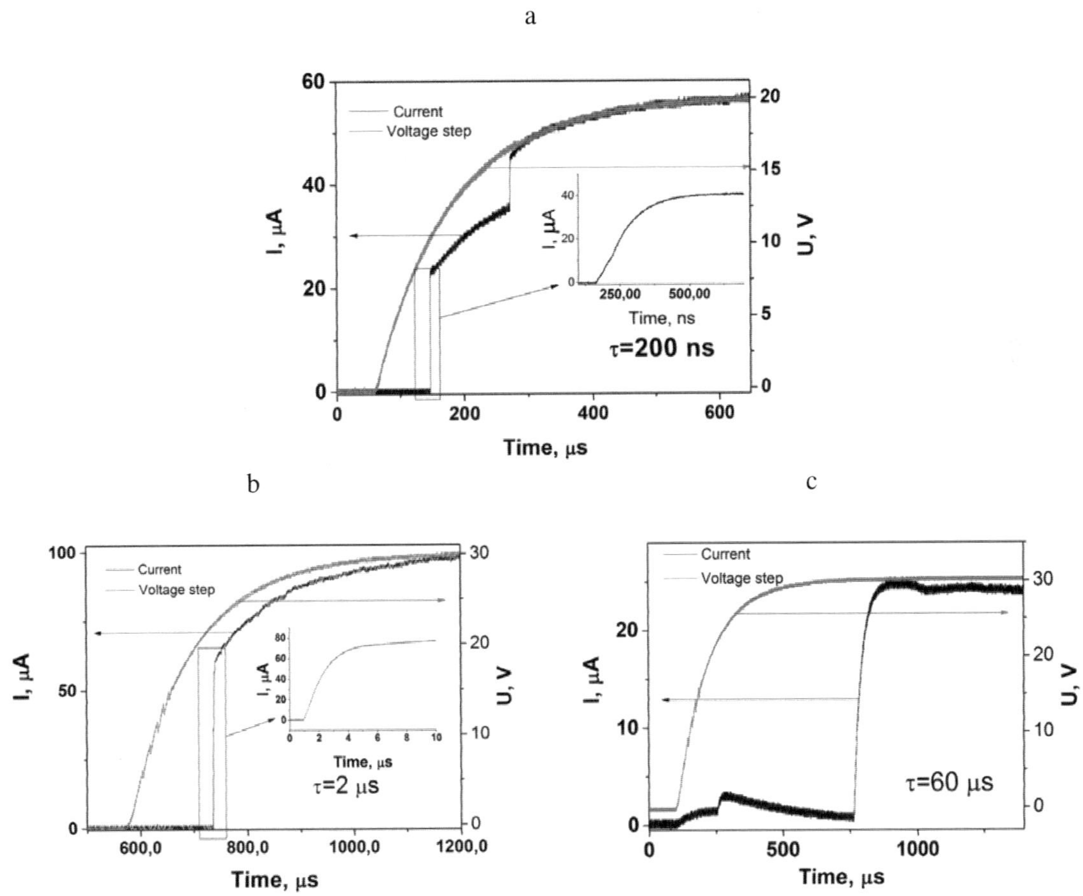

Figure 5. Switching kinetics between the low-conductivity and high conductivity states of silver (a), gold (b) and copper (c) films brought to the percolation threshold by the procedures described in the text. The inserts show the transition regions on a still finer time scale.

4. Conclusions

The two step method consisting of the conductivity controlled physical vapor deposition followed by the conductivity controlled thermal annealing leads to the robust production of the metal films at the percolation threshold. The films obtained in this way exist in two stable states with low and high conductivity. The conductivity difference of these states reaches seven orders of magnitude. The films may be switched between these states by the applied voltage, the threshold value being 8 V for silver and slightly larger for gold and copper. The voltage cycling leads to the slight reduction of the threshold voltage. The switching time for silver is 200 ns, while for gold is 2 μs and copper is 60 μs, it is much longer. A plausible explanation of the switching mechanism may be related to the fine mechanical deformations of the metal nanoparticles induced by the applied voltage. This deformation may greatly enhance the probability of electron tunneling between almost touching nanoparticles. Despite the applied voltage is rather small the electric field in the nanogaps between the particles can exceed 10^3 V/cm. This value is large enough to stimulate tiny deformations of nanoparticles and cause the resistance drop. After the voltage switch off the particles restore their original shape and the conductivity reduces as well.

The results of this study can be used in the future development of the resistive memory devices.

5. Acknowledgements

We are grateful to V. A. Polishchuk and to P.S. Parfenov for SEM and AFM study of samples.
This study was done in the framework of the state assignment of the Ministry of education and science of Russian Federation #3.4903.2017/6.7 and was supported by the Russian Foundation for Basic Research (project nos. 16-32-60028 mol_a_dk and 16-32-00165 mol_a) and by the Government of Russian Federation (Grant 074-U01).

6. References

[1] Kiesow A, Morris J E, Radehaus C and Heilmann A 2003 *J. Appl. Phys.* **94** 6988-90
[2] Sarkar D K, Cloutier F and El Khakani M A 2005 *J. Appl. Phys.* **97** 084302
[3] Terabe K, Hasegawa T, Nakayama T and Aono M 2005 *Nat.* **433** 47-50
[4] Waser R and Aono M 2007 *Nat. Mat.* **6** 833-40
[5] Vartanyan T, Gladskikh I, Leonov N and Przhibel'skii S 2014 *Phys. Solid State* **56** 816-22
[6] Wei H and Eilers H 2009 *J. Phys. Chem. Solids* **70** 459-65
[7] Wagner S, Pundt A 2008 *Phys. Rev.* **78** 155131
[8] Gladskikh I A, Leonov N B, Przhibelskii S G and Vartanyan T A 2014 *J. Opt. Tech.* **81** 280-4
[9] Leonov N B, Gladskikh I A, Polishchuk V A and Vartanyan T A 2015 *Opt. Spectr.* **119** 450-5
[10] Kim H C, Alford T L and Allee D R 2002 *Appl. Phys. Lett.* **81** 4287-9
[11] Aslan K, Leonenko Z, Lakowicz J R and Gedde C D 2005 *J. Fluor.* **15** 643

IOP Conf. Series: Journal of Physics: Conf. Series **951** (2017) 012002 doi:10.1088/1742-6596/951/1/012002

High extinction ratio integrated optical modulator for quantum telecommunication systems

A Tronev[1,2], M Parfenov[3], P Agruzov[1], I Ilichev[1], A Shamray[1,2,3]
[1]Ioffe Institute, Saint Petersburg 194021, Russia
[2]ITMO University, Saint Petersburg 197101, Russia
[3]Peter the Great St. Petersburg Polytechnic University, Saint Petersburg 195251, Russia
e-mail: achamrai@mail.ioffe.ru

Abstract. A method for increasing the extinction ratio of integrated optical Mach-Zehnder modulators based on $LiNbO_3$ via the photorefractive effect is proposed. The influence of the photorefractive effect on the X- and Y-splitters of intensity modulators is experimentally studied. An increase in the modulator extinction ratio by 17 dB (from 30 to 47 dB) is obtained. It is shown that fabricated modulators with a high extinction ratio are important for quantum key distribution systems.

1. Introduction

One of the most widely spread materials in modern integrated optics is lithium niobate ($LiNbO_3$) [1, 2], which is used as a wafer for different devices because of its transparency in the visible and near infrared spectral regions [3], the presence of the electro-optical Pockels effect [4], and the applicability in nonlinear optics [5]. Moreover, there are technologies of fabrication of stable optical waveguides on lithium niobate with low losses and the mode sizes suitable for coupling to standard single-mode optical fibers [1, 2].

After the birth of quantum information science, lithium niobate waveguide architectures emerged as one of key platforms for enabling photonics quantum technologies. However, new photonic systems place new demands on characteristics of integrated optical devices. This paper is devoted to the Mach-Zehnder (MZ) intensity modulator with an improved extinction ratio which is important for quantum key distribution systems [6].

2. Mach-Zehnder intensity modulator

The Mach-Zehnder intensity modulator is based on a Mach-Zehnder waveguide interferometer [4]. This device consists of an input waveguide and an X- or Y-splitter, which divides the input optical power into two interferometer arms. An electric field applied to the waveguide arms through electrodes induces a phase difference between the waves in the arms via the Pockels effect. The second X- or Y-splitter brings the optical waves together and the result of interference exits through the output waveguide (figure 1).

In the ideal case, the modulator transmits the maximal power if the signals from the interferometer arms are in phase and transmits no power if the signals are out of phase. However, in a real MZ modulator the difference between the maximal and minimal transmitted power is smaller because of finite tolerances of fabrication processes, and, first of all, because of a finite photolithography resolution. A typical extinction ratio (i.e., the ratio between the maximal and minimal transmitted powers) of commercially available MZ intensity modulators lies in the range 20–30 dB. Since the precision of fabrication cannot be increased, the ways for correcting the waveguide elements that have already been fabricated should be investigated.

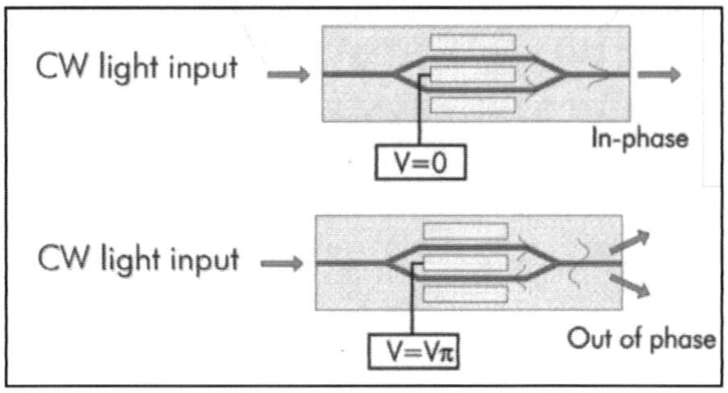

Figure 1. Integrated MZ intensity modulator

One of the ways is to use photorefractive properties of lithium niobate. Due to the photorefractive effect, a local irradiation of its surface can cause a decrease in the refractive index. The maximal value of such a decrease can be about 10^{-3} [7], while the refractive index increase of the waveguides on $LiNbO_3$ is about 10^{-2}-10^{-3} [1], so the photorefractive effect can be used for correction of waveguide elements. Such a possibility was demonstrated experimentally.

3. Experiment

The experimental setup used for the photorefractive waveguide correction is presented in figure 2. A local irradiation of the lithium niobate wafer was performed with the help of a He-Ne laser (633 nm, 4.5 mW), the light from which was directed through a standard single-mode optical fiber. The light spot positioning on the integrated optical scheme was performed by a 3-axis translation stage (figure 2).

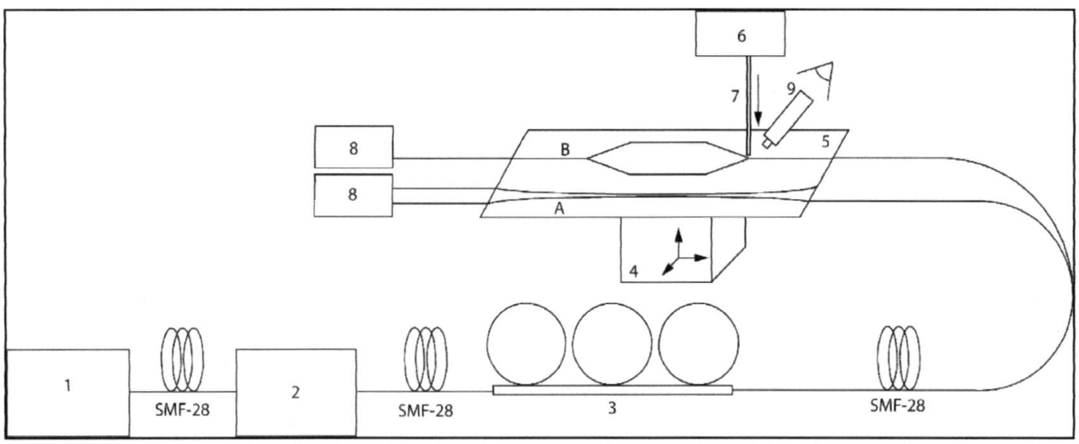

Figure 2. Experimental setup. 1 – DFB-laser diode, 2 – fiber optical attenuator, 3 – polarization controller, 4 – 3-axis translation stage, 5 – integrated optical chip, 6 – He-Ne laser, 7 – single mode optical fiber, 8 – optical detector, 9 – microscope.

The correction of optical waveguide splitters of two types (X and Y) was investigated. The dependence of the change in the splitting ratio on the light spot position was measured (figures 3, 4).

During the correction of the Y-splitter the light spot was placed on the arms of the splitter, for the X-splitter the light spot was on the lines perpendicular to the waveguides.

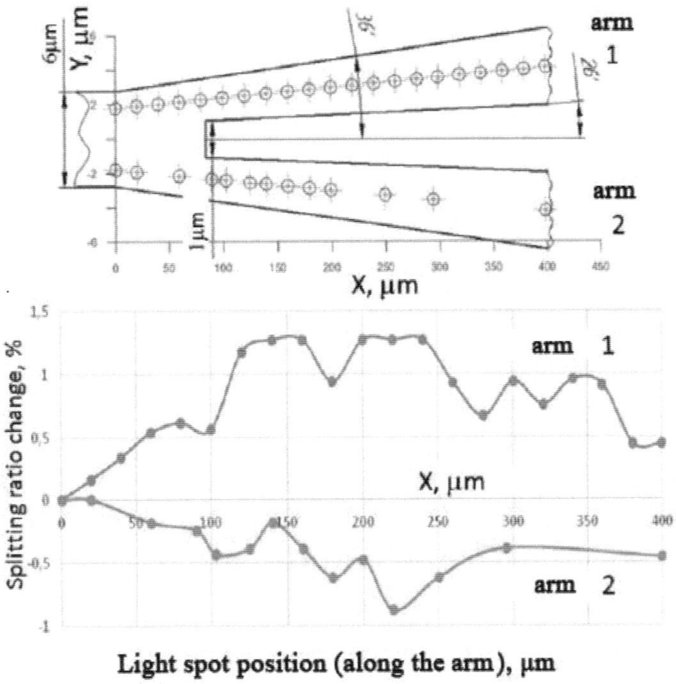

Figure 3. Splitting ratio change versus light spot position (Y-coupler).

A typical time of irradiation for reaching the steady state condition was about 2 minutes and was in agreement with the photorefractive response time. A homogeneous illumination of the chip with an intense visible or UV light erased the local refractive index change and returned splitters to the initial states.

The measured dependencies presented in figures 3 and 4 show that there are regions of the splitters which are sensitive to the photorefractive correction. The light spot position corresponding to the maximum splitting ratio change depended on the light distribution between coupled waveguides in the X-coupler. For the parallel section of the X-coupler the spot was in the mid position between the waveguides and shifted towards the waveguide with a higher intensity in the input (converging) and output (diverging) sections of the X-splitter. For the Y-splitter the influence of the photorefractive correction on the arms lead to opposite changes in the splitting ratio (see figure 3).

After the experiments with individual X and Y power splitters, the possibility of photorefractive correction was studied for the Y-couplers included into a MZ modulator. The extinction ratio measurements for such a corrected modulator showed an increase of 17 dB (from 30 to 47 dB).

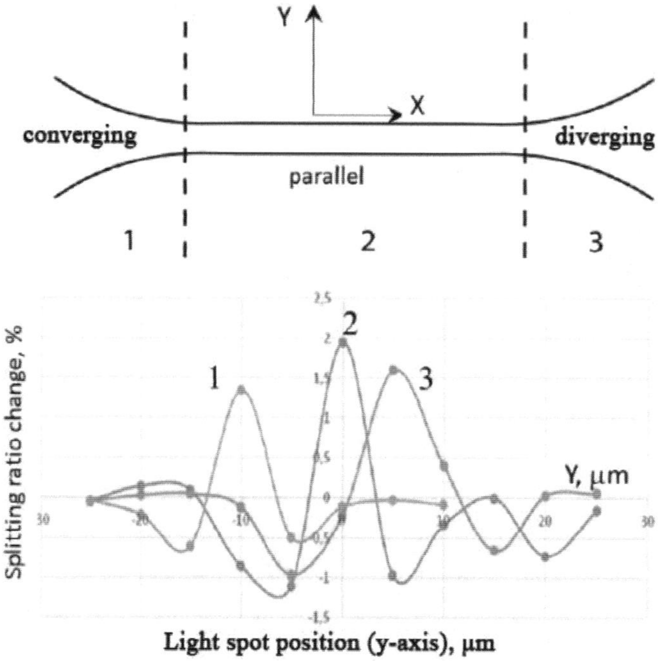

Figure 4..Splitting ratio change versus light spot position (X-coupler).

4. Conclusions

The results obtained show that the photorefractive effect can be used for correcting the splitting ratio of X- and Y-power splitters and, therefore, for increasing the MZ modulator extinction ratio. Modulators with an increased extinction ratio are promising for applications in quantum key distribution systems [6].

The dependences measured in the experiments show the regions of the power splitters most sensitive to the correction and can be used in further investigations of other methods of correction, such as photorefractive correction with thermal fixing [8] or intense femtosecond laser writing [1].

References

[1] Bazzan M, Sada C 2015 *Appl. Phys. Rev.* **2(4)** 040603
[2] Wooten Ed L, Kissa K M, Yi-Yan Al, Murphy Ed J et al. 2000 *IEEE J. Sel. Top. Quantum Electron.* **6(1)** 69-82
[3] Arizmendi L 2004 *Phys. Status Solidi (a)* **201(2)** 253-283
[4] Toney J E 2015 *Lithium Niobate Photonics* (Norwood: Artech House)
[5] Weis R S, Gaylord T K 1985 *Appl. Phys. A* **37** 191-203
[6] Wang X, Liu J, Li X, and Li Y 2015 *IEEE J. Quantum Electron.* **51(6)** 1–6
[7] Chen F 1969 *J. Appl. Phys.* **40(8)** 3389-3396
[8] Grachev A I, Chamrai A V and Petrov M P 2001 *OSA TOPS* **62** 203

IOP Conf. Series: Journal of Physics: Conf. Series **951** (2017) 012003 doi:10.1088/1742-6596/951/1/012003

Optical properties of β-BBO and potential for THz applications

N. A. Nikolaev[1, 2, a], Yu. M. Andreev[3, 4], V. D. Antsygin[2], T. B. Bekker[5], D. M. Ezhov[1], A. E. Kokh[5], K. A. Kokh[5], G. V. Lanskii[3, 4], A. A. Mamrashev[1, 2], V. A. Svetlichnyi[1, 4]

[1]High Current Electronics Institute, SB RAS, 2/3 Akademicheskii Ave., 634055, Tomsk, Russia;
[2]Institute of Automation and Electrometry, SB RAS, 1 Academician Koptyug Ave., 630090, Novosibirsk, Russia;
[3]Institute of Monitoring of Climatic and Ecological Systems, SB RAS, 10/3 Akademicheskii Ave., 634055, Tomsk, Russia;
[4]Siberian Physical Technical Institute of Tomsk State University, 1 Novosobornaya Sq., 634050, Tomsk, Russia;
[5]Institute of Geology and Mineralogy, SB RAS, 3 Academician Koptyug Ave., 630090, Novosibirsk, Russia;

[a]Corresponding author's e-mail: nazar@iae.nsk.su

Abstract. The anisotropy of optical properties of high quality beta barium borate crystal (β-BaB_2O_4, β-BBO) was studied in the main transparency window by using classic spectroscopic methods and in the range of 0.2 – 2 THz by using THz time-domain spectroscopy. β-BBO crystals were grown by the top-seeded solution technique in a highly resistive furnace with a heat field of 3-fold axis symmetry. At room temperature (RT), absorption coefficient in the maximal transparency window in grown crystals did not exceed 0.05 cm^{-1}. Strong absorption anisotropy was observed in 3 – 5 μm and the THz range. At 1 THz absorption coefficients for *e* and *o* wave were, respectively, 7 cm^{-1} and 21 cm^{-1} at RT; 2 cm^{-1} and 10 cm^{-1} at 81 K. At the most attractive for out-of-door applications range < 0.4 THz the absorption coefficient is found to be very low: below 0.2 cm^{-1} at RT and 1 cm^{-1} at 81 K. Refractive indices dispersions measured by THz-TDS were approximated in the form of Sellmeier equations. Birefringence is found quite large for phase matched difference frequency generation (DFG) or down-conversion into the THz range (THz-DFG) under near IR pump at RT and 81 K. Type II (*oe-o* and *eo-o*), and type I (*ee-e*) three wave interactions can be realized at RT. THz-DFG of Nd:YAG laser and KTP OPO can be realized by type II (*oe-o*) three-wave interaction. For selected spectral ranges of femtosecond Ti:Sapphire laser efficient phase matched and group velocity matched optical rectification can be realized by another two types of three wave interactions. Accounting other well-known attractive physical properties of β-BBO crystal, wide application in THz technique can be forecasted.

1. Introduction

The interest to terahertz (THz) sources has been growing significantly because of promising applications in the fields of spectroscopy of biomolecules and rotation lines of gases, nondestructive imaging, communication [1] and environmental monitoring systems [2]. Progress in the THz technology developments has, however, been hampered both by a lack of intense THz sources and operational difficulties in using available nonlinear crystals for down-converters of IR lasers as THz wave generators [2, 3]. Disadvantages of widely used nonlinear crystals for THz applications stimulate a growing interest in searching for new or known, but not efficiently utilized, highly-reliable anisotropic crystals.

In this study, anisotropy of optical properties of high quality β-BBO crystal discovered in 1985 by Chen et al. [4] was studied in the main transparency window by using classic spectroscopic methods and, for the first time, in the range of 0.2 – 2 THz by the methods of THz time-domain spectroscopy to find out possibility of efficient down-conversion of near IR lasers into the THz range.

2. Experimental results

The High quality β-BBO crystals were grown by the modified top-seeded solution technique in a highly resistive furnace with a heat field of 3-fold axis symmetry [5] (Fig. 1).

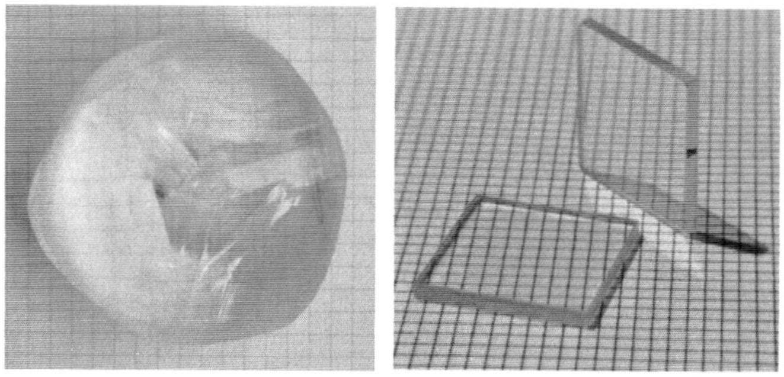

Fig. 1. As-grown single crystal BBO boule (a) and test samples (b).

Crystals were grown in BaB_2O_4-$NaBaBO_3$-$Ba_2Na_3[B_3O_6]_2F$ system. After determining the equilibrium temperature, the seed was allowed to grow with continuous rotation 1 rpm. The cooling and pulling rates varied, respectively, from 0.4 to 2 °C/day and 0.5 to 0.1 mm/day. Test samples of $10 \times 10 \times 2$ mm^3 cut at $\theta = 0°$ and 90° to study optical properties for *o*- and *e*-wave were manufactured (Fig. 2b). Test samples were cooled by mounting in the bath cryostat with liquid nitrogen.

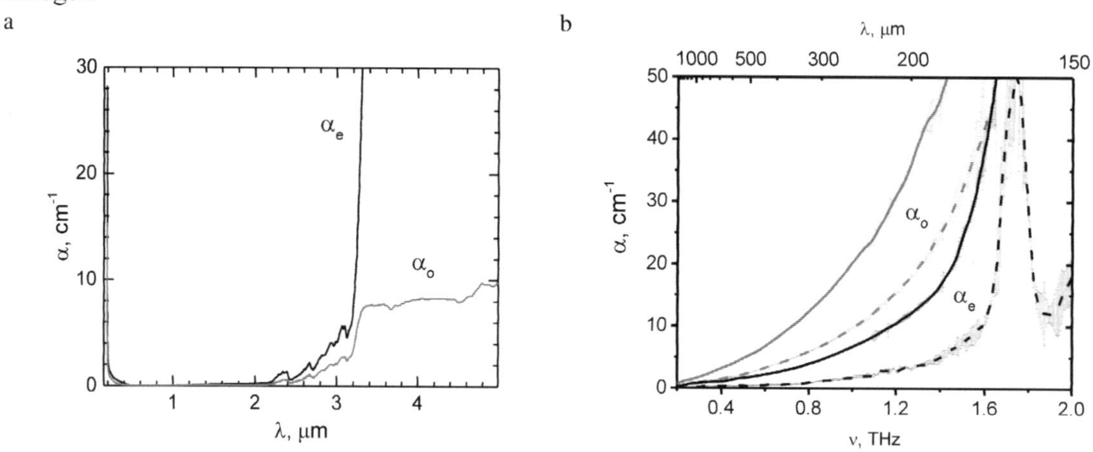

Fig. 2. Absorption spectra (a) in the maximal transparency window and (b) in the THz range at RT (solid lines) and 81 K (dashed lines) in BBO crystal.

At room temperature (RT), absorption coefficient in the maximal transparency window in grown BBO

IOP Conf. Series: Journal of Physics: Conf. Series **951** (2017) 012003 doi:10.1088/1742-6596/951/1/012003

crystals did not exceed 0.05 cm^{-1} (Fig. 2a). Strong absorption anisotropy was observed in 3 – 5 μm and the THz range (Fig. 2b). At 1 THz absorption coefficients for *e* and *o* wave were, respectively, 7 cm^{-1} and 21 cm^{-1} at RT; 2 cm^{-1} and 10 cm^{-1} at 81 K. At the most attractive for out-of-door applications range < 0.4 THz, absorption coefficient is found to be very low: below 1 cm^{-1} at RT and 0.2 cm^{-1} at 81 K that is favorable for long THz wave generation.

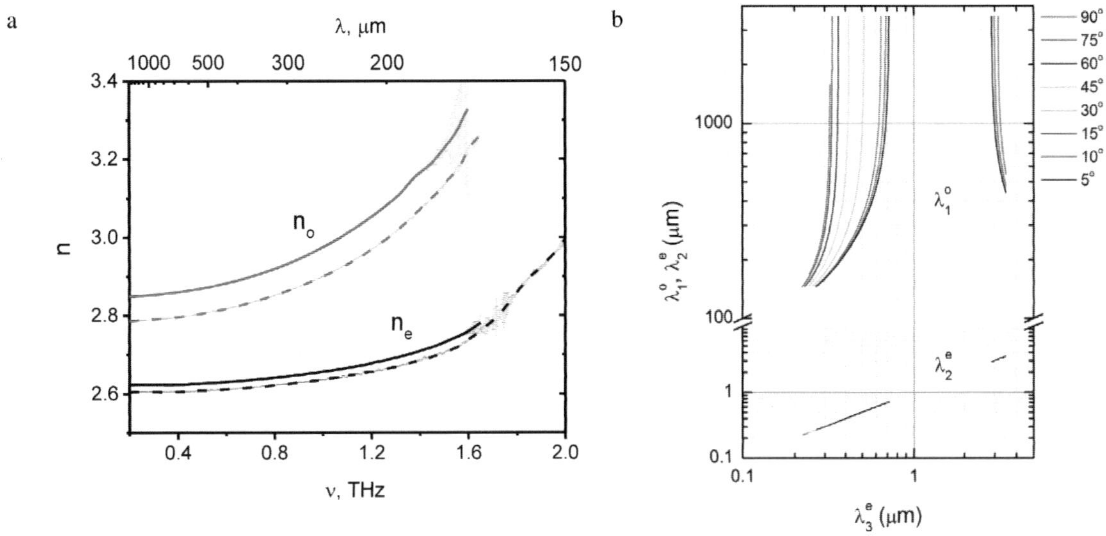

Fig. 3. (a) Dispersions of refractive indices in BBO crystal in the THz range and (b) and RT *oe-o* type phase matching curves.

Dispersions of refractive indices were approximated in the form of Sellmeier equations for RT (1, 2) and 81 K (3, 4) with validity range 140 – 3500 μm.

$$n_o^2 = 2.040 + \frac{0.816\,\lambda^2}{\lambda^2 - 12815}, \tag{1}$$

$$n_e^2 = 2.478 + \frac{0.160\,\lambda^2}{\lambda^2 - 15598} \tag{2}$$

$$n_o^2 = 4.612 + \frac{3.117\,\lambda^2}{\lambda^2 - 16175}, \tag{3}$$

$$n_e^2 = 5.970 + \frac{0.800\,\lambda^2}{\lambda^2 - 16659} \tag{4}$$

Birefringence is found quite large for phase matched difference frequency generation (DFG) or down-conversion into the THz range (THz-DFG) under near IR pump at RT and 81 K (Fig. 3b). It was established that type II (*oe-o* and *eo-o*), and type I (*ee-e*) of three wave interactions can be realized at room temperature. THz-DFG of Nd:YAG laser and KTP OPO can be realized by type II (*oe-o*) three-wave interaction. For selected spectral ranges of femtosecond Ti:Sapphire laser efficient phase matched and group velocity matched optical rectification can be realized by another two types of three wave interactions.

3. Conclusion

Optical properties of high optical quality BBO crystals were studied in the THz range. It was found that at long wavelength range (> 650 μm) BBO possesses low optical absorption coefficient down to below 1 cm^{-1} and it is phase matchable for down conversion of near IR lasers into this range. Accounting other well-known attractive physical properties of β-BBO crystal, wide application in THz technique can be forecasted.

Acknowledgements
The work has been supported by the Russian Science Fund, Project 15-19-10021. The THz-TDS spectrometer provided by "Spectroscopy and Optics" Shared Equipment Center of the Institute of Automation and Electrometry, SB RAS.

References

[1] Saeedkia D (Ed.) 2013 *Handbook of terahertz technology for imaging, sensing and communications* Elsevier
[2] Stepanov A, Henin S, Petit Y, Bonacina L, Kasparian J and Wolf J 2010 *Applied Physics B* **101**(1-2) 11-14
[3] Guo J, Xie J, Li D, Yang G, Chen F, Wang C et al 2015 *Light: Science & Applications* **4**(12) e362
[4] Chen C, Sasaki T, Li R, Wu Y, Lin Z et al. 2012 *Nonlinear optical borate crystals: Principals and applications* John Wiley & Sons
[5] Kokh A, Bekker T, Vlezko V and Kokh K 2011 *J. Cryst. Growth* **318**(1) 602-605

Quantum treatment of laser cooling on weak transitions: multipeaks and bimodal momentum distributions.

R Ya Ilenkov[1,2], A V Taichenachev[1,2], V I Yudin[1,2,3] and O N Prudnikov[1,2]

[1] Institute of Laser Physics SB RAS, 13/3 Acad. Lavrentyev avenue, Novosibirsk 630090, Russia
[2] Novosibirsk State University, 2 Pirogova street, Novosibirsk 630090, Russia
[3] State Technical University, 20 K. Marx avenue, Novosibirsk 630073, Russia

e-mail: ilenkov.roman@gmail.com

Abstract. The work is devoted to the study of the features and parameters of the momentum distributions of atoms laser cooled on weak optical transitions. It was shown that atoms distributions are described by a bimodal momentum distribution whose characteristics depends on the parameters of the light field. In a strong field a velocity selective coherent population trapping effect is observed.

1. Introduction

The laser is a powerful and precise tool for effectively controlling the translational degrees of freedom of atoms. Currently laser cooling is rapidly developing field of science at the intersection of laser physics and atomic optics [1-3], and has a plurality fundamental and applied applications. In particular, laser cooling is used to obtain and study the Bose-Einstein condensate [4], in the field of quantum computer science [5], atomic nanolithography [6] and interferometry [7]. The combination of laser cooling and modern methods of precision spectroscopy allows the creation of standards for frequency and time, the accuracy of which reaches values of the order of 10^{-17} - 10^{-18} [8,9]. To date, various methods for localizing and cooling atoms (and magneto dipole traps, optical lattices, etc.) have become an integral part of modern basic and applied science. This progress would not have been possible without theoretical analysis and consideration of the processes arising from the interaction of atoms with the electromagnetic field.

However, the theoretical description, taking into account the multilevel structure of the atom, the degeneracy of the atomic levels, spontaneous decay, the recoil effect, and the effect of field polarization, is an extremely difficult task. The beginning of its solution was laid in the 70's - 80's by the study of the simplest system: a two-level atom in a resonant light field [1,2,10-12]. An analytical study of this model using the quasiclassical approach (see, for example, [13]), which considers cooling in terms of light-induced forces and their fluctuations (diffusion in momentum space), made it possible to understand many of the basic cooling mechanisms. But, the quasi-classical approach describes only those cases where single photon recoil frequency ($w_r / \gamma = \hbar k^2 / 2m$, where k – wave vector, m m – mass of atom) is small compared with γ - natural linewidth cooling transition, and is not suitable for

IOP Conf. Series: Journal of Physics: Conf. Series **951** (2017) 012004 doi:10.1088/1742-6596/951/1/012004

consideration forbidden transitions such as intercombination transitions Yb, Mg, Sr, etc used in experiments for ultradeep laser cooling [14-17].

In this paper we will investigate the laser cooling of atoms on weak optical transitions where the recoil w_r / γ is not small parameter. The study will be conducted using an accurate quantum calculation [18,19], because it is in contrast to previously developed methods, allows to obtain the density matrix of atoms with full consideration of recoil effects due to the interaction of atoms with the field photons. Density matrix containing full information about both the internal and external degrees of freedom. This method allows us to correctly take into account both the cooling of atoms localized in the optical potential, and the more heat fraction of atoms that perform above-barrier motion.

Particular attention will be paid to the features of impulse distributions of cold atoms as a function of the intensity of the light field.

2. Formulation of the problem

The field of a standing monochromatic wave has the form:

$$E = 2E_0 \exp(-i\omega t)\cos(kx), \tag{1}$$

where E_0 the amplitude of the light field, and ω ω is the frequency of the light field. The interaction of a two-level atom with the field of a given configuration is considered on the basis of the quantum kinetic equation for the density matrix $\widehat{\rho}(x_1, x_2, t)$ in the coordinate representation (see, for example, [2]):

$$\frac{\partial}{\partial t}\widehat{\rho}(x_1, x_2, t) = -\frac{i}{\hbar}\left[\widehat{H}(x_1, x_2, t)\widehat{\rho}(x_1, x_2, t) - \widehat{\rho}(x_1, x_2, t)\widehat{H}(x_1, x_2, t)\right] + \widehat{\Gamma}\{\widehat{\rho}(x_1, x_2, t)\}, \tag{2}$$

where $\widehat{H}(x_1, x_2, t)$ - Hamiltonian of atom, $\widehat{\Gamma}\{\widehat{\rho}(x_1, x_2, t)\}$ - Operational functional that describes the processes of spontaneous relaxation. In subsequent calculations, it is convenient to go from the coordinates x_1 x_1 and x_2 to the new (Wigner) coordinates $x = (x_1 + x_2)/2$ 2 and $q = (x_1 - x_2)$. This transition is dictated by the convenience of numerical calculations.

In the resonance approximation and in the transition to the basis of the rotating wave the Hamiltonian of the atom becomes independent of time:

$$\widehat{H}(x, q) = -\frac{\hbar^2}{2m}\frac{\partial}{\partial x}\frac{\partial}{\partial q} + \widehat{H}_0 + \widehat{V}. \tag{3}$$

The first term is the operator of the kinetic energy of the atom, the second describes the internal structure of the free atom and depends only on the internal degrees of freedom, and the latter describes the dipole interaction of the atom with the field of the standing light wave (1). In the one-dimensional case, the interaction operator can be represented in the form:

$$\widehat{V} = -\hbar\Omega(x)\begin{pmatrix} 0 & 1 \\ 1 & 0 \end{pmatrix}, \tag{4}$$

where $\Omega(x) = 2\Omega\cos(kx)$, $\Omega = dE_0/\hbar$ is the Rabi frequency, and d is the matrix element of the dipole moment operator of the atom. The operator $\widehat{H}_0 = -\hbar\delta\,|e><e|$ can be represented as:

$$\widehat{H}_0 = -\hbar\delta\begin{pmatrix} 0 & 0 \\ 0 & 1 \end{pmatrix}, \tag{5}$$

Here $\delta = \omega - \omega_0$ - detuning the frequency of the field from the frequency of the atomic transition. The operator $\widehat{\Gamma}$ in (2) describes spontaneous relaxation taking into account the recoil effect:

$$\widehat{\Gamma} = -\frac{\gamma}{2}\begin{pmatrix} -2f(q) & 1 \\ 1 & 2 \end{pmatrix}. \tag{6}$$

In the absence of recoil $f(q \to 0) = 1$, which corresponds to the conservation of normalization for spontaneous transitions.

As a result, the system of equations to be solved for the elements of a two-point density matrix has the form:

$$\begin{cases} \left(-\frac{i\hbar}{m}\frac{\partial}{\partial q}\frac{\partial}{\partial x}+\frac{\gamma}{2}-i\delta\right)\rho_{21}(x_1,x_2)=-\frac{i}{\hbar}\left(V(1)\rho_{11}(x_1,x_2)-\rho_{22}(x_1,x_2)V(2)\right), \\ \left(-\frac{i\hbar}{m}\frac{\partial}{\partial q}\frac{\partial}{\partial x}+\frac{\gamma}{2}-i\delta\right)\rho_{12}(x_1,x_2)=-\frac{i}{\hbar}\left(V^*(1)\rho_{22}(x_1,x_2)-\rho_{11}(x_1,x_2)V^*(2)\right), \\ \left(-\frac{i\hbar}{m}\frac{\partial}{\partial q}\frac{\partial}{\partial x}+\gamma\right)\rho_{22}(x_1,x_2)=-\frac{i}{\hbar}\left(V(1)\rho_{12}(x_1,x_2)-\rho_{21}(x_1,x_2)V^*(2)\right), \\ \left(-\frac{i\hbar}{m}\frac{\partial}{\partial q}\frac{\partial}{\partial x}\right)\rho_{11}(x_1,x_2)-\gamma f(q)\rho_{22}(x_1,x_2)=-\frac{i}{\hbar}\left(V^*(1)\rho_{21}(x_1,x_2)-\rho_{12}(x_1,x_2)V(2)\right), \end{cases} \tag{7}$$

where ρ_{22}, ρ_{11} are the populations of the excited and unexcited levels, and ρ_{12}, ρ_{21} - optical coherences between levels.

A method of solving this equation system with full consideration of recoil and localization of atoms was developed by us in [18,19].

3. Results

Various quantum regimes of laser cooling of two-level atoms were studied for different values of the intensity of the light field of a standing light wave with spatial homogeneous polarization. For these parameters, as already noted, the quasiclassical approach is inapplicable the recoil frequency is large enough. It should be noted that the impulse distribution is essentially nonequilibrium and is not described by the Gaussian distribution, therefore, cannot be characterized in terms of the temperature dependence. The momentum distributions obtained in the results of calculations can be roughly divided into two large groups: multi-peak structures with narrow peaks $\Delta p < \hbar k$ (Fig. 1a) and narrow distributions with wide wings (Fig. 1b).

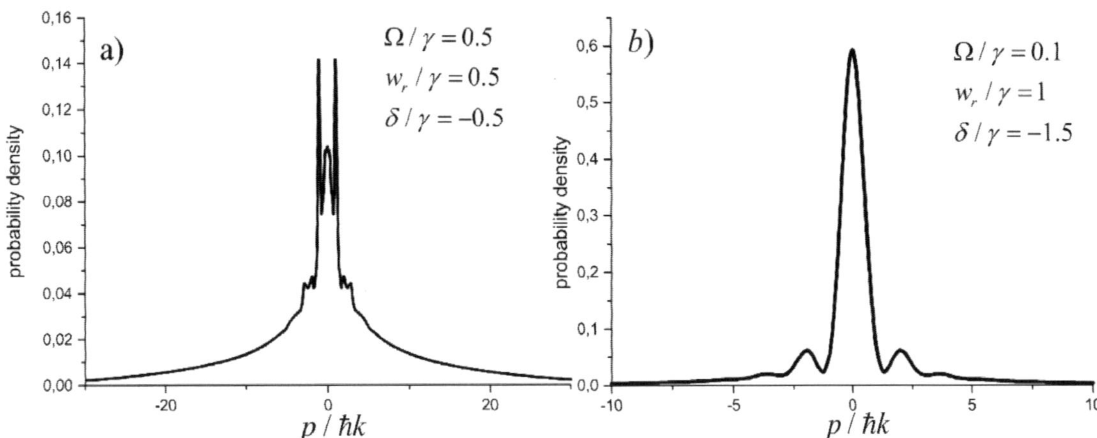

Figure 1. Two-structure momentum distributions of cold atoms in quantum regimes: a) The distribution has a clearly expressed narrow peaks which width of order a single photon momentum. b) The distribution has narrow structure of cold atoms and wide wings. Parameters of the problem are located on graphs.

To begin with, let us consider the second case, which is more important from the point of view of the obtained ultra-low temperatures of the atomic ensemble. It is seen that momentum distribution of atoms has a complex structure - a narrow peak is located on a wide substrate. Such a uniquely bimodal structure can be approximated with good accuracy by two Gaussian functions (Fig.2) corresponding to two velocity groups of atoms: for parameters ($\delta / \gamma = -1$, $\Omega / \gamma = 0.1$, $w_r / \gamma = 0.64$), the "cold" fraction atoms has a characteristic temperature $T_1 = 0.67\hbar\gamma$, and the "hot" fraction of atoms with temperature $T_1 = 26.2\hbar\gamma$.

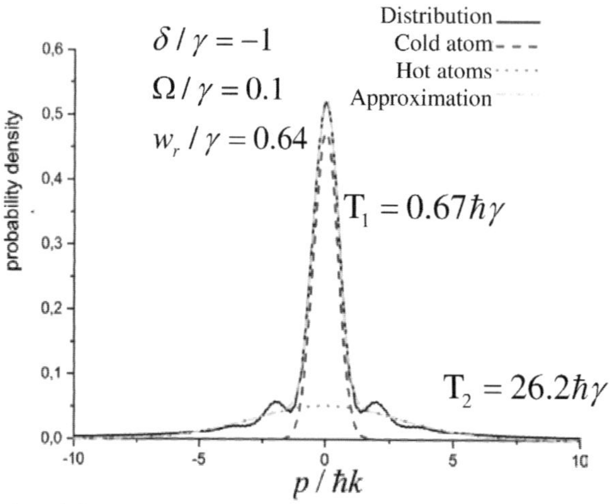

Figure 2. Bimodal momentum distribution of cold atoms, approximated by two Gaussian functions. Parameters of the problem are located on graph.

Similar bimodal distributions arise at different parameters of the light fields for the investigated weak optical transitions. In this case, the proportion of hot and cold atoms, as well as their temperatures depend on the parameters of the light fields (detuning and Rabi frequency).

Depending temperature distributions of cold and hot fractions atoms on the intensity of the light field is shown in Fig. 3a. The temperature of the cold fraction is much lower than the average temperature, defined as $< p^2 > / m$ ($< p^2 >$ is the average square of momentum) and reaches the limiting values $< \hbar\gamma$.

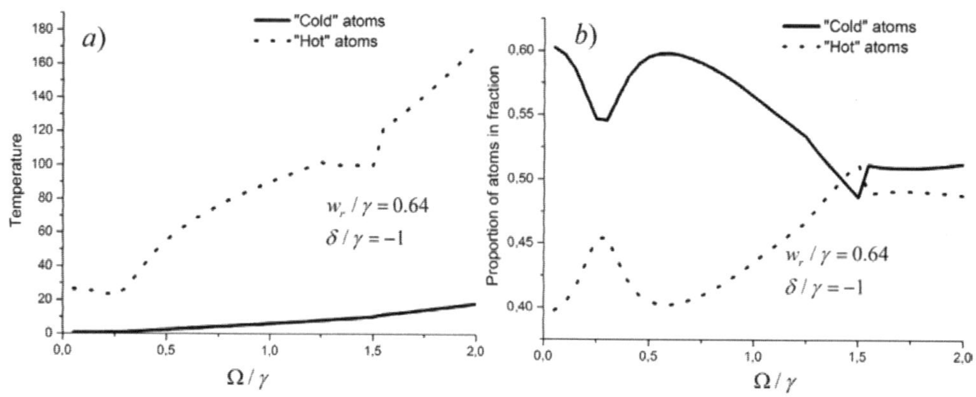

Figure 3. Dependences of temperature a) and proportion b) of atoms for "cold" and "hot" fractions of atoms of Rabi frequency. Parameters of the problem are located on graphs.

The main tendency consists in the general heating of the impulse distribution with increasing intensity of the light field, both the temperature of the cold fraction and the temperature of the hot fraction increase. At the same time, the temperature of the hot fraction has two peculiarities of behavior that disrupt the monotony of growth. Both can be explained by the distortions of the shape of the momentum distributions of the atoms. For the region $\Omega / \gamma = 0.2 - 0.4$ a major role in the distribution begin to play narrow pulse-order structure of a single photon, accompanying the central peak, and in the area of strong magnetic fields $\Omega / \gamma = 1.2 - 1.6$ the distribution is distorted so that the most probable impulses for atoms are not zero (Fig. 4) , but the corresponding $p = \pm\hbar k$, which somewhat worsens the accuracy of the approximation, because Approximation averages the contribution of emissions.

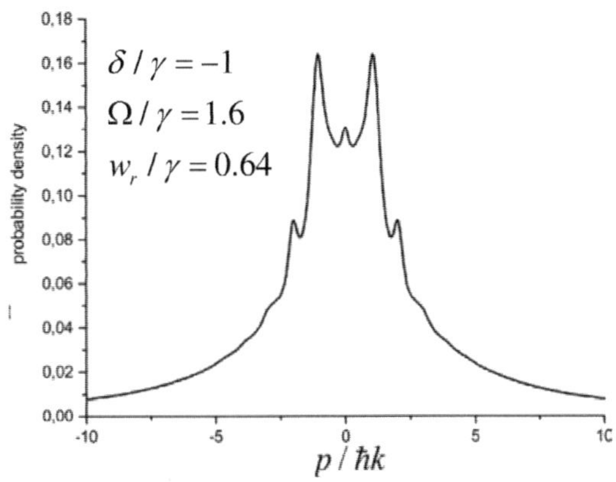

Figure 4. Multipeak momentum distribution of cold atoms. Parameters of the problem are located on graph.

The fraction of fractions of cold and hot atoms as a function of the intensity is shown in Fig. 3b. The number of atoms in the cold distribution is maximal in a weak field, and reaches 60%. In strong fields, the contribution of both fractions is equalized and amounts to about 50%. Features of behavior manifest at the same intensity as the temperature graphs.

Such complex multi-peak structures are the second basic mode of cooling atoms on weak clock transitions. This structure is associated with the manifestation of the effect of velocity selective coherent population trapping [20,21] and Fig. 5 illustrates this mechanism. For a two-level atom, this structure exists only in an essentially quantum mode.

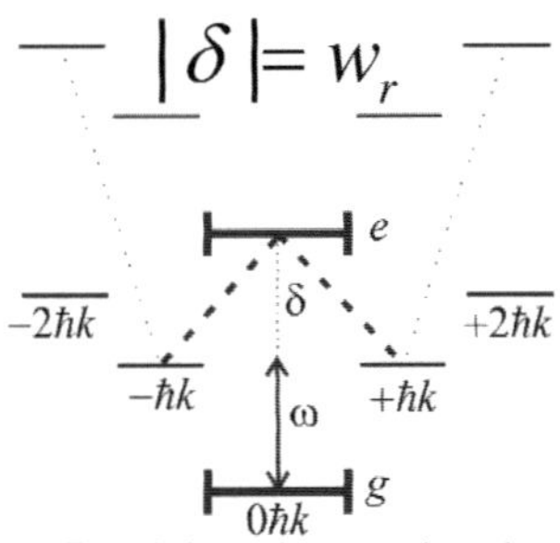

Figure 5. VSCPT in two-level system

The intensity of VSCPT peaks becomes maximum when the detuning and the recoil frequency are equal (in modulus), this leads to the most efficient excitation of the lambda circuit inside which most of the atoms are located. The atoms captured in the Lambda diagram form two distinct narrow peaks located at points with impulses $\pm\hbar k$ (Fig. 6a).

With increasing intensity, an increasing number of such peaks are observed, but along with this, the temperature of the hot fraction of atoms also increases, which makes it impossible to obtain fractions of the cold-atom fractionally distributed over the $\pm n\hbar k$ separated velocities (Fig. 6b).

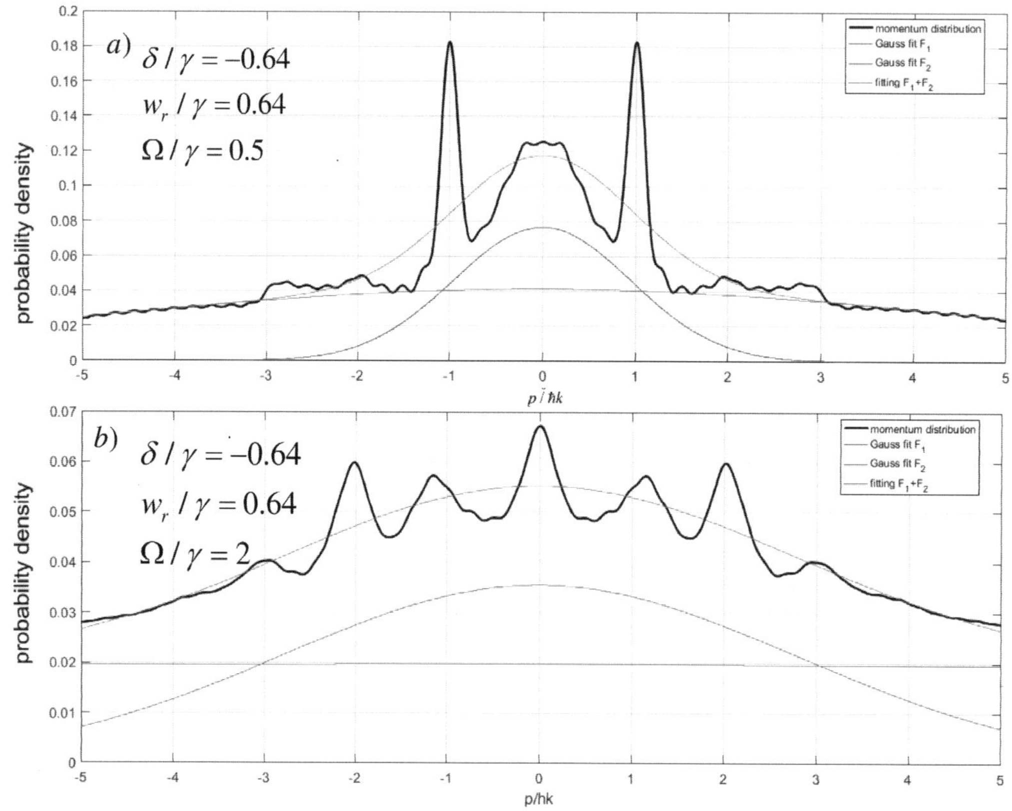

Figure 6. Momentum distribution contains narrow peaks. a) medium field; b) strong field. Parameters of the problem are located on graphs.

4. Conclusion

Laser cooling of two-level atoms in quantum regimes was studied. It is shown that there are two main modes - bimodal distribution, and a multi-peak distribution. The multi-peak distribution is caused by the manifestation of VSCPT and is particularly pronounced when the detuning and the recoil frequency are equal in a strong light field, when 7 or more narrow peaks in the momentum distribution are resolved. The bimodal distribution makes it possible to obtain a cold fraction of atoms whose temperature is $< \hbar\gamma$ (which can reach less than 1 millikelvin on researched clock transitions), with up to 60% of atoms in the cold fraction.

Acknowledges

The work of R. Ya. Ilenkov was personally supported by Russian Science Foundation (project № 17-72-10139). V.I.Yudin acknowledges the support of the Ministry of Education and Science (3.1326.2017/4.6) and RFBR (grant 17-02-00570).

References

[1] Minogin V G, Letochow W.S. 1986 *Laser Light Pressure on Atoms* (Moscow: Nauka)
[2] Kazantsev A P, Surdutovich G I, Yakovlev V P 1991 *Mechanical Action of Light on Atoms* (Moscow: Nauka)
[3] Cohen-Tannoudji C. 1992 *Atomic motion in laser light* (Paris: Elsevier Science Publishers B.V.)
[4] Anderson M H, Ensher J R , Matthews M R, Wieman C E , Cornell E A 1995 *Science* **269** 198
[5] Noguchi A, Eto Y, Ueda M, Kozuma M 2011 *Phys. Rev. A.* **84** 030301
[6] Fischer J, Wegener M 2013 *Laser & Photonics Reviews* **7** 22
[7] Müller H, Chiow S, Herrmann S, Chu S, Chung K 2008 *Phys. Rev. Lett.*. 031101
[8] Bloom B J, Nicholson T L, Williams J R, Campbell S L, Bishof M, Zhang X, Zhang W, Bromley S L, Ye J 2014 *Nature* **506** 71
[9] Hinkley N, Sherman J A, Phillips N B, Schioppo M , Lemke N D , Beloy K , Pizzocaro M , Oates C W , Ludlow A_D 2013 *Science*. **341** 1215
[10] Dalibard J, Cohen-Tannoudji C 1985 *JOSA B.* **2** 1707
[11] Cook R J 1979 *Phys. Rev. A.* **20** 224
[12] Gordon J P, Ashkin A. 1980 *Phys. Rev. A* **21** 1606
[13] *Minogin V G* 1980 *JETP* **79** 2044
[14] Barber Z W, Stalnaker J E, Lemke N D, Poli N, Oates C W, Fortier T M, Diddams S A, Hollberg L, Hoyt C W, Taichenachev A V, Yudin V I 2008 *Phys. Rev. Lett.* **100** 103002
[15] Poli N, Barber Z W, Lemke N D, Oates C W, Ma L S, Stalnaker J E, Fortier T M, Diddams S A, Hollberg L, Bergquist J C, Brusch A, Jefferts S, Heavner T, Parker T 2008 *Phys. Rev. A.* **77** 050501
[16] Poli N, Drullinger R E, Ferrari G, L´eonard J, Sorrentino F, Tino G M 2005 *Phys. Rev. A.* **71** 061403
[17] Brazhnikov D V, Bonert A E, Goncharov A N, Taichenachev A V, Yudin V I 2014 *Laser Phys.* **24** 074011
[18] Prudnikov O N, Taichenachev A V, Tumaikin A M, Yudin V I 2008 *JETP* **133** 963
[19] Prudnikov O N, Ilenkov R Ya, Taichenachev A V, Tumaikin A M, Yudin V I 2011 *JETP* **139** 1074-1080
[20] Wilkowski D, Chalony M, Kaiser R, Kastberg A 2009 *EPL.* **86** 53001
[21] Widmer M, Doery M, Bellanca M. et al. 1996 *Phys. Rev. A.* **53** 946

IOP Conf. Series: Journal of Physics: Conf. Series **951** (2017) 012005 doi:10.1088/1742-6596/951/1/012005

Temperature dependence of terahertz optical properties of LBO and perspectives of applications in down-converters

N. A. Nikolaev[1, 2, a], Yu. M. Andreev[3, 4], N. G. Kononova[5], G. V. Lanskii[3, 4], A. A. Mamrashev[1, 2], V. D. Antsygin[2], K. A. Kokh[5], A. E. Kokh[5]

[1]High Current Electronics Institute, SB RAS, 2/3 Akademicheskii Ave., 634055, Tomsk, Russia;
[2]Institute of Automation and Electrometry, SB RAS, 1 Academician Koptyug Ave., 630090, Novosibirsk, Russia;
[3]Institute of Monitoring of Climatic and Ecological Systems, SB RAS, 10/3 Akademicheskii Ave., 634055, Tomsk, Russia;
[4]Siberian Physical Technical Institute of Tomsk State University, 1 Novosobornaya Sq., 634050, Tomsk, Russia;
[5]Institute of Geology and Mineralogy, SB RAS, 3 Academician Koptyug Ave., 630090, Novosibirsk, Russia

[a]Corresponding author's e-mail: nazar@iae.nsk.su

Abstract. Lithium triborate LiB_3O_5 (LBO) crystals are widely used for frequency conversion of the near-IR lasers within main transparency windows. Their optical properties at these wavelengths are well studied. However, very little work has been published on the properties in the terahertz (THz) range. There was a lack of data on the refractive indices, the absorption coefficients spectra and their temperature dispersions. There are no reports of THz applications. Present work reveals all these topics including the prospects for use LBO crystals as down-converters of the near-IR lasers radiation. Optically finished samples of flux-grown LBO crystals were studied by THz-TDS. The refractive index dispersions were recorded and then approximated in the form of Sellmeier equations for the temperatures of 300 and 81 K. The phase-matching curves for the IR-THz and THz-THz frequency conversions were calculated. It was found that the absorption coefficients of LBO decrease significantly with cooling to cryogenic temperatures, but the overall character of optical properties changes is intricated. Experimental results are discussed in detail considering potential characteristics of THz down-converters.

1. Experimental setup and methods

The LBO crystals were grown from MoO_3 flux by modified Kyropoulos method at Institute of Geology and Mineralogy, Siberian Branch of Russian Academy of Science (SB RAS) [1]. LBO samples of thicknesses 2068 ± 5 mkm and faces of 10×10 mm^2 were cut along directions perpendicular to the crystallographic axes **a**, **b**, **c**. The correspondence between the optical and crystallographic axes was determined to be $\boldsymbol{X}, \boldsymbol{Y}, \boldsymbol{Z} \leftrightarrow$ **a**, **c**, **b** [2]. The studies were carried out using Time-Domain Terahertz Spectrometer (THz-TDS) which description and signal processing methods were described elsewhere [3]. The fine-tuning of the spectrometer polarization optics was performed before the studies, which is a prerequisite for measurements of birefringent media [4]. The bath cryostat was used for cooling down the samples to the liquid nitrogen temperature.

IOP Conf. Series: Journal of Physics: Conf. Series **951** (2017) 012005 doi:10.1088/1742-6596/951/1/012005

2. Results

LBO crystal shows significant changes in their optical properties being cooled down to the liquid nitrogen temperature. The absorption coefficients α_x and α_y decrease vastly to an immeasurably small level (Figure 1a) while α_z remains almost unchanged and keeps the value of about 1.5 cm^{-1} at 1 THz (Figure 1b). The refractive indices n_x and n_y decrease by ~0.37 and ~0.47, respectively, (Figure 2a) whereas n_z increases by ~0.07 (Fig.2 b). Thus, the thermo-optical coefficients in the THz frequency range are $dn_x/dT \approx 1{,}7\cdot10^{-3}$ K^{-1}, $dn_y/dT \approx 2{,}2\cdot10^{-3}$ K^{-1} and $dn_z/dT \approx -3{,}2\cdot10^{-4}$ K^{-1}. They are higher at least by two orders of magnitude than thermo-optical coefficients in the optical range.

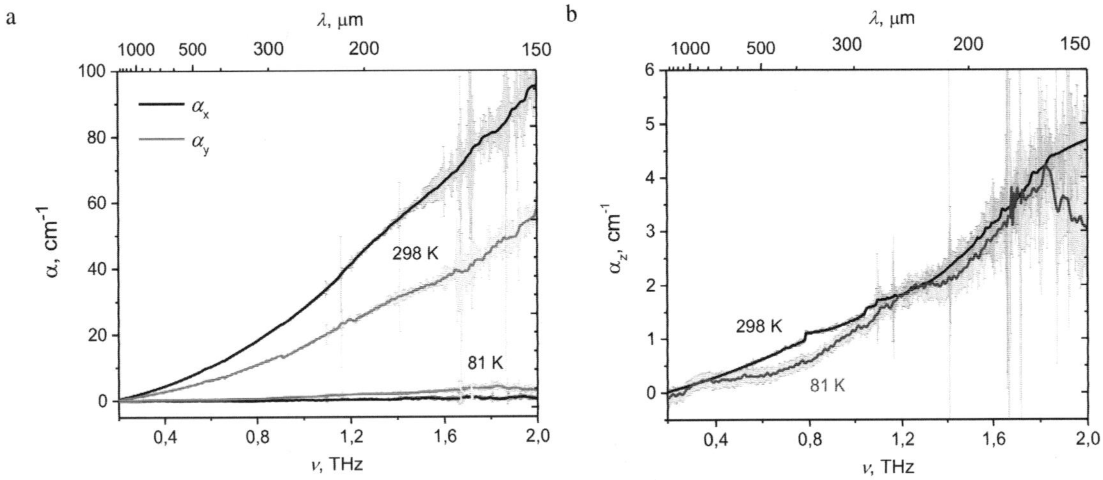

Figure 1. LBO Absorption coefficients α_x, α_y (a) and α_z (b).

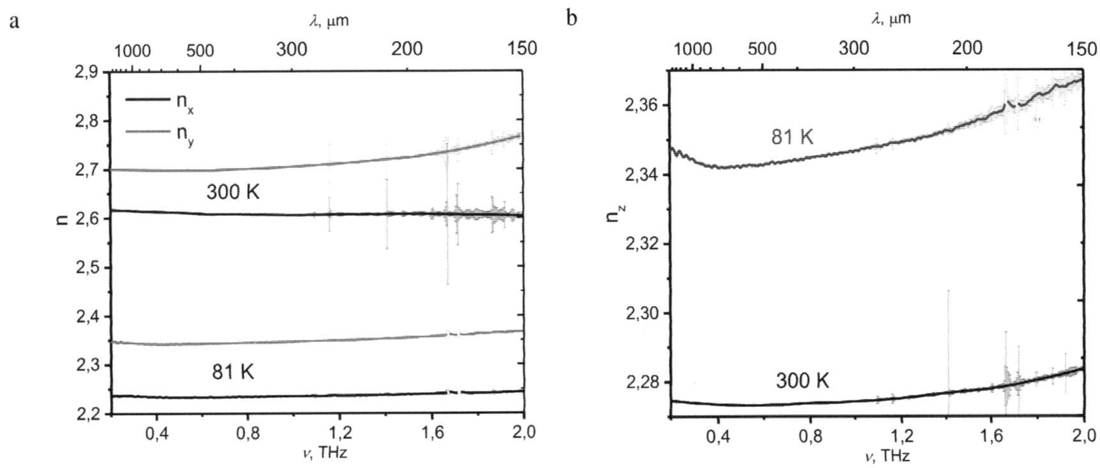

Figure 2. LBO refractive indices n_x, n_y (a) and n_z (b).

The dispersion equations based on measured data were approximated for the range of 0.2 – 2 THz (Table 1). Overall results show that the LBO becomes quasi isotropic at the temperature of liquid nitrogen. It means that the phase-matching conditions for Difference-Frequency Generation (DFG) cannot be fulfilled at 81 K. The phase-matching curves in ***XZ*** plane were calculated for the room temperature by using approximated dispersion equations (Figure 3).

Table 1. LBO refractive indices dispersion equations approximation for the range of 0.2 – 2 THz.

Room temperature	81 K
$n_x^2 = 6.82232 + \frac{0.03648\,\lambda^2}{\lambda^2 - 11676}$	$n_x^2 = 4.86951 + \frac{0.11663\,\lambda^2}{\lambda^2 - 6516}$
$n_y^2 = 6.84070 + \frac{0.45596\,\lambda^2}{\lambda^2 - 9418}$	$n_y^2 = 4.79318 + \frac{0.68735\,\lambda^2}{\lambda^2 - 3316}$
$n_z^2 = 5.06773 + \frac{0.10363\,\lambda^2}{\lambda^2 - 7012}$	$n_z^2 = 4.81255 + \frac{0.66883\,\lambda^2}{\lambda^2 - 3329}$

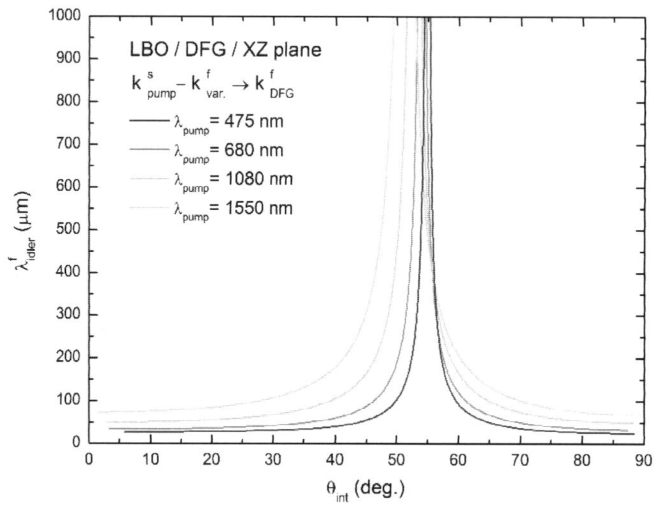

Figure 3. THz-DFG phase-matching curves for the typical wavelengths of telecom lasers (1550 nm), Ti:Sa lasers (680 – 1080 nm) and multi-terawatt system based on XeF(C-A) amplifier (475 nm) [5].

3. Conclusion

Despite having very low absorption coefficients in the terahertz range LBO can't be used for the down-conversion of visible and IR lasers radiation due to unfulfillment of phase-matching conditions. However, LBO could be perspective for the THz generation via optical rectification of femtosecond laser pulses. Moreover, the small changes of α_z and n_z with the cooling makes LBO perspective material for the manufacturing of temperature insensitive optical elements.

Acknowledgements

The work has been supported by the Russian Science Fund, Project 15-19-10021. The THz-TDS spectrometer provided by "Spectroscopy and Optics" Shared Equipment Center of the Institute of Automation and Electrometry, SB RAS.

References

[1] Kokh A, Vlezko V, Kokh K, Kononova N, Villeval P and Lupinski D 2012 *J. Cryst. Growth* **360** 158-161

[2] Nikogosyan D 2006 *Nonlinear optical crystals: a complete survey* Springer Science & Business Media

[3] Antsygin V, Kaplun A, Mamrashev A, Nikolaev N and Potaturkin 2014 *Optics express* **22**(21) 25436-25443

[4] Antsygin V, Losev V, Mamrashev A, Nikolaev N and Potaturkin O 2016 *Optoelectronics, Instrumentation and Data Processing* **52**(4) 374-380

[5] Alekseev S, Aristov A, Ivanov N et al. 2013 *Laser and Particle Beams* **31**(1) 17-21

IOP Conf. Series: Journal of Physics: Conf. Series **951** (2017) 012006 doi:10.1088/1742-6596/951/1/012006

Determination of the local field in the nuclear spin system of n-type GaAs

V M Litvyak[1*], R V Cherbunin[1], K V Kavokin[1,2] and V K Kalevich[1,2]

[1] Spin Optics Laboratory, 1 Ulianovskaya, 198504 St. Petersburg, Russia

[2]Ioffe Institute, 26 Polytekhnicheskaya, 194021 St. Petersburg, Russia

* e-mail: valiok.ok@gmail.com

Abstract. The nuclear spin system of a semiconductor can be cooled down to the microKelvin range of spin temperatures using optical pumping with subsequent adiabatic demagnetization. The efficiency of the nuclear spin cooling depends on the local field B_L, which is created by dipole – dipole interaction between nuclear spins. In this work we develop a method of experimental determination of the local field in semiconductor structures by using optical cooling and adiabatic demagnetization of the nuclear spin system, and apply it to measure B_L in n-GaAs.

1. Introduction

Electron spins in semiconductors can be optically oriented by illuminating the semiconductor crystal with circularly polarized light [1]. Recombination of such spin-polarized electrons with holes results in circularly polarized photoluminescence (PL). The degree of polarization of the PL is proportional to the electron spin, and can be used as an indicator of the spin states of electrons. At the same time, optical polarization of electron spins can give rise to nuclear polarization in semiconductors. The reason for this is a hyperfine interaction between electrons and nuclear spins. Such interaction provides a transfer of angular momentum from electrons to nuclei, via flip-flop transitions between electron and nuclear spins. The polarization of nuclear spins leads to appearance of the effective magnetic field created by nuclei upon electron spins. The nuclear field (Overhauser field) B_N can reach several Tesla ($B_{N\,max}$~ 5.3 T for GaAs [1]), strongly affecting the electron spin state [1, 2]. So, due to hyperfine interaction and, in particular, due to the existence of nuclear field, processes in the nuclear spin system can be studied by measuring the degree of polarization of the PL.

Nuclear spins are also affected by interactions within the nuclear spin system (NSS). In particular, the dipole-dipole interaction creates local fields in which nuclear spins precess with Larmor periods of order of 10^{-4} s. These magnetic fields, created by magnetic moments of nuclei, fluctuate in time due to directional fluctuations of the nuclear spins. The time average of these fields is zero, but the mean squared local field, B_L^2, is not zero. The dipole-dipole interaction leads to relaxation of nuclear angular momentum with the characteristic time T_2. During this time a thermodynamic equilibrium is established, so that the NSS can be characterized by a nuclear spin temperature Θ (β=1/ Θ – is the inverse spin temperature). This temperature can be significantly different from the lattice temperature, towards which it would relax with the spin-lattice relaxation time $T_1 >> T_2$. The NSS can be effectively cooled by circularly polarized optical pumping in an external magnetic field, when the energy flow from optically polarized electrons into the NSS can change nuclear spin temperature by orders of magnitude [1].

Apart of kinetic properties of the NSS, the local field B_L characterizes also its thermodynamic properties, being the measure of the NSS heat capacity [3, 4]. In particular, B_L determines the limit of lowering the temperature of the NSS by adiabatic demagnetization. This

IOP Conf. Series: Journal of Physics: Conf. Series **951** (2017) 012006 doi:10.1088/1742-6596/951/1/012006

method allows one to obtain extremely low nuclear spin temperatures, reaching 10^{-6} K at the lattice temperature of 4 K [5, 6]. In external fields $B< B_L$ the spin temperature does not depend on B, and the nuclear magnetization is proportional to B and inversely proportional to the spin temperature (Curie law). The Overhauser field created by polarized nuclear spins enhances the external field by hundreds times [1], allowing one to manipulate electron spin polarization in semiconductor structures by applying very weak (microTesla) external fields. So, experimental determination of the magnitude of B_L is vital for planning experiments on deep cooling of the NSS and studying its interaction with electron spins.

2. Experimental results

We study n-type GaAs doped by silicon with concentration n ~ 10^{15}cm^{-3} at the liquid helium temperature. Samples under study were pumped by circularly polarized light from a semiconductor laser, and we measured the degree of circular polarization of the luminescence ρ which is proportional to electronic spin $< S_z >$, where z is the direction along the excitation beam. For our sample, the maximum degree of the PL circular polarization was $\rho_{max} = 4\%$. We realized a precise (limited by the photon statistics) method for measuring the PL polarization, using a photoelastic modulator combined with a two-channel photon counter [5]. PL was detected by an avalanche photodiode. The geomagnetic fields were compensated to the level below 0.005 G by three pairs of Helmholtz coils. Our experiments were conducted in three stages. First, the NSS was prepared in the equilibrium with lattice temperature (optical excitation and magnetic fields were turned off). Secondly, the NSS was cooled by circularly polarized light in the longitudinal magnetic field $B_z = 100$ G during the time $t_{cooling}$ = 10 s. Flip-flop transitions with optically spin-polarized electrons provided the energy flow into the NSS that resulted in decreasing the absolute value of the nuclear spin temperature. Depending on mutual orientation of $\vec{B}_z$ and $\vec{S}_z$, positive (at $B_z \uparrow\uparrow S_z$) or negative (at $B_z \uparrow\downarrow S_z$), nuclear spin temperatures could be obtained. At the third stage, longitudinal magnetic field B_z was adiabatically slowly turned off and the transverse magnetic field was switched on. The cooled NSS was adiabatically re-polarized in the transverse field, with nuclear spin polarization obeying the dependence [5]:

$$p_N = \langle I \rangle / I = \frac{I+1}{3k_B\theta_N(0)} \hbar\gamma_N B_\perp \sqrt{\frac{B_L^2}{B_\perp^2 + B_L^2}} \quad , \tag{1}$$

where I is the nuclear spin, $<I>$ is the mean nuclear spin, γ_N is the nuclear gyromagnetic ratio averaged over all the isotopes (^{71}Ga, ^{69}Ga, ^{75}As) present in GaAs, k_B is the Boltzmann constant, $B_\perp$ is the transverse magnetic field, and $\theta_N(0)$ is the nuclear spin temperature immediately after adiabatic demagnetization.

The nuclear spin polarization then relaxed with the spin-lattice relaxation time T_1. We could trace the dynamics of nuclear polarization by measuring the Overhauser field [1], acting upon electron spins:

$$B_N = b_N \cdot \langle I \rangle / I \quad , \tag{2}$$

where b_N is the maximum value of the Overhauser field corresponding to complete polarization of nuclei; for GaAs, $b_N = 5.3$ T [1].

To this end, we continued to pump the sample with circular-polarized light, and detected the PL polarization ρ. The value of ρ was affected by electron spin depolarization in the effective transverse magnetic field B_{eff} (Hanle effect [1]). As in our case, the time dependence of reflected the dynamics of nuclear spin polarization. A typical experimental curve is presented in figure 1 (blue line).

We fit the experimental time dependence of ρ(t) by the following Eq. (3) (figure 1, red dashed line):

$$\rho(t) = \rho(0)\cdot\frac{B_{1/2}^2}{B_{1/2}^2+(B_\perp+B_N(t))^2}\quad, \tag{3}$$

where $\rho(0) = \rho_{max}$, $B_{1/2}$ is the half-width at half maximum of the Hanle curve [1] and $B_N(t) = B_N(0)\cdot e^{-\frac{t}{T_1}}$. The value of parameter $\rho(0) = 0.04$ is the maximum polarization degree of the PL for our sample, and $B_{1/2} = 100$ G were taken from the experiment on depolarization of electron spin in transverse magnetic field (Hanle curve) in absence of nuclear effects.

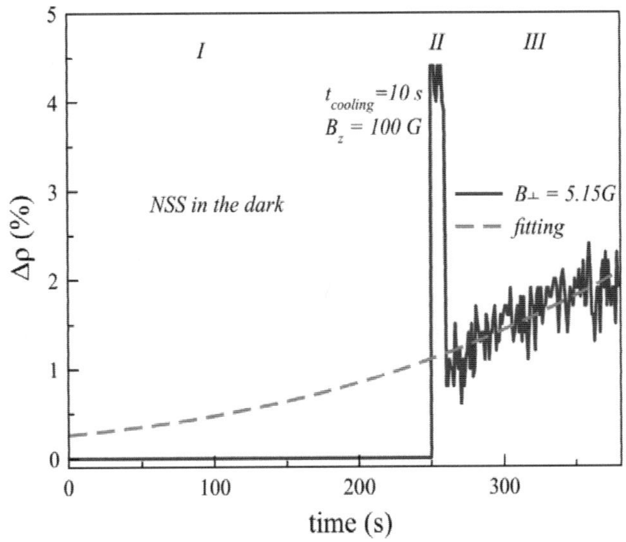

Figure 1. Time dependence of the PL circular polarization in n-GaAs (blue line): I) the NSS in the dark during 250 s; II) optical cooling of the NSS in the longitudinal magnetic field B_z = 100 G during 10 s; III) nuclear warm up at transverse magnetic field $B_\perp$ = 5.15 G. The red dashed curve is the fit by Eq. (1).

Using $B_N(0)$ as the adjustable parameter, we fit experimental curves for each transverse magnetic field, and obtain values of $B_N(0)$ for different values of transverse magnetic field $B_\perp$.The nuclear field $B_N(0)$ just after adiabatic cooling depends on the external magnetic field $B_\perp$, local field B_L and on the inverse spin temperature $\beta(0)$:

$$B_N(0) = \frac{b_N\hbar\gamma_N I(I+1)}{3k_B}B_\perp\beta(0)\sqrt{\frac{B_L^2}{B_L^2+B_\perp^2}}\quad. \tag{4}$$

Based on experimental data and fitting of warm up of the NSS, we obtain the experimental dependence $B_N(B_\perp)$ shown by circles in figure 2. After fitting experimental curve in figure 2 by Eq. (4) with the adjustable parameter B_L (figure 2, solid lines), we obtained the value of the local field B_L = (6±2) G.

It is worth to note that the value of the squared local field determined in our experiments is larger than the value B_L^2 = (2.1±0.1) G^2 calculated for GaAs by D. Paget [2]. This difference may be a result of the quadrupole splitting of nuclear spin levels due to small residual strains in the structure; increased local nuclear fields in GaAs structures were earlier inferred from the experimental magnetic field dependences of the spin-lattice relaxation time T_1 [7, 8].

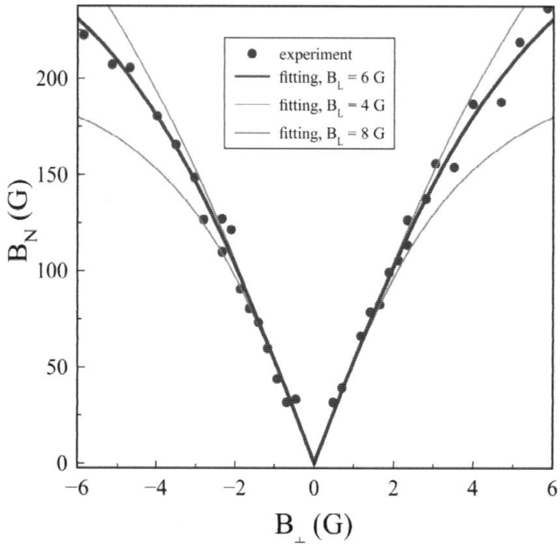

Figure 2. Experimental dependence $B_N(B_\perp)$ for cooling time equal to 10 s (circles). Solid red, blue and green lines are calculated using Eq. (4) with the adjustable parameter B_L = 4, 6 and 8 G, respectively.

3. Conclusion

To conclude, in this work we presented an experimental method for determination of the local field B_L in the nuclear spin system of a semiconductor, exploiting optical cooling combined with adiabatic demagnetization of the NSS, and apply it to n-GaAs. The measured value of B_L in *n*-GaAs is larger than expected from purely dipole-dipole nuclear spin interaction, and most likely originates from strain-induced quadrupole splitting in our sample.

4. Acknowledgment

This work was carried out in the framework of the RFBR grant No 16-52-150008.

References

[1] Optical Orientation, ed. by B.Zakharchenya and F.Meier, North-Holland, 1984.
[2] D. Paget, G. Lampel, B. Sapoval, V.I. Safarov. Low-field electron-nuclear spin coupling in gallium arsenide under optical pumping conditions. Phys. Rev. B **15**, 12 (1977).
[3] A. Abragam, W.G. Proctor, Phys. Rev. **109**, 1441 (1958)
[4] V.K. Kalevich, K.V. Kavokin, Igor Merkulov and M.R. Vladimirova. Dynamic Nuclear Polarization and Nuclear Fields, Chapter 12 in Spin Physics in Semiconductors ed. by M.I.Dyakonov, 2nd edition, Springer, 2017.
[5] V.K. Kalevich, V.D. Kulkov, V.G. Fleisher. Optical cooling of the nuclear spin system of a semiconductor in conjunction with adiabatic demagnetization. Bulletin of Academy of Science of the USSR, Physical Series, **46**, 70 (1982).
[6] V.K. Kalevich, V.D. Kulkov, V.G. Fleisher. Onset of a nuclear polarization front due to optical spin orientation in a semiconductor. JETP Lett. **35**, 20 (1982).

[7] M. Kotur, R.I. Dzhioev, M. Vladimirova, B. Jouault, V.L. Korenev, and K.V. Kavokin. Nuclear spin warm up in bulk n-GaAs. Phys. Rev. B **94**, 081201 (2016).
[8] M. Vladimirova, S. Cronenberger, D. Scalbert, M. Kotur, R. I. Dzhioev, I. I. Ryzhov, G. G. Kozlov, V. S. Zapasskii, A. Lematre, and K. V. Kavokin. Nuclear spin relaxation in n-GaAs: From insulating to metallic regime. Phys. Rev. B **95**, 125312 (2017).

Study of properties and development of sensors based on graphene films grown on SiC (0001) by thermal destruction method

A A Lebedev[1,2], V Y Davydov[1], D Y Usachov[3], S P Lebedev[1], A N Smirnov[1,2], V S Levitskii[1], I A Eliseyev[1,3], P A Alekseev[1], M S Dunaevskiy[1], A G Rybkin[3], S N Novikov[4] and Yu N Makarov[5,6]

[1] Ioffe Institute, Russian Academy of Sciences, St. Petersburg, 194021 Russia
[2] ITMO University, St. Petersburg, 197101 Russia
[3] Saint Petersburg State University, St. Petersburg, 199034 Russia
[4] Aalto University, Tietotie 3, 02150, Espoo, Finland
[5] GK «Nitride Crystals», 194156, St. Petersburg, Russia
[6] Nitride Crystals Inc., 181 E Industry Court, Suite B, Deer Park, NY 11729, USA

e-mail: shura.lebe@mail.ioffe.ru

Abstract. The structural, chemical, and electronic properties of epitaxial graphene films grown by thermal decomposition of the Si-face of a semi-insulating 6H-SiC substrate in an argon environment are studied by Raman spectroscopy, X-ray photoelectron spectroscopy and angle-resolved photoemission. It was demonstrated the possibility of fabrication of the gas and biosensors that is based on grown graphene films. The gas sensors are sufficiently sensitive to NO_2 at low concentrations. The biosensor operation was checked using an immunochemical system comprising fluorescein dye and monoclonal anti fluorescein antibodies. The sensor detects fluorescein concentration on a level of 1–10 ng/mL and bovine serum albumin–fluorescein conjugate on a level of 1–5 ng/mL. The proposed device has good prospects for use for early diagnostics of various diseases.

1. Introduction

Graphene (single layer of graphite) is presently the object of extensive studies due to its unique physical properties and huge potential in the development of new-generation devices that implement the principles of ballistic electronics, spintronics, optoelectronics, nanoplasmonics, and other promising alternatives to conventional semiconductor electronics [1]. For graphene to be economically feasible and attractive for numerous applications, it is necessary to have largesize wafers of high-quality homogeneous graphene. The sublimation of silicon from a semi-insulating single-crystal silicon carbide (SiC) substrate is one of the most promising ways to epitaxially grow graphene [2, 3]. A great advantage of the given technology is that there is no need to transfer the resulting film onto an insulator substrate, as this is done, e.g., in the synthesis of graphene on metals. In addition, this growth technique can grow graphene on the surface of commercial SiC substrates with a diameter of up to 4 in., the industrial manufacture of which has been mastered. Structures of this kind can be used in the standard technological line for the fabrication of semiconductor devices.

The structural characteristics of graphene films strongly depend on which face of SiC crystals they are grown [2, 4]. Films of graphene grown on the C face are commonly constituted by several

IOP Conf. Series: Journal of Physics: Conf. Series **951** (2017) 012007 doi:10.1088/1742-6596/951/1/012007

mutually misoriented single layers, with their surface topography having the form of separate flakes. By contrast, films grown on the Si face have a more uniform surface structure, which has the form of long terraces with width of ~1–2 μm. As a rule, films of this kind are constituted by a single layer, or a small number of these. Thus, the morphological and structural characteristics of graphene films grown on the Si face of SiC are more favorable for the development of devices on their basis.

The goal of our study is the systematic integrated diagnostics of graphene films produced by thermal decomposition of the Si face of a 6H-SiC substrate in order to relate the structural, chemical, and electronic properties of the films to the technological growth parameters.

2. Samples and experimental

To obtain graphene films, we used the method of the thermal decomposition of a SiC surface. In contrast to [5, 6], where graphene was grown in a high vacuum, we used the modified growth technique in an inert gas (argon), which enables more precise control over the sublimation of silicon-carbide components. As a consequence, it is possible to control with high precision both the growth process of a graphene film and the uniformity of coating a substrate with graphene.

As substrates served high-resistivity 6H-SiC wafers (manufactured by Svetlana OAO) with (0001) ± 0.25° orientation (Si face). Graphene was grown in a graphite crucible inductively heated by a high-frequency generator. To determine the optimal conditions of graphene synthesis, we varied the main technological parameters: the growth temperature was in the range from 1750 to 1900°C; the growth duration was 10 to 30 min; the argon pressure in the growth chamber was 600 to 800 Torr; and the sample heating rate was 100 to 350°C/min.

The structural, chemical, and electronic characteristics of the graphene being grown were monitored by Raman spectroscopy, X-ray photoelectron spectroscopy (XPS), and angle-resolved photoemission spectroscopy (ARPES).

Raman measurements were performed at room temperature in the backscattering configuration using a LabRam HR 800 spectrometer installation equipped with a confocal microscope, which could provide information from a part of the graphene film ~1 μm in diameter. Together with performing local diagnostics, we analyzed large sample areas (10 × 10 to 30 × 30 μm), with the subsequent plotting of Raman maps of the spectral line parameters. This procedure made it possible to carry out one of the main tasks of the present study, which consisted in assessing the uniformity of the structural characteristics over the sample area. We used in these measurements an excitation light wavelength of 532 nm and a laser power of 2.0 mW on a sample in a spot ~1 μm in diameter at a spectrometer resolution of 2.5 cm^{-1}.

ARPES, and XPS data with AlK_α radiation were obtained at the Resource Centre "Physical Methods of Surface Investigation" of the Research Park of Saint Petersburg State University (SPbU). The XPS spectra at various photon energies were measured with synchrotron radiation at the Russian-German beam line of the BESSY II synchrotron radiation facility.

Below results that are typical of samples grown in the final stage of optimization of the technological parameters are presented.

3. Graphen films structural characterization

3.1. Raman data

Figure 1a shows an array of Raman spectra measured in the spectral range 1300–2800 cm^{-1} on a sample area of 12.5×12.5 μm^2. The spectra show features appearing upon light scattering from the graphene film: *G* and *2D* lines and a weak *D* line [7]. The Raman diagnostics of graphene layers is based on an analysis of the frequency position, width, and intensity ratio of the *G*, *2D*, and *D* lines in the spectra. Analysis of the *G* line intensity map, which was obtained by processing this array, revealed a rather uniform distribution of the line intensity over the sample area. This is indicative of good thickness uniformity of the graphene film in the region being analyzed. It was found that the *2D* line is symmetric in the majority of spectra and is well fitted by a single Lorentzian, which is a fingerprint of a single-layer graphene [7]. The shape of the *2D* line which can be approximated by an envelope of four Lorentzians is observed in no more than 10% of the total number of spectra in the array. It follows from these data that the sample being analyzed is mostly formed by single-layer graphene with a small number of double-layer inclusions. The ratio between the integral intensities of D and G lines (I_D/I_G) can be used to evaluate the structural perfection of a graphene film. It was found that the I_D/I_G ratio distribution has a maximum at around 0.08, which corresponds to a defect concentration of $N_d < 10^{10}$ cm^{-2} [7]. The compression strain in the graphene layer estimated from the correlation between the positions of the G and $2D$ lines is $\varepsilon_{\|} = (0.35\pm0.03)\%$.

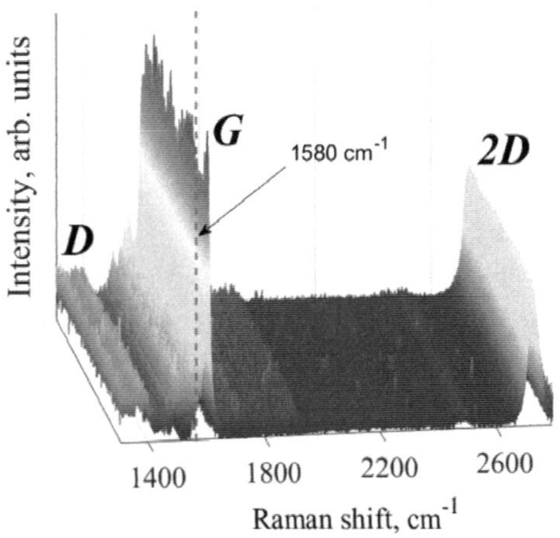

(a)

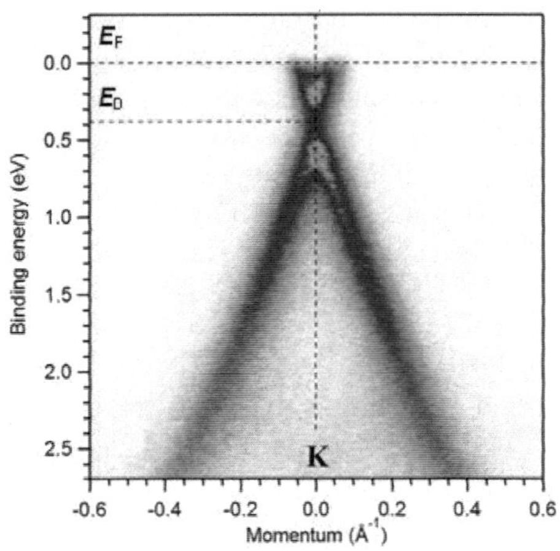

(c)

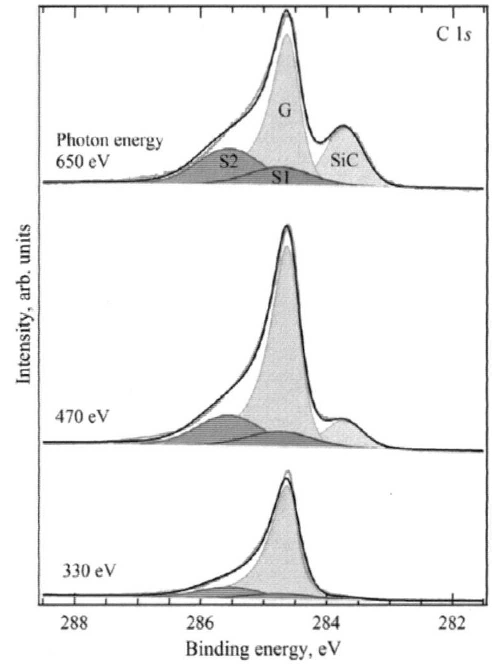

(b)

Figure 1. (a) Array of Raman spectra of a sample grown on the Si face of 6H-SiC. (b) XPS spectra measured in the C 1*s* region at various photon energies. (c) Electronic structure of the valence band in the vicinity of point *K* of the Brillouin zone. The ARPES data were obtained with He II radiation (40.8 eV).

3.2. XPS and ARPES data and Results of Synchrotron Measurements

To determine the chemical composition of the near-surface region, a survey XPS spectrum was measured upon removing molecules adsorbed in air via heating in ultrahigh vacuum at a temperature of 500°C. Only a small amount of oxygen was found in the spectrum in addition to lines related to carbon and silicon and peaks associated with the characteristic loss of energy by electrons. The content of oxygen in the upper atomic layers of SiC is approximately 1 at %. A detailed analysis of the shape of the XPS data in the region of the C 1*s* line makes it possible to estimate the average graphene thickness. To reliably determine the thickness of graphene, we analyzed XPS spectra obtained at four photon energies providing different depths of analysis within the range 5–23 Å (Figure 1b). The layer thicknesses were determined by choosing those thicknesses of the graphene and buffer layer that provided the best coincidence between the calculated and measured intensities of separate components of C 1*s* spectra.

It is known that the electronic structure of the valence band varies with the number of graphene layers. Single-layer graphene is characterized by a Dirac cone of electronic states at point *K* of the Brillouin zone, whereas the formation of a second layer leads to energy splitting of the cone and to doubling of the number of states. The ARPES data for the electronic structure of the valence band of the graphene/SiC(0001) system are shown in Figure 1c. An unsplit Dirac cone is seen at point *K*. This indicates that a single-layer graphene coating is predominant on the surface. This conclusion agrees with the Raman, AFM, and XPS data, according to which a sample grows as predominantly single-layer graphene with a small amount (~10%) of inclusions of double-layer islands with submicrometer dimensions. It is also seen from the ARPES data that the Dirac point lies below the Fermi level and has a binding energy of $E \approx 0.4$ eV. This unambiguously points to *n*-type doping with an electron concentration in the conduction band of $n = E^2/\pi(\hbar v_F)^2 \approx 10^{13}$ cm^{-2} [8].

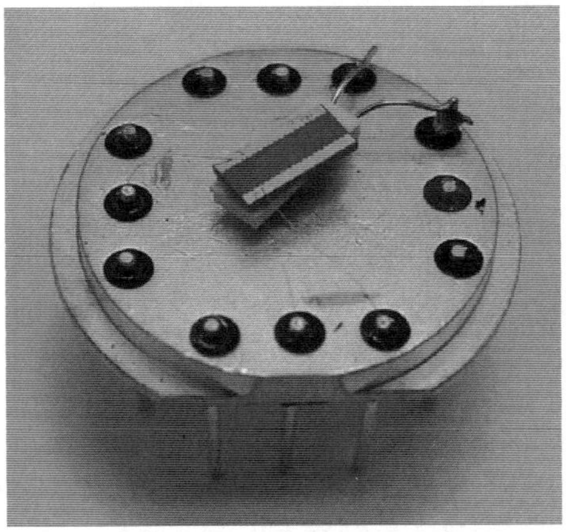

Figure 2. Graphene sensor chip on the holder.

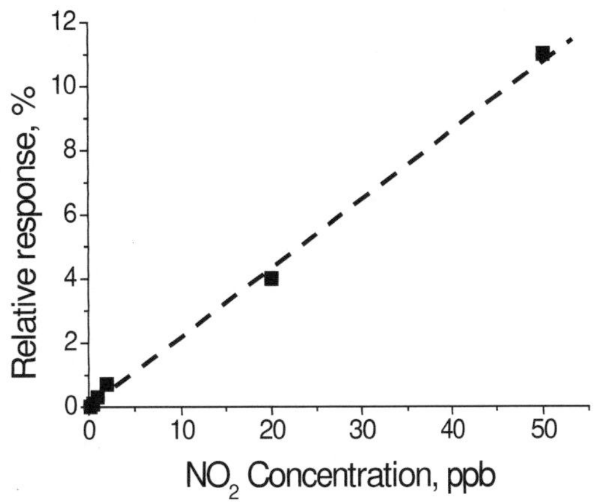

Figure 3. Plot of the response of the graphene-based gas sensor *vs.* NO_2 concentration in gas mixture at a temperature of 20°C.

4. Sensors developing and characterization

We fabricated the sensor structure on the grapheme film using laser photolithography and an AZ5214 photoresist [9,10] (Figure 2). Excess graphene is removed from the substrate surface with the aid of etching in the oxygen–argon plasma. Ti/Au (5/50 nm) ohmic contacts are fabricated using lift-off photolithography after deposition of metal on the surface of photoresist using electron-beam evaporation. The chip of the sensor is fixed on a holder with two Pt100 resistors that are used for temperature measurement and heating, respectively. In the measurements of the sensor sensitivity, we use the existing system for mixing and supply of gas mixtures [9,10]. Such a system allows a variation in the dilution coefficient from 1 : 1 to 1 : 10^5, so that the output concentration of the detected gas ranges from 0.1 ppb to 10 ppm. Purified air serves as a carrier gas. Sensor sensitivity *r* is measured in percents and determined as a relative variation in the resistance of the sample in the presence of the gas mixture containing the detected gas:

$$r = (R - R_0)/R_0,$$

here, R is the resistance of the sensor in the presence of the gas and R_0 is the original resistance in the absence of the detected gas in the air flow.

Figure 3 shows relative variations in the resistance of the graphene sensor in the presence of NO_2 in the gas mixture at a temperature of 20°C. The desorption rate of NO_2 is extremely low at room temperature. Therefore, annealing at a temperature of 110°C is used to refresh the sensor [9].

It should be noted that a significant disadvantage of graphene-based gas sensors is the lack of selectivity. Indeed, a change in the conductivity does not indicate which particular molecule is adsorbed on the graphene surface. Moreover, some molecules contribute to conductivity with opposite signs, so that the total change in resistance can be nearly zero. The problem of selectivity in graphene-based sensors can be solved by using the antigen–antibody reaction. The components of such a pair can only interact with each other and not with any other protein. As is known, certain stages of many diseases involve the appearance of antigen–markers that are specific of one disorder or a group of several disorders. These antigens may bind to specific antibodies preliminarily immobilized on the surface of a graphene sensor. Then, the antigen–antibody reaction will lead, by analogy with a gas sensor, to a change in resistivity of the graphene film.

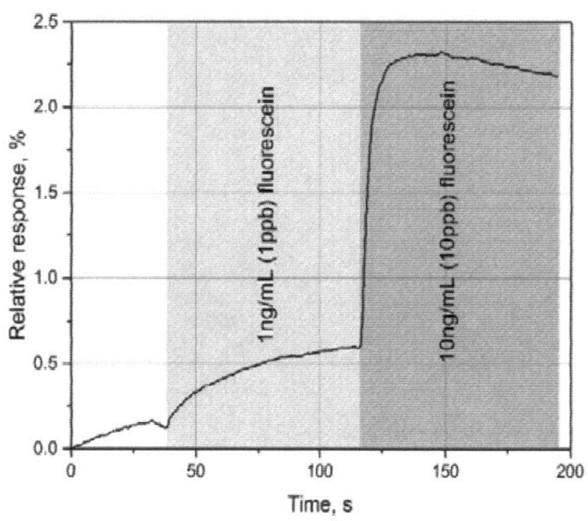

Figure 4. Variation of graphene-based sensor resistance during contact with solutions containing free fluorescein (at indicated concentrations).

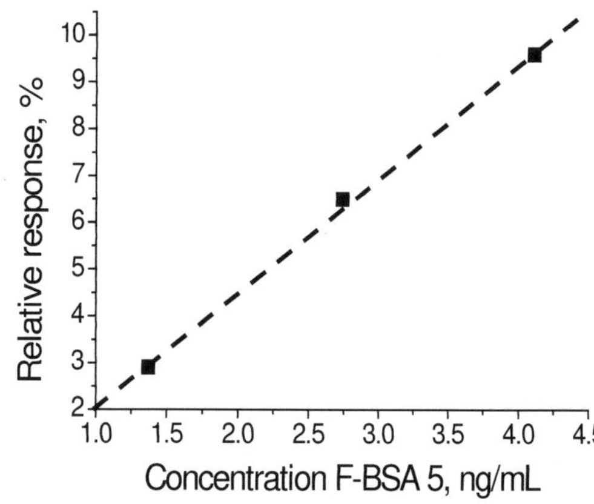

Figure 5. Variation of the graphene-based sensor resistance during contact with solutions containing BSA–fluorescein conjugate (F-BSA-5)

The technology of graphene-based biosensors was developed with respect to an immunochemical system comprising fluorescein dye and monoclonal anti-fluorescein antibodies (MAbs), in which the dye can be present both in the form of free molecules and as those chemically conjugated to protein carriers. The molecular mass of fluorescein (0.322 kDa) is comparable to that of some biologically significant marker molecules, such as hormones, nucleotides, and short peptides.

The measurements were carried out in borate buffer solution. The results of testing graphene-based sensors with anti-FITC MAbs grafted on the surface showed that the sensor resistance exhibited clearly detected changes upon contact with solutions containing low-molecular-mass ligands at concentrations of 1 and 10 ng/mL (Figure 4). Tests with high-molecular-mass ligands were performed using specially prepared bovine serum albumin (BSA) conjugated to fluorescein (F-BSA-5). The molecular mass of F-BSA-5 conjugate (69 kDa) is comparable to those of proteins used as markers of human diseases. In the F-BSA-5 conjugate, each fluorescein group accounts for about 5% of the total molecular mass. According to spectrophotometric data, the conjugate used in this work contained three fluorescein groups per F-BSA-5 molecule. When in contact with the solution of high-molecular-mass ligands (F-BSA-5), changes in the electric resistance of the graphene-based sensor were detected at conjugate concentrations within 1.37–3.11 ng/mL (Figure 5).

5. Conclusions

The results of an integrated study related the structural, chemical, and electronic characteristics of graphene films grown by the method of thermal decomposition of the SiC surface to the technological growth modes. This made it possible to optimize the technological parameters and develop a reproducible growth technology of single-layer graphene films with a small fraction of fine inclusions of double-layer graphene on the Si face of 6H-SiC substrates. It was found that the best structural characteristics are observed for graphene films grown with the following technological parameters: growth temperature of 1850°C; growth duration of 10 min; argon pressure in the growth chamber of 750 Torr; sample heating rate of 100–150°C/min. On the whole, the samples can be characterized as high-quality graphene, which enables the development of devices on the basis of these samples.

Also, we have demonstrated the possibility of fabrication of a gas sensor that is based on grapheme grown with the aid of sublimation in vacuum. The sensors are sufficiently sensitive to NO_2 at low concentrations. The results can be used in commercial production of graphene sensors.

At present, the diagnostics of human diseases is based to a significant degree on determining the concentrations and/or presence of certain marker molecules in the blood, which are specific to one or a group of disorders. The detection of diagnostically significant amounts of marker macromolecules is now performed using laborious, time-consuming procedures of enzyme immunoassay. The use of graphene-based biosensors may provide an accelerated and technically advanced approach to diagnostics of human diseases.

Acknowledgments

This work was financially supported by the Ministry of Education and Science of the Russian Federation (agreement No 14.575.21.0148, unique identifier of applied scientific research RFMEFI57517X0148).

References

[1] Ferrari A C *et al.* 2015 *Nanoscale* **7** 4598
[2] Emtsev K V *et al.* 2009 *Nat. Mater.* **8** 203
[3] Yazdi G R, Iakimov T and Yakimova R 2016 *Crystals* **6** 53
[4] Tedesco J L, VanMil B, Myers-Ward R, McCrate J M, Kitt S A, Campbell P M, Jernigan G G, Culbertson J C, Eddy Jr. C R and Gaskill D K 2009 *Appl. Phys. Lett.* **95** 122102
[5] Agrinskaya N V, Berezovets V A, Kozub V I, Kotousova I S, Lebedev A A, Lebedev S P, and Sitnikova A A 2013 *Semiconductors* **47** 301
[6] Lebedev A A, Lebedev S P, Novikov S N, Davydov V Yu, Smirnov A N, Litvin D P, Makarov Yu N and Levitskii V S 2016 *Tech. Phys.* **61** 453
[7] Ferrari A C and Basko D M 2013 *Nature Nanotech.* **8** 235
[8] Davydov V Yu, Usachov D Yu, Lebedev S P, Smirnov A N, Levitskii V S, Eliseyev I A, Alekseev P A, Dunaevskiy M S, Vilkov O Yu, Rybkin A G and Lebedev A A 2017 *Semiconductors* **51** 1072
[9] Novikov S, Lebedeva N and Satrapinski A 2015 *J. Sensors* **2015** 108581
[10] Novikov S, Lebedeva N, Satrapinski A, Walden J, Davydov V and Lebedev A 2016 *Sens. Actuators B: Chem.* **236** 1054

Morphology, granulometric and structural phase composition of mechanically synthesized composite powder Al-Mg+Al/MWCNTs

A V Aborkin[1], A V Sobol`kov[1], A V Kireev[1], A T Volochko[2], A Yu Izobello[2], N V Sachkova[3] and A E Sytschev[3]

[1]Vladimir State University named after Alexander and Nikolay Stoletovs, Vladimir, Russia

[2]Physical-Engineering Institute, National Academy of Sciences of Belarus, Minsk, Belarus

[3]Merzhanov Institute of Structural Macrokinetics and Materials Science, Russian Academy of Sciences, Chernogolovka, Russia

E-mail: aborkin@vlsu.ru

Abstract. A method of mechanical synthesis in a planetary ball mill produced composite powders based on a nanocrystalline Al-Mg matrix reinforced with micro additives of Al / MWCNTs. The granulometric composition of the obtained composite powders was studied. It has been established that the concentration of Al / MWCNTs does not significantly affect the average particle size of the composite powder formed in the process of mechanical synthesis. The resulting composite powders were studied by X-ray diffraction analysis. The influence of the Al / MWCNTs concentration on the size of the coherent scattering regions of the synthesized powder is determined. The morphology of the resulting composite powders was studied by high-resolution scanning microscopy. The resulting powders can be used in the creation of bulk composite materials and functional coatings.

1. Introduction

Creation and study of composite materials reinforced with hybrid carbon structures is a promising trend of scientific researches especially with manufacturing industry demands in parts with improved physical and mechanical properties.

The main problem faced by most researchers in the production of aluminum-matrix composite materials, hardened MWCNTs, is the agglomeration of nanotubes, a weak bond between MWCNTs and the matrix.

One of the solutions to this problem is the use of physicochemical treatment of MWCNTs, which leads, in particular, to the production of metal-containing nanoparticles or coatings on the surface of MWCNTs, which can be used as a boundary phase to improve interaction with the matrix material. This treatment can be carried out by deposition of vapor-phase coatings of organometallic compounds (MOCVD technology). During processing, hybrid nanostructures are formed, for example, Cu/MWCNTs, Al/MWCNTs, etc.

Works [1,2] show that using hybrid carbon structures for reinforcement helps gradually improve tribological behavior of frictionproof babbit coatings.

The observed increases in the mechanical properties of composite materials modified by hybrid nanostructures are certainly related to the uniformity of the distribution of nanostructures in the

volume of matrix materials. At the same time, the improvement of properties is also due to the modifying effect of nanostructures on the interphase boundaries. From previous studies it follows that the introduction of carbon nanostructures is accompanied by the chemisorption of atoms or metal clusters on their surface that causes a change in the electronic structure of the metal in the contact region and, ultimately, the redistribution of electronic interactions in the composite structure.
The aim of this work is to research the influence of microadditions of hybrid carbon structures to morphology, granulometric and structural phase composition of mechanically synthesized composite powders Al-Mg+Al/MWCNTs (Al-Mg+Al/MWCNTs ration changes between 0.05 and 0.3 wt.%).
Initial mixture used consisted of globular granules 1-2 mm in diameter from aluminium alloy AlMg2 with addition of 0.05, 0.15 and 0.3 wt.% of Al/MWCNTs.

2. Experimental methods and equipment

Mechanical treatment of the initial mixture was conducted in a globular planetary mill FRITSCH PULVERISETTE 6 using surface-active materials (stearic acid 0.75 wt.%). Milling was conducted using steel balls 8 mm in diameter at the ratio of mixture components to milling bodies 1:20. Mechanical synthesis was conducted at rotation rate 600 rpm during 4.5 hours.
Particle size distribution of synthesized composite powder was measured using MICROSIZER-201C. Structural phase composition of composite powder was performed using X-ray structural analysis. For that purpose a diffractometer D8 ADVANCE was used. Composite powder morphology was examined using ultra high resolution field-emission scanning election microscope ZEISS ULTRA PLUS based on ULTRA 55.

3. Results

Study of granulometric size composition of composite shows (see fig. 1a) the increase of Al/MWCNTs volume in the mixture from 0.05 to 0.3 wt.% leads to decrease of medium particle size from 18.7 to 17.6 μm. It should be noted that in all compositions fraction size is less that 10 μm.
Results study of X-ray structural analysis (see fig. 1b) shows that diffraction patterns have similar characteristics.

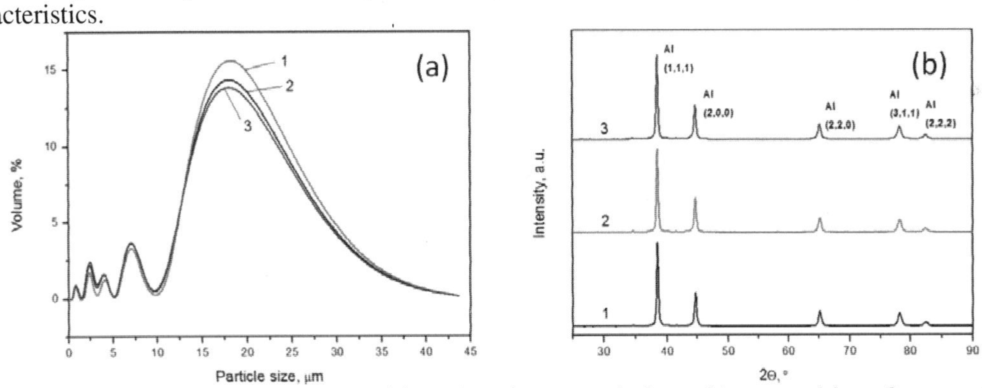

Figure 1. Granulometric size composition (a) and structural phase (b) composition of composite powders. Numbers identify: 1 – 0.05 wt.%; 2 – 0.15 wt.%; 3 – 0.3 wt.%

Peaks that correspond to aluminum were noted. At the same time there are no peaks that correspond to Al/MWCNTs; this is due to small sensibility of X-ray structural method to identifying phases with volume less that 3%.
Selyakov-Scherrer equation was used to calculate the size of coherent-scattering region. Calculation results show that increase of Al/MWCNTs volume in the mixture leads to decrease the size of coherent-scattering region from 65 to 57 nm.

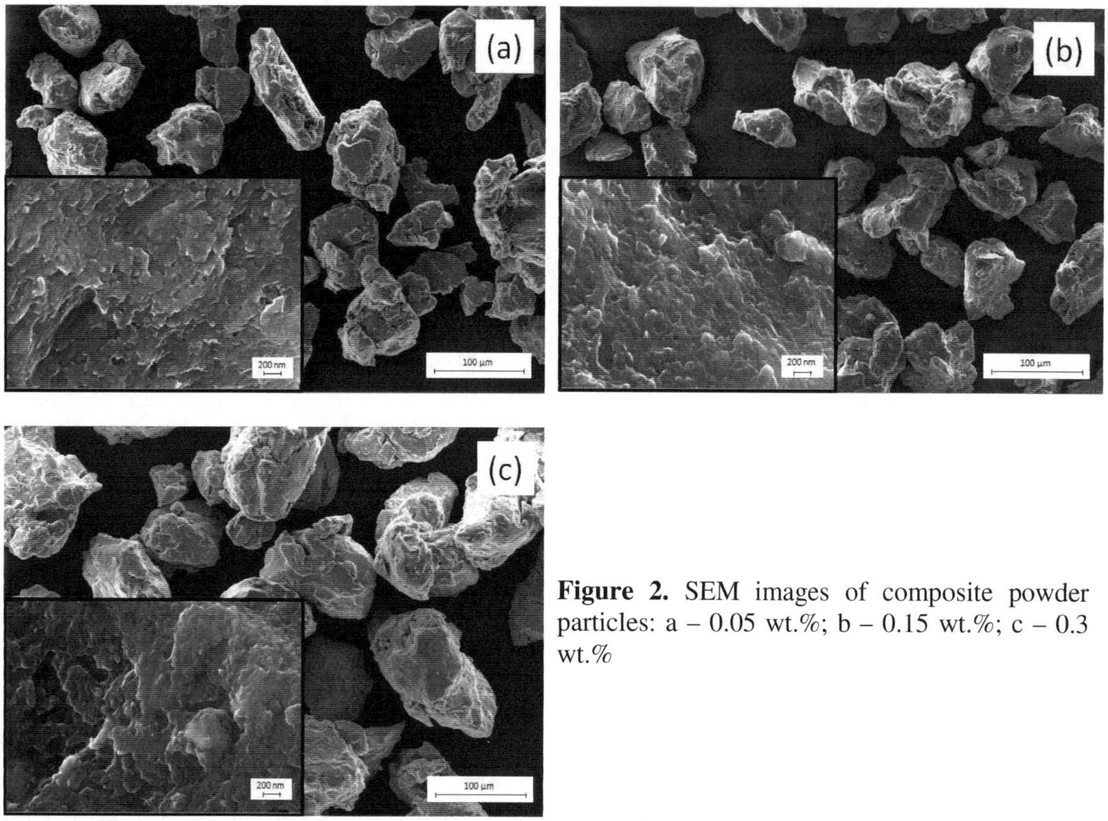

Figure 2. SEM images of composite powder particles: a – 0.05 wt.%; b – 0.15 wt.%; c – 0.3 wt.%

Fig. 2 shows SEM images of powder particles that characterize it's morphology. At higher resolution on the fracture surface shows micro-crystals which sizes correspond to calculated values of coherent-scattering region. On the surface of powder particles no hybrid carbon structures were detected. This is related to its small concentration and specific formations of powder particles during mechanical synthesis [3].

The powders obtained can be used for the preparation of semi-finished products or finished products by the method of thermal deformation processing, and also for the creation of protective coatings by the method of gas-dynamic spraying.

Acknowledgments

The reported study was funded by RFBR according to the research project № 17-58-04048

References

[1] Kobernik N V, Miheev R S, Vaganov V E and Aborkin A V *Voprosy materialovedeniya* 2016 **85** 60–66.

[2] Aleshin N P, Kobernik N V, Mikheev R S, Vaganov V E, Reshetnyak V V and Aborkin A V *Rus. Eng. Res.* 2016 **36** 46–52.

[3] Aborkin A V, Evdokimov I A, Vaganov V E, Alymov M I, Abramov D V and Khor'kov K S *Nanotechnologies in Russia* 2016 **11** 297–304.

Control of propagation of spatially localized polariton wave packets in a Bragg mirror with embedded quantum wells

I.E. Sedova[1], I.Yu. Chestnov[1], S.M. Arakelian[1], A.V. Kavokin[2,3,4] and E.S. Sedov[1,3]

[1]Department of Physics and Applied Mathematics, Vladimir State University named after A. G. and N. G. Stoletovs, Gorky Street 87, 600000, Vladimir, Russia
[2]CNR-SPIN, Viale del Politecnico 1, I-00133, Rome, Italy
[3]School of Physics and Astronomy, University of Southampton, SO17 1NJ Southampton, United Kingdom
[4]Spin Optics Laboratory, St. Petersburg State University, Ul'anovskaya 1, Peterhof, St. Petersburg 198504, Russia

evgeny_sedov@mail.ru

Abstract. We considered the nonlinear dynamics of Bragg polaritons in a specially designed stratified semiconductor structure with embedded quantum wells, which possesses a convex dispersion. The model for the ensemble of single periodically arranged quantum wells coupled with the Bragg photon fields has been developed. In particular, the generalized Gross-Pitaevskii equation with the non-parabolic dispersion has been obtained for the Bragg polariton wave function. We revealed a number of dynamical regimes for polariton wave packets resulting from competition of the convex dispersion and the repulsive nonlinearity effects. Among the regimes are spreading, breathing and soliton propagation. When the control parameters including the exciton-photon detuning, the matter-field coupling and the nonlinearity are manipulated, the dynamical regimes switch between themselves.

1. Introduction

Our nowdays' life is largely built on optoelectronic devices. The most of already existing devices operate separately with their optical and electronic parts. We believe that joining together the photonic and electronic parts will substantially change the face of modern optoelectronics, make new devices more efficient, targeted and successful. To answer this technological challenge, we resort to methods and approaches of Polaritonics, the interdisciplinary research area which is at the heart of modern optoelectronics and quantum light wave technologies.

Among the most demanded and promising optoelectronic devices are the devices for information storage and processing based on macroscopic manipulation by coherent coupled matter-light states [1]. Cavity quantum electrodynamics, the study of strong coherent interactions of quantized cavity electromagnetic field with matter, provides a versatile platform for this goal. Cavity-QED arrays composed of engineered optical cavity modes and few-level oscillators, e.g. excitons in quantum wells (QWs), may serve as many-body systems for the light manipulation [2]. However, the strength of light-matter coupling in microcavities is limited by the number of QWs embedded in each cavity. To overcome this technological limitation, the concept of Bragg polaritons has been proposed [3,4].

IOP Conf. Series: Journal of Physics: Conf. Series **951** (2017) 012009 doi:10.1088/1742-6596/951/1/012009

The Bragg polaritons are eigenmodes of the structure created by incorporation of QWs that are holders of excitons periodically throughout a Bragg mirror representing, in fact, a 1D binary photonic crystal.

Recently it has been shown that the modified Bragg mirror structures can act as hyperbolic dispersion materials for Bragg polaritons [5–7]. The effects of reduction of the group velocity and the strong negative refraction have been predicted in them. Control of the dispersion properties of the structure through manipulation by the radiative properties of excitons has been shown in the linear regime. However, the considered structure can demonstrate strong nonlinear properties in comparison with pure optical nonlinearities achieved [8]. The nonlinearity originated from the repulsive Coulomb interaction of excitons has a defocusing character. The combination of the defocusing nonlinearity and the anomalous dispersion allows one to expect formation of spatially localized Bragg polariton wave packets in the considered structure. Furthermore, the tunability of the dispersion and nonlinear properties of the structure opens the way to the control of dynamics of the polariton wave packets. These issues are addresses in our work.

2. The model of the resonant Bragg polaritons structure

The structure under consideration has been first proposed in [5,6]. It represents a 1D binary photonic crystal composed of alternating semiconductor layers of two types, A and B. The layers differ from each other by their widths and refractive indices, d_A, n_A and d_B, n_B. The structure is periodic with the period $d = d_A + d_B$. Single QWs that are holders of excitons are embedded in the centers of the layers of type A. Schematically variation of the refractive index in the structure growth direction is shown in figure 1. In the mean-field approximations the evolution of the coupled exciton-photon system can be described by the coupled Schrödinger and Gross-Pitaevskii equations for the photon, $\psi \equiv \psi(t,z)$, and the exciton, $\chi \equiv \chi(t,z)$, wave functions

$$i\hbar\partial_t\psi = \hat{H}_{\mathrm{C}}\psi + \hbar\Omega_{\mathrm{R}}\chi, \tag{1}$$

$$i\hbar\partial_t\chi = \left[\hat{H}_{\mathrm{X}} + \alpha n_{\mathrm{X}}|\chi|^2\right]\chi + \hbar\Omega_{\mathrm{R}}\psi \tag{2}$$

with the linear operators $\hat{H}_{\mathrm{C}}$ and $\hat{H}_{\mathrm{X}}$ given by

$$\hat{H}_{\mathrm{C,X}} = -\frac{\hbar^2}{2m_{\mathrm{C,X}}}\frac{\partial^2}{\partial z^2} + V_{\mathrm{C,X}}(z), \tag{3}$$

where m_{C} and m_{X} are the effective masses of photons and bulk excitons, respectively. $V_{\mathrm{C,X}}(z)$ describe potentials for photons and excitons that are periodic in the growth direction, $V_{\mathrm{C,X}}(z) = V_{\mathrm{C,X}}(z+d)$. The constant α is the strength of the exciton-exciton interactions that are responsible for nonlinear processes, n_{X} is the line density of excitons in the growth direction. The parameter Ω_{R} is the exciton-photon coupling constant.

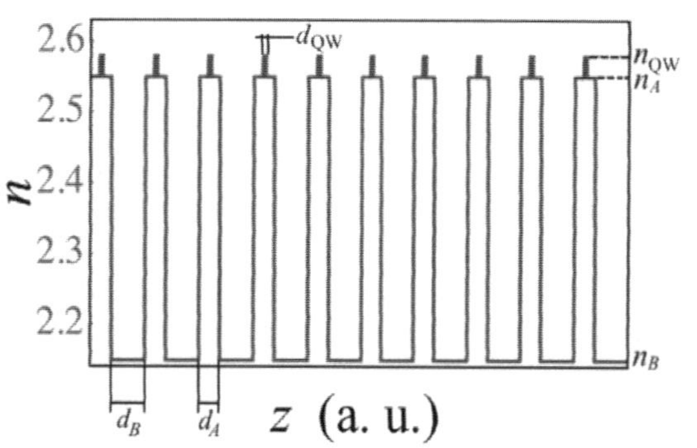

Figure 1. Schematic picture of the refractive index variation along the growth axis (labelled as z axis) of the structure of the Bragg mirror with embedded QWs.

To study the 1D macroscopic dynamics of light under the coupling with the QW excitons in the considered structure, we use one of the perturbation methods that is the method of multiple scales [9]. Following [10–12], we reduce the coupled photon and exciton equations (1) and (2) to the 1D equation for the slowly varying Bragg polariton envelope function. To this end, we introduce different time and space coordinate scales, $t_j = \varepsilon^j t$ and $z_j = \varepsilon^j z$, where ε is a dimensionless fictitious small parameter, $\varepsilon \ll 1$. Importantly, the new variables, t_j and z_j, are now considered independently.

We expand the solution of Eqs. (1) and (2) in the series

$$\psi = \sum_{j=1} \varepsilon^j \psi_j, \quad \chi = \sum_{j=1} \varepsilon^j \chi_j. \tag{4}$$

We now substitute the expansions (4) to Eqs. (1) and (2) provided with the introduction of the new variables, t_j and z_j. The further computational routine is connected with the subsequent analysis of the equations appearing at different orders of the perturbation parameter ε.

We search the first-order photon and exciton WFs in the following form:

$$\psi_1 = \zeta_1 \Xi(\mathbf{T},\mathbf{Z}) e^{-i\Omega_{\nu n}(k)t} \phi_{\nu k}(z), \tag{5}$$

$$\chi_1 = \zeta_2 \Xi(\mathbf{T},\mathbf{Z}) e^{-i\Omega_{\nu n}(k)t} f_{nk}(z), \tag{6}$$

where $\phi_{\nu k}(z)$ and $f_{nk}(z)$ are the eigenfunctions of the linear eigenproblems

$$\hat{H}_C \phi_{\nu k}(z) = \hbar\omega_{C\nu}(k)\phi_{\nu k}(z), \tag{7}$$

$$\hat{H}_X f_{nk}(z) = \hbar\omega_{Xn}(k) f_{nk}(z). \tag{8}$$

The variable k is the quasimomentum of particles that is actually the Bloch wave vector. The functions $\phi_{\nu k}(z)$ and $f_{nk}(z)$ obey the orthonormality conditions

$$\int_d \phi^*_{\nu k}\phi_{\nu' k'} dz = \frac{d}{2\pi}\delta_{\nu\nu'}\delta(k-k'), \quad \int_d \phi^*_{\nu k}\phi_{\nu' k'} dz = \frac{d}{2\pi}\delta_{\nu\nu'}\delta(k-k'). \tag{9}$$

In (5)–(8), ν and n numerate photonic and excitonic energy bands, $\omega_{C\nu}(k)$ and $\omega_{Xn}(k)$ describe dispersions for photons and excitons in the corresponding bands, respectively. $\Omega_{\nu n}(k)$ describes the dispersion of the eigenmodes of the modified Bragg mirror appearing as a result of coupling of photons in the band ν and excitons in the band n. Two modes are supported possessing the dispersion in the form:

$$\Omega_{\nu n}(k) = 0.5\left(\omega_{C\nu}(k) + \omega_{Xn}(k) \pm \sqrt{\delta^2_{\nu n}(k) + 4\omega^2_R(k)}\right), \tag{10}$$

where $\delta_{\nu n}(k) = \omega_{C\nu}(k) - \omega_{Xn}(k)$ is the exciton-photon detuning, $\omega_R(k)$ is the k-dependent splitting of the structure eigenmodes, found as

$$\omega_R(k) = \frac{2\pi\Omega_R}{d}\int_d \phi^*_{\nu k} f_{nk} dz. \tag{11}$$

The function $\omega_R(k)$ is convex and symmetric with respect to $k=0$ that means $\omega_R(0) = \mathrm{Max}\left[\omega_R(k)\right]$. For the simplicity, we make the notation $\omega_R = \omega_R(0)$.

The eigenmodes of the structure referred to as Bragg polaritons [4] belonging to different dispersion branches (11) can be considered independently of each other in different experimental conditions, e.g. in the case of resonant excitation of the lower branch. Below we pay our attention only to the lower-branch polaritons characterized by the eigenfrequency (10) with "–" at the last term.

The amplitude $\Xi(\mathbf{T},\mathbf{Z})$ is a function of slow variables $\mathbf{T} = (\varepsilon t, \varepsilon^2 t)$ and $\mathbf{Z} = (\varepsilon z, \varepsilon^2 z)$. It describes the envelope function of the Bragg polariton wave packet. For the lower-branch polaritons the weighting coefficients $\zeta_{1,2}$ in (5) and (6) are given by

$$\zeta_{1,2} = \mp\frac{1}{\sqrt{2}}\left(1\mp\frac{\delta_{vn}(k)}{\sqrt{\delta_{vn}^2(k)+4\omega_R^2(k)}}\right)^{1/2}. \tag{12}$$

As a result of applying the method of multiple scales, we finally arrive at the nonlinear equation for the Bragg polariton envelope function

$$i\hbar\left(\frac{\partial\Psi}{\partial t}+v_g\frac{\partial\Psi}{\partial z}\right)=\hbar\hat{\Omega}\Psi+g\,|\Psi|^2\,\Psi, \tag{13}$$

where we made the redefinition $\Psi=\varepsilon\Xi$, $\int_{-\infty}^{\infty}|\Psi|^2\,dz=1$. The operator $\hat{\Omega}$ describes the non-parabolic dispersion of polaritons. It is obtained by the change $k\to\hat{k}$ in (10), where $\hat{k}=-i\partial_z$ is the momentum operator. g is the nonlinearity coefficient

$$g=\frac{2\pi\alpha n_X\,|\zeta_2|^4}{d}\int_d|f_{nk}(z)|^4\,dz. \tag{14}$$

3. Bragg polaritons with convex dispersion

We start our consideration with the QW-free structure which represents a conventional 1D photonic crystal. To reveal its dispersion properties, we follow the references [5–7] and use the convenient transfer matrix method [13].

Figure 2a demonstrates dispersions of the three lower photon eigenmodes of the infinite 1D photonic crystal. As a model structure, we consider GaN/$Al_{0.3}Ga_{0.7}N$ structure. We choose the following parameters of the layers: $n_A=2.55$, $d_A=64.8$ nm and $n_B=2.15$, $d_B=115.3$ nm. Since the Bragg reflection condition is not fulfilled for our structure, i.e. $n_Ad_A\neq n_Bd_B$, the second photonic band gap opens. The centre, $\omega_{2\mathrm{Br}}$, and the half-width, $\Delta\omega_{2\mathrm{Br}}$, of the gap are estimated as $\omega_{2\mathrm{Br}}=2\pi c/(n_Ad_A+n_Bd_B)$ and $\Delta\omega_{2\mathrm{Br}}\approx0.5(n_B-n_A)(1-n_Ad_A/n_Bd_B)/(n_A+n_B)$, respectively — see Supplementary material to reference [5]. The second photonic band is convex that provides anomalous dispersion even in the center of the 1st BZ. Keeping the convex shape of the dispersion of eigenmodes of the structure in combination with the defocusing nonlinearity one possible to expect that the structure supports propagation of spatially localized optical wave packets.

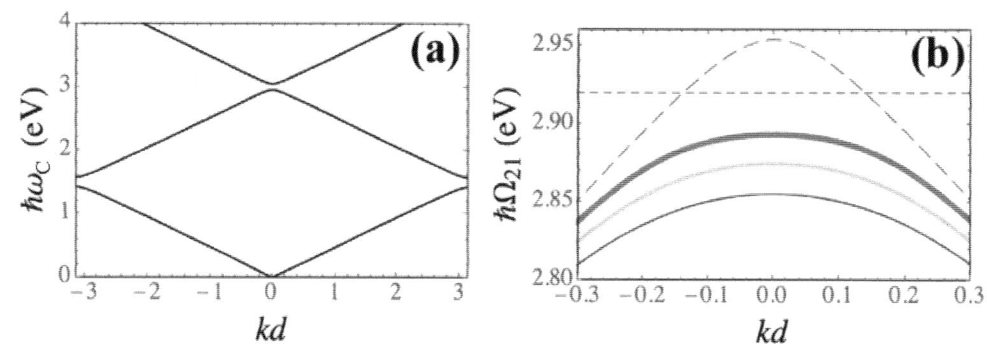

Figure 2. (a) Energy band gap structure of the QW-free photonic crystal. In the considered structure (parameters are given in the text) the second photonic gap opens. (b) Dispersion of the lower branch Bragg polaritons for different values of the exciton-photon coupling strength (from the thickest to the thinnest curves): $\hbar\omega_R$ = 40 meV, 60 meV and 80 meV. The dashed curve and the dashed horizontal line correspond to the dispersions of non-interacting photons and excitons, respectively. The exciton-photon detuning is taken Δ = 35 meV.

Now let us turn to the considered structure with embedded QWs. In order to obtain the structure possessing the convex dispersion of polariton eigenmodes, we choose QWs such that the ground-state ($n=1$) exciton energy is tuned to the vicinity of the top of the second photonic band, $\omega_{X1}=\omega_{C2}(0)-\Delta$, where Δ is the small k-independent exciton-photon detuning, $|\Delta| \ll \omega_{X1}, \omega_{C2}(0)$. Due to a strong confinement of excitons in QWs, ω_{X1} can be considered constant. In our model structure we consider $In_{0.12}Ga_{0.88}As$ QWs. Figure 2b demonstrates dispersion of the lower Bragg polariton mode resulting from coupling of the ϕ_{2k} photon mode and the f_{1k} exciton mode at different exciton-photon coupling strength ω_R.

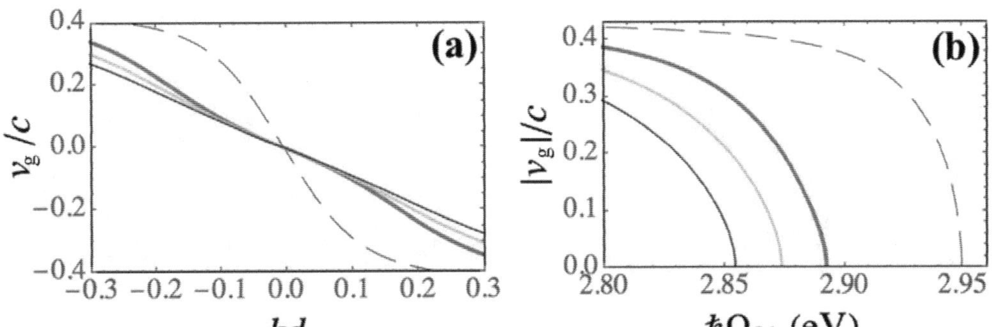

Figure 3. (a) The group velocity of the lower-branch Bragg polaritons in dependence of the polariton quasimomentum and (b) the parametric dependence of the group velocity on the polariton energy for different values of the exciton-photon coupling strength. Notations and values of the parameters are the same as those in figure 2b.

According to the definition, the group velocity of the Bragg polaritons can be found as $v_g=\partial\Omega_{21}/\partial k$ that gives

$$v_g=-i\frac{2\pi\hbar}{d\,m_C m_X}\left(m_X\zeta_1^2\int_d \phi_{2k}^*\partial_z\phi_{2k}dz+m_C\zeta_2^2\int_d f_{1k}^*\partial_z f_{1k}dz\right). \quad (15)$$

Noteworthy, the expression (15) obtained from the definition coincides with that found at the second order of the method of multiple scales. Figure 3a shows the group velocity of the lower-branch Bragg polaritons in dependence of their wave vector for different exciton-photon coupling strengths. Even at a small change of the wave vector the group velocity can change by tens of percent. It is noteworthy that the eigenwaves with positive wave vectors possess a negative group velocity while the eigenwaves with the negative wave vector propagate with the positive group velocity, i.e. the group and phase velocities in the considered structure are anti-parallel [14–15]. The optical waves with parallel group and phase velocities entering the structure from vacuum excite polariton Bloch waves with negative wave vectors that propagate with the positive group velocity. Figure 3b demonstrates the parametric dependence of the absolute value of the group velocity on the polariton energy for different coupling strengths. The variables v_g and Ω_{21} are linked through the polariton quasimomentum k.

4. Nonlinear dynamics of polariton wave packets in the convex-dispersion structure

The dynamics of Bragg polariton wave packets in the considered structure is determined by competition of the dispersion and nonlinearity effects. Summarizing the discussion above, the following options allow one to control the dispersion of Bragg polaritons (see (10)): the exciton-photon detuning Δ and the matter-field coupling strength ω_R. Herewith, if the former can be chosen only at the stage of growth of the structure, the latter can be tuned at any stage by external impact through the tuning of the radiation properties of excitons, see Appendix to reference [7].

IOP Conf. Series: Journal of Physics: Conf. Series **951** (2017) 012009 doi:10.1088/1742-6596/951/1/012009

The nonlinear effects result in the considered structure from the Coulomb repulsion of excitons within each QW. According to (14), the nonlinearity is k-dependent. To estimate this dependence, we use the assumption given above: we consider excitons perfectly localized within each QW and neglect interactions between QWs because of large distance between them. Neglecting the k-dependence of the exciton WF, we combine the coefficients in (14) in a fitting constant $G = 2\pi\alpha n_X d^{-1}\int_d |f_1(z)|^4\, dz$ and re-write (14) in the reduced form

$$g = G|\zeta_2|^4. \tag{16}$$

We also have two options to control the nonlinearity in the considered structure. The first one is the exciton-photon coupling enclosed in $|\zeta_2|^4$ together with the dependence on k. The second option is the intensity of the wave packet proportional to the line density of excitons in the coefficient G.

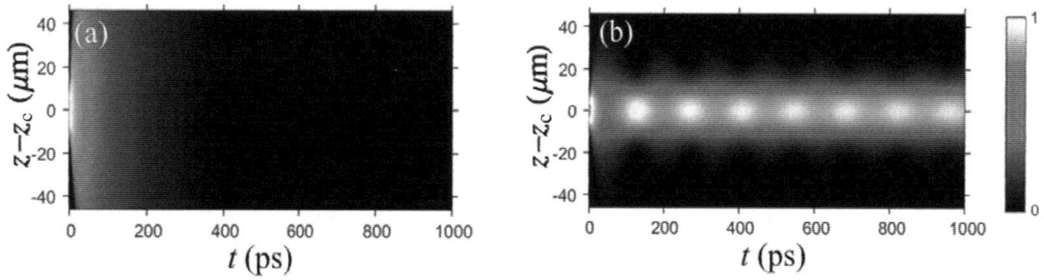

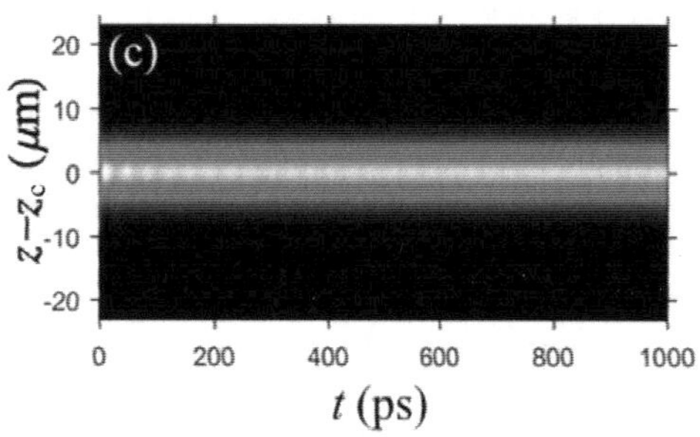

Figure 4. Propagation of the Bragg polariton wave packets in the growth direction of the Bragg mirror with embedded QWs in three dynamical regimes: (a) spreading, (b) breathing and (c) soliton propagation. The coordinate z is dynamically shifted by the position of the centre of the wave packet z_c. The parameters taken for modelling are $G = 3.1$ meV·μm, $\Delta = 35$ meV, and $\omega_R = 32$ meV, 35 meV and 40 meV for (a), (b) and (c), respectively.

In the considered structure, we revealed three dynamical regimes for Bragg polariton wave packets that are spreading, breathing and soliton propagation. Such a set of regimes is typical for dynamical systems. For example, similar regimes have been found in the discrete systems of ultracold atoms [16,17] and conventional atomic polaritons [18,19] with normal dispersion and attractive nonlinearity. Examples of the Bragg polariton wave packet evolution in the considered structure are presented in figure 4. The first regime is one of spreading, when the polariton wave packet blurs in time, see figure 4a. Two other regimes are spatially localized that means that the wave packet width changes in finite limits during propagation. In the breathing regime presented in figure 4b the wave packet moves with an oscillating width. In the soliton regime, the wave packet propagates with unchanged shape. Propagation of the polariton wave packet in the regime close to solitonic is presented in figure 4c. All pictures were obtained for the fixed detuning Δ and the nonlinearity coefficient G. The matter-field coupling constant ω_R is the only control parameter different for all pictures.

5. Conclusions

We have considered the nonlinear dynamics of the Bragg exciton-polaritons in the modified Bragg mirror with embedded periodically arranged QWs. Staring with the Schrödinger and Gross-Pitaevskii equations for photon and exciton wave functions in a spatially periodic potential, we have obtained the generalized nonlinear Gross-Pitaevskii equation with the convex non-parabolic dispersion for Bragg polaritons. We have shown the possibility of formation of localized polariton wave packets under the competition of the effects of the convex dispersion and defocusing nonlinearity. We have revealed three dynamical regimes for polaritonic wave packets including spreading, breathing and soliton propagation. Switching between the regimes can be carried out using the control parameters that are the exciton-photon detuning, the matter-field coupling strength and the intensity of the polariton wave packet. Our results provide a platform for the studies of collective properties of light in stratified nonlinear media.

Acknowledgements
IES acknowledges support from the grant of the President of Russian Federation for state support of young Russian scientists No. MK-8031.2016.2. IYuCh acknowledges funding from RFBR Grant No. 16-32-60102 and from the grant of the President of Russian Federation for state support of young Russian scientists No. MK-2988.2017.2. Work of SMA was supported by the RFBR grant No. 15-59-30406 and the Ministry of Education and Science of the Russian Federation project No. 16.1123.2017/4.6. AVK acknowledges the partial support from the EPSRC Programme grant on Hybrid Polaritonics No. EP/M025330/1 and the HORIZON 2020 RISE project CoExAn (Grant No. 644076). ESS acknowledges support from the RFBR grant No. 16-32-60104.

References
[1] Ed Bruß D and Leuchs G 2007 *Lectures on Quantum Information* (Weinheim: Wiley-VCH Verlag)
[2] Hartmann M J, Brandao F G S L and Plenio M B *Effective Spin Systems in Coupled Microcavities* 2007 *Phys. Rev. Lett.* **99** 160501
[3] Faure S *et al. Relaxation and emission of Bragg-mode and cavity-mode polaritons in a ZnO microcavity at room temperature* 2009 *Appl. Phys. Lett.* **95** 121102
[4] Askitopoulos A, Mouchliadis I, Iorsh I, Christmann G, Baumberg J J, Kaliteevski M A, Hatzopoulos Z and Savvidis P G 2011 *Phys. Rev. Lett.* **106** 076401
[5] Sedov E S, Iorsh I V, Arakelian S M, Alodjants A P and Kavokin A 2015 *Phys. Rev. Lett.* **114** 237402
[6] Sedov E S, Charukhchyan M V, Arakelyan S M, Alodzhants A P, Lee R-K and Kavokin A V 2016 *JETP Letters* **104** 62–67
[7] Sedov E S, Cherotchenko E D, Arakelian S M and Kavokin A V 2016 *Phys. Rev.* B **94** 125309
[8] Barland S *et al.* 2002 *Nature* **419** 699
[9] De Sterke C M and Sipe J E 1988 *Phys. Rev.* A **38** 5149–5165
[10] Paternostro M, Agarwal G S and Kim M S 2009 *New J. Phys.* **11** 013059.
[11] Chen I-H, Lin Y Y, Lai Y-C, Sedov E S, Alodjants A P, Arakelian S M and Lee R-K 2012 *Phys. Rev.* A **86** 023829
[12] Konotop V V and Salerno M 2002 *Phys. Rev.* A **65** 021602
[13] Kavokin A, Baumberg J, Malpuech G and Laussy F 2017 *Microcavities, 2nd ed., Series on Semiconductor Science and Technology* (Oxford: Oxford University Press)
[14] Boardman A D, King N and Velasco L 2005 *Electromagnetics* **2**5, 365
[15] Jiang P, Yang H and Xie K 2014 *Optica Applicata* **44** 17
[16] Trombettoni A and Smerzi A 2001 *Phys. Rev. Lett.* **86** 2353
[17] Wang J-J, Zhang A-X, Zhang K-Z, Ma J and Xue J-K 2010 *Phys. Rev.* A **81** 033607
[18] Sedov E S, Alodjants A P, Arakelian S M, Chuang Y-L, Lin Y, Yang W-X and Lee R-K 2014 *Phys. Rev.* A **89** 033828
[19] Sedov E S, Arakelian S M and Alodjants A P 2014 *Nanosystems: Physics, Chemistry, Mathematics* **5** 234–248

Structural phase composition and effectiveness of gas-dynamic spraying of hybrid coatings based on AlMg2 nanocrystalline matrix reinforced with graphene-like structures and micro-size corundum

A V Aborkin[1], A V Sobol`kov[1], A I Elkin[1] and V E Arkhipov[2]
[1]Vladimir State University named after Alexander and Nikolay Stoletovs, Vladimir, Russia
[2] Mechanical Engineering Research Institute, Russian Academy of Sciences, Moscow, Russia
E-mail: aborkin@vlsu.ru

Abstract. The method of cold gas-dynamic spraying of mechanically synthesized powders based on a nanocrystalline AlMg2 matrix reinforced with graphene-like structures and micro-sized corundum particles was used for obtaining hybrid coatings. A feature of the formed coatings is the presence of a two-level micro- and nanocomposite structure. It was found that an increase in the content of corundum microdimensional particles in the mixture from 10 to 30% by weight contributes to an increase in the thickness of the coating obtained at the same time by a factor of 2 from 140 to 310 μm. Further increase in the content of a mixture of micron-sized corundum particles to 50% by weight leads to a decrease in the thickness of the coating formed to 40 μm. The resulting coatings correspond to a high microhardness, varying depending on the composition in the range from 1.7 GPa to 3.2 GPa. The high hardness of the coatings is due to the increase in the hardness of the matrix material due to the creation of a nanocomposite structure, which increases the strength of fixing micro-sized corundum particles therein, improving the characteristics of the heterogeneous coating as a whole.

1. Introduction

Composite coatings based on aluminum [1-5], taking into account their high corrosion resistance, are in demand in engineering when creating the specified functional properties of the working surfaces of structural elements and machine parts. The main disadvantage of composite coatings based on aluminum and its alloys, which suppresses their wider use, are low wear resistance and mechanical properties.

Performance increase possibilities of functional properties of aluminum matrix based composite coating reinforced with only one type of filler are by now almost used up. Creation of hybrid nanocrystalline matrix based coatings reinforced with nano- and micro-sized fillers opens new opportunities of increasing and forming desired complex of working properties of element surfaces of constructions and machine parts.

IOP Conf. Series: Journal of Physics: Conf. Series **951** (2017) 012010 doi:10.1088/1742-6596/951/1/012010

2. Results and discussion

The objective of this work is to research on the structural phase composition and gas-dynamic spray coating of hybrid AlMg2 nanocrystalline matrix based coatings reinforced with graphene-like structures and micro-sized corundum.

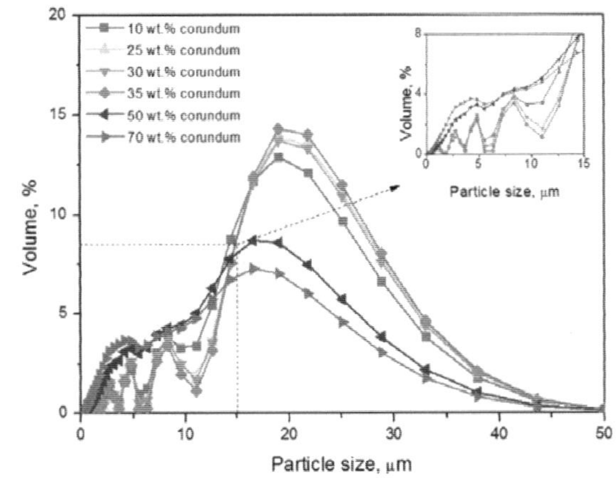

Figure 1. The granulometric composition of the hybrid powder

Powder for gas-dynamic spray coating was manufactured by mechanic synthesis in globular planetary mill during two technological conversions. During the first one a nanocomposite powder AlMg2+1 wt.% graphene-like structures was produced [6]. During the second conversion 10-70 wt.% of corundum was added to the produced nanocomposite powder and treatment was continued under same parameters [7].

A study of the granulometric composition of the powder obtained in the first redistribution shows that the particle size of the mechanically synthesized powder does not exceed 300 μm. The average particle size of the powder is 60 μm. The powder particles have a nanocomposite structure (a nanocrystalline matrix reinforced with graphene-like structures located between the crystallites of the matrix material).

After the second redistribution, the granulometric composition of the hybrid powder changes. Fig. 1 presents data on the granulometric composition of a mechanically synthesized hybrid powder containing 10-70% by weight of corundum.

Figure 2. SEM image of a hybrid powder containing: a - 10% by weight of corundum; b - 30% by weight of corundum; c -50% by weight of corundum. The numbers indicate: 1 - agglomerates; 2 - particles of corundum.

IOP Conf. Series: Journal of Physics: Conf. Series **951** (2017) 012010 doi:10.1088/1742-6596/951/1/012010

An analysis of the data obtained shows that the introduction of 10-35% by weight of corundum into the composite powder obtained in the first repartition, and mechanical treatment even for 15 minutes exert a considerable influence on the average size of the formed particles, which decreases by more than 3 times and amounts to ~ 17 μm. A further increase in the corundum content from 50 to 70% by weight leads to a decrease in the average particle size, which is 11 and 8 μm, respectively.
In Fig. 2 shows the SEM image of a hybrid powder with different corundum content. The study of the morphology of particles of a mechanically synthesized hybrid powder showed that the powder obtained is a mechanical mixture.
The mechanical mixture consists of agglomerates and separately located corundum particles. The agglomerates are particles of the matrix material AlMg2, containing graphene-like structures with corundum particles embedded in them, as well as surface particles. It should be noted that the use of powders in the form of agglomerates for gas-dynamic spraying can ensure the production of hybrid coatings with uniformly distributed nano- and micro-sized reinforcing particles.
The analysis of X-ray diffraction study results of different powder compositions were conducted on Bruker D8 Advance X-ray diffractometer and showed that diffractograms have similar characters. Peaks corresponding to aluminum and corundum were noticed. Calculation of coherent-scattering region performed using Selyakov-Scherrer equation showed that matrix material has a nanocomposite structure (medium size of crystals ~60 nm).
Manufactured powders were sprayed on mounts from sheets of AlMg2 (120±15 HV) that had a 50 μm thick cladding layer of clean aluminum (45±10 HV) on its surface. For gas-dynamic spraying of manufactured powders a DIMET-404 machine was used. Spraying mode was stationary (point spraying) at the air flow temperature 270°C, exposing time 15 seconds.
After comparing results of powder and manufactured hybrid coatings X-ray diffraction study it can be stated that during gasdynamic splaying phase composition doesn't change. At the same time there is a 10–20% decrease of half-widths of peaks that correspond to matrix material after spraying; this proves size increase of coherent-scattering region meaning coarsening of fine structure. Median size of manufactured coatings crystallites was ~85 nm.

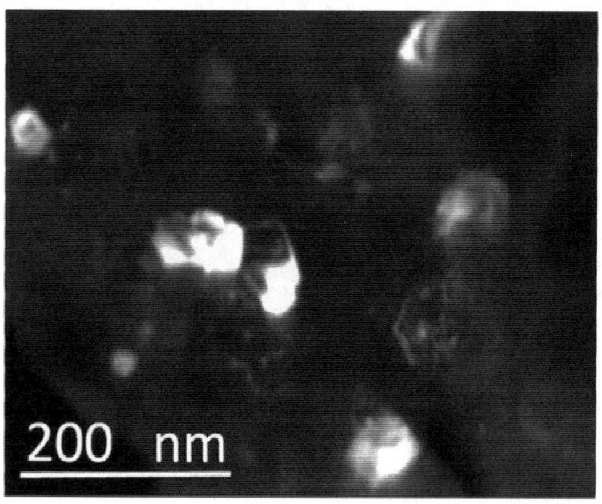

Figure 3. Dark-field TEM image of hybrid coating structure AlMg2+1 wt.% graphene-like structures + 30 wt.% corundum

Study of hybrid coating manufactured using straying powder consisting 30% volume of corundum that was conducted using electric microscopes (JEM-2010) shows that the coating has nanocrystalline structure.
Typical sizes of crystallites visible on TEM images are less that 100 nm (see fig. 3). The analysis of those images proves absence of preferred orientation of aluminum particles and it's oxide to each other and orientation dependencies between them. Thin structure analysis shows presence of areas between aluminum crystals that correspond to graphene-like structures (sizes no more than 20 nm) with strongly distorted layers.
Effectiveness of spray coating was based on incremental value of the surface and calculated as a difference of mount mass before and after spraying. The analysis of results shows that the lost effective spraying is at 30% of corundum in the powder mixture. Further increase of corundum leads to decrease of effectiveness of gas-dynamic spraying.
Studying sections of hybrid coatings shows that during gas-dynamic spraying a considerable change of microgeometry of mount's cladding layer is happening; it is determined by deformational and erosive impact of sprayed powder (see fig. 4).

Increasing the volume of micro-sized particles of corundum in the mixture from 10 to 30% helps increasing the coating thickness produced during the same amount of time 2 times, from 140 to 310 μm. Further increase of corundum volume in the mixture from 35 to 50% leads to decrease of the thickness down to 150 and 40% respectively. With 70 wt.% of corundum in the mixture surface thickness does not exceed 20 μm, in addition a solid coating is not formed since erosive processes become predominate.

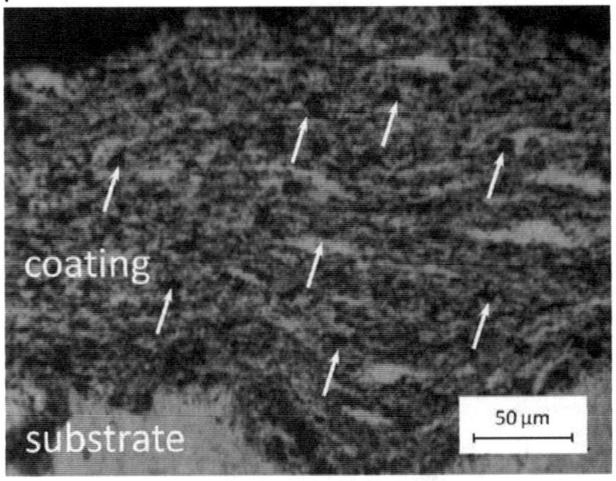

Figure 4. Microstructure of hybrid coating AlMg2+1 wt.% graphene-like structures + 30 wt.% corundum. Arrows indicate some micro-sized particles of corundum.

Metallographic examination of sections shows that independently from the corundum composition the formed surfaces are pretty solid with equally distributed particles of corundum (median size 3-5 μm) on the surface thickness. Nevertheless compacting (pressing) cracks can be noticed on the surface; it's lengths decreases with increase of corundum composition from couple hundreds to ten micrometers. Study of "coating – cladding plate" interface shows increase of corundum concentration in that zone; it's related to specific character of surface growth process initiation during gas-dynamic spraying on mounts with low firmness and high plasticity.

Acknowledgments
The reported study was funded by RFBR according to the research project № 16-48-330156.

References

[1] Alkhimov A P, Kosarev V F and Papyrin A N 1990 *Sov. Phy. Dokl.* **35** 1047–49.
[2] Arkhipov V E, Londarskii A F, Moskvitin G V, Mel'shanov A F and Pugachev M S 2013 *Journal of Machinery Manufacture and Reliability* 2013 **42** 495–99.
[3] Arkhipov V E, Londarskii A F, Moskvitin G V, Pugachev M S and Shirokova N V *Rus. Eng. Res.* 2017 **37** 321–325.
[4] Oryshchenko A S and Gerashchenkov D *Inorganic Materials: Applied Research* 2016 **7** 863–67.
[5] Shkodkin A, Kashirin A, Klyuev O and Buzdygar T *J. Therm. Spray Techn.* 2006 **15** 382–86.
[6] Aborkin A V, Evdokimov I A, Vaganov V E, Alymov M I, Abramov D V and Khor'kov K S *Nanotechnologies in Russia* 2016 **11** 297–304.
[7] Khrenov D S, Sobol'kov A V, Elkin A I, Aborkin A V, Sytschev A E *Explosive Production of New Materials: Science, Technology, Business, and Innovations* 2016 74-5.

Analysis of the temperature dependence of the luminescence spectra of liquid-crystal nanocomposites with a cadmium selenide quantum dots

K R Karimullin[1,2], M A Mikhailov[2], M G Georgieva[2], K A Magaryan[1,2] and I A Vasilieva[2]

[1] Molecular Spectroscopy Dept., Institute for Spectroscopy RAS, 5 Fizicheskaya Str., Troitsk, Moscow, 108840, Russia.
[2] Shpol'sky Chair of Theoretical Physics, Moscow Pedagogical State University, 29/7 Malaya Pirogovskaya Str., Moscow, 119435, Russia.

E-mail: kamil@isan.troitsk.ru

Abstract. A theoretical model is proposed for describing the interaction of excitons with phonons and defect states in semiconductor nanocrystals. On the basis of this model, numerical simulation was performed and experimental data on the temperature dependence of the absorption and fluorescence spectra of cadmium selenide quantum dots in liquid crystal solid solutions of cadmium caprylate was discussed.

1. Introduction

Nanocomposites which combine the properties of semiconductor nanocrystals and the material of a solid matrix are widely used in various fields of modern science and technology. Using materials with semiconductor colloidal quantum dots, efficient and economical light-emitting devices are produced (e.g., Samsung's QD-LED technology), as well as various luminescent coatings and labels, and spectral devices [1]. Among the current developments, one can emphasized the search for new gain media for laser generation, sources of nonclassical light for quantum optics and informatics, elements for photovoltaic devices and nanoelectronics. For example, the modeling of collective processes in a dense ensemble of preliminary excited quantum dots demonstrates the possibility of forming short pulses of an electromagnetic field with an extra wide spectrum [2]. In addition, fluorescent labels and markers based on semiconductor quantum dots are highly demanded in biology, medicine, and various diagnostic methods as multiparameter nanosensors, since their spectral properties (e.g., luminescence kinetics) are essentially determined by the parameters of the local field, depend on the thermodynamic parameters of the system and external fields (see, e.g., [3]). In particular, quantum dots are used as temperature sensors [4]. For practical use of nanocomposite materials, a detailed study of their photophysical properties and spectral-luminescent characteristics is necessary. One of the important problems in this area is the investigation of the nature of the fluorescence blinking of single quantum dots [5, 6]. To use semiconductor nanocrystals as temperature sensors, a fundamental study of the dynamical processes that determine the photophysical properties of quantum dots and their relationship to sample temperature is necessary.

The temperature dependences of the absorption and fluorescence spectra of various quantum dots, including CdSe, CdS, InAs, etc., placed in crystalline solutions, have been investigated experimentally in a number of papers (see, e.g., [7-10]). Therefore, it is of interest to obtain a theoretical model describing the interaction of excitons with phonons and defect states in nanocrystals that would allow one to analyze the redistribution of the intensity in the fluorescence spectra of cadmium selenide (CdSe) quantum dots in liquid crystal solutions of cadmium caprylate (CdC_8) with a temperature change from 77 K to 300 K.

2. Theoretical model

Conjugate absorption and luminescence spectra of liquid crystal solutions of cadmium caprylate with 1.8 nm and 2.3 nm cadmium selenide quantum dots in a wide range of temperatures have been experimentally investigated in [10-12]. Two exciton maxima were observed in the absorption spectra of samples with 2.3 nm cadmium selenide quantum dots (see Fig. 1).

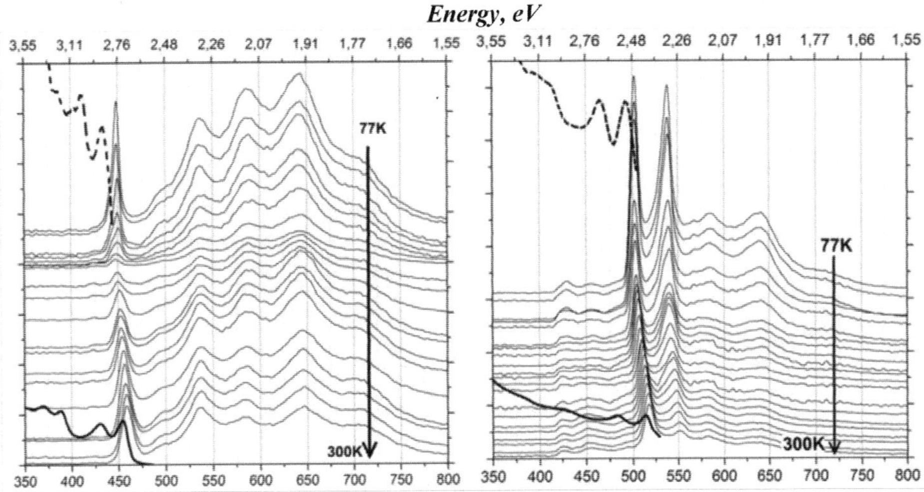

Figure 1. Fluorescence (solid curve), absorption (bold curve) and fluorescence excitation (dotted curve) spectra of samples with 1.8 nm (left image) and 2.3 nm (right image) CdSe quantum dots in the temperature range 77-300 K.

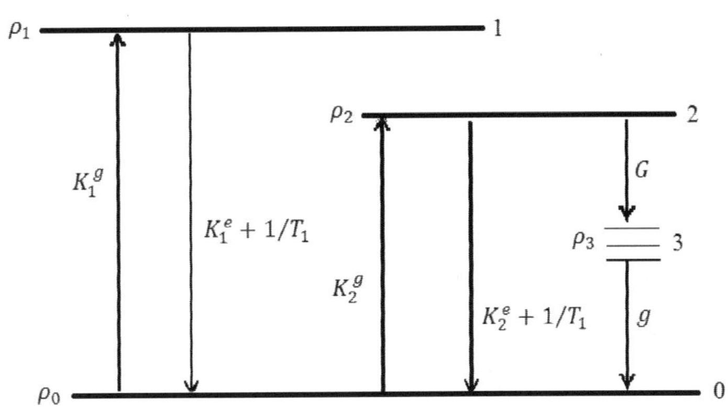

Figure 2. Scheme of energy levels and quantum transitions for the system under study.

To describe the quantum states and transitions in the samples, the energy scheme shown in Fig. 2 was proposed. In the scheme, *0* denotes the ground state of the crystal with a population ρ_0 corresponding to the Fermi energy. The figure shows only the electron states, the states of the holes correspond to energy levels are symmetric to the energy levels of the electrons relative to the ground level. In the absence of excitation $\rho_0 = 1$. *1* and *2* denote two exciton states with populations ρ_1, ρ_2 correspondingly. The absorption of external electromagnetic radiation originates from the ground state *0* into exciton states *1*, *2* with absorption coefficients (transition rates) K_1^g, K_2^g which explains the occurrence of a broad inhomogeneously broadened absorption band with two exciton maxima.

The radiation from states *1* and *2* is due to spontaneous transitions with a rate T_1^{-1} (T_1 is the eigen lifetime of the exciton state), and induced transitions with the rates K_1^g, K_2^g correspondingly. These transitions form a luminescence band with two maxima. Such absorption and luminescence bands are observed for a sample with 2.3 nm quantum dots. For a sample with 1.8 nm quantum dots only one maximum corresponding to the interband recombination of the *e-h* pair in the luminescence band [12] and four exciton maxima in the absorption band were detected. An explanation of this effect will be given below.

Transitions from state *2* to the ground state are also possible through a series of intermediate defect states. Such transitions lead to an additional decay of level *2* with a total rate *G*. They form wide intense bands in the red region of the fluorescence spectrum. In the model, we neglect transitions between the exciton states *1* and *2*. Such transitions was not experimentally observed in the spectra and are unlikely because of low volume concentration of excitons in the sample. These transitions can be further taken into account in the theory, but they have virtually no effect on the final conclusions.

The system of balance equations for the scheme shown in Fig. 2, can be written as:

$$\begin{aligned}\dot{\rho}_0 &= -(K_1^g + K_2^g)\rho_0 + \left(K_1^e + \frac{1}{T_1}\right)\rho_1 + \left(K_2^e + \frac{1}{T_1}\right)\rho_2 + g\rho_3,\\ \dot{\rho}_1 &= K_1^g\rho_0 - \left(K_1^e + \frac{1}{T_1}\right)\rho_1,\\ \dot{\rho}_2 &= K_2^g\rho_0 - \left(K_2^e + \frac{1}{T_1}\right)\rho_2 - G\rho_2\\ \dot{\rho}_3 &= G\rho_2 - g\rho_3.\end{aligned} \tag{1}$$

Putting all derivatives with respect to time in the (1) equal to zero, one can find the steady-state solution:

$$\rho_0 = \frac{1}{1 + \frac{K_1^g}{\Gamma_1} + \frac{K_2^g}{\Gamma_2}\left(1 + \frac{G}{g}\right)} \text{(a)}, \quad \rho_1 = \frac{K_1^g/\Gamma_1}{1 + \frac{K_1^g}{\Gamma_1} + \frac{K_2^g}{\Gamma_2}\left(1 + \frac{G}{g}\right)} (b),$$

$$\rho_2 = \frac{K_2^g/\Gamma_2}{1 + \frac{K_1^g}{\Gamma_1} + \frac{K_2^g}{\Gamma_2}\left(1 + \frac{G}{g}\right)} (c), \quad \rho_3 = \frac{\frac{G}{g}(K_2^g/\Gamma_2)}{1 + \frac{K_1^g}{\Gamma_1} + \frac{K_2^g}{\Gamma_2}\left(1 + \frac{G}{g}\right)} \quad (d), \tag{2}$$

here $\Gamma_1 = K_1^e + \frac{1}{T_1}$, $\Gamma_2 = K_2^e + \frac{1}{T_1} + G$. (3)

If pumping is performed only at the *0*↔*1* transition, i.e. $K_2^g = 0$, then we obtain a model of a two-level atom. The interaction of a two-level atom with external electromagnetic radiation can be described using a system of optical Bloch equations [13, 14]. From such equations one can proceed to the balance equations for the populations of states of a two-level atom [14]. The system (1) is a generalization of such a system of equations for the case of the presence of two exciton states and defect levels. In this case the zero-phonon line of the Lorentz profile in Bloch's theory is an inhomogeneously broadened exciton maximum. Comparison with a two-level atom is justified, since for quantum dots spectroscopic phenomena specific of a two-level atom are observed [15-17]. In particular, exciton resonance absorption lines with the Lorentz profile are observed [15]. But there are differences. As can be seen from Fig. 1, the emission and absorption bands are not resonant. Therefore, for the transition rate, the upper indices *e* (radiation) and *g* (absorption) are introduced.

Suppose that the pump is only on the *0*↔*1* transition, and, as is done in the Bloch theory

$$K_1^g = K_1^e = K_1 = 2\chi^2\frac{\gamma_1}{{\Delta_1}^2 + (\gamma_1)^2}, \tag{4}$$

we find that the shape of the absorption band is described by the Lorentz profile

$$\rho_1(\Delta_1, t \to \infty) = I_1^g(\Delta_1) = 2\chi^2\frac{T_1\gamma_1}{{\Delta_1}^2 + (\gamma_1)^2 + 4\chi^2(T_1\gamma_1)} \tag{5}$$

with a half-width $\Delta_{1/2} = 2\gamma_1$. (6)

In formulas (5-6) $\Delta_1 = \Omega_1 - \omega_l$ – is the difference between the frequency of excitonic excitation and the frequency of electromagnetic radiation (detuning), $\chi = (\vec{d}\vec{E})/\hbar$ – is the Rabi frequency, $\vec{d}$ – is the transition dipole moment, $\vec{E}$ – is the electric field. The value of γ_1 determines the half-width of the inhomogeneously broadened absorption maximum. Similar results one can obtain for the case of pumping at the $0 \rightarrow 2$ transition and $G = 0$, i.e. there are no intermediate defect states. In the case of simultaneous pumping of two transitions, two exciton maxima appear in the absorption spectrum. Thus, if $K_1^g \neq 0$ ($K_2^g = 0$) and $K_1^g = K_1^e = K_1$, then we have a model of a two-level atom. In this case, the absorption and luminescence bands are resonant to each other and have a Lorentz profile. In our case, the absorption and luminescence lines are not resonant to each other, i.e. there is a Stokes shift [18].

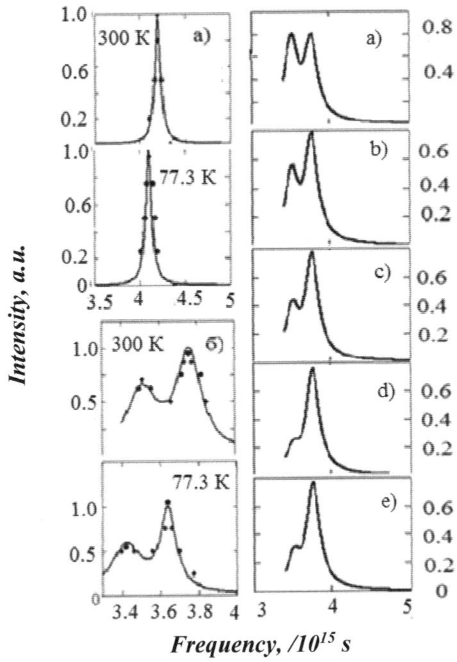

Figure 3. Calculated luminescence bands (solid curve) and experimental data (points) [10-12] for a sample with a) 1.8 nm and b) 2.3 nm quantum dots at 77.3 K and 300 K (on the left). Modeling the redistribution of intensity in the luminescence spectrum for a sample with 2.3 nm quantum dots for different values of GT_1 and G/g: a) 0, 0; b) 0.5, 0.5; c) 1, 1; d) 3, 3; e) 3, 1 (on the right).

The results of a numerical calculation of the absorption and luminescence bands for 1.8 nm and 2.3 nm CdSe quantum dots at two different temperatures are shown in Fig. 3. The half-width of the inhomogeneous luminescence band for a sample with 1.8 nm quantum dots was 0.5×10^{14} s^{-1} and 0.4×10^{14} s^{-1} at temperatures of 77.3 K and 300 K correspondingly. The luminescence bands for a sample with 2.3 nm quantum dots of size at given temperatures are represented by the sum of two Lorentzians with half-widths $(0.5\div1)\times10^{14}$ s^{-1}. Calculation for a sample with 1.8 nm quantum dots at 77.3 K indicates that the luminescence band contains one exciton maximum, while the absorption band contains two exciton maxima. The absorption band is described by Lorentz profiles with a half-width of 0.5×10^{14} s^{-1} and 0.8×10^{14} s^{-1}.

The obtained values indicate a strong inhomogeneous broadening of spectra, which is practically independent on temperature. We associate it with the distribution of quantum dots sizes. The width of homogeneous luminescence bands depends strongly on temperature, but against a background of strong inhomogeneous broadening, this effect is practically not manifested. In this case, the integral intensity of the luminescence bands and the redistribution of the intensity between the luminescence maxima are noticeably manifested with a simultaneous increase in the intensity of broad bands in the red spectral region, which are associated with the defect states of the nanocrystals. These effects can be explained on the basis of the proposed balance equations (1) due to the temperature variation of the corresponding transition rates and changes in the level populations. The temperature increase leads to a significant increase in the value of *G* and an increase in the probability of transitions from the exciton level to the defect states. Consequently this leads to a decrease of the stationary values of the population of the exciton states and an increase of the population of the defect states.

Consider the effects of redistribution of intensity in the lines of luminescence. The luminescence band is proportional to the sum of the populations of the exciton states:

$$I^e(\omega) \propto \rho_1 + \rho_2 = \frac{K_1^g/\Gamma_1 + K_2^g/\Gamma_2}{1+\frac{K_1^g}{\Gamma_1}+\frac{K_2^g}{\Gamma_2}\left(1+\frac{G}{g}\right)}. \qquad (7)$$

In accordance with (7) and the scheme in Fig. 2, two exciton maxima should be observed in the luminescence band. Since $K_1^g \cong K_2^g$ the ratio of the intensities of the luminescence maxima

$$I_2^e(\omega)/I_1^e(\omega) \sim \rho_2/\rho_1 \sim \Gamma_1/\Gamma_2 \qquad (8)$$

Consider the case when fast transitions from the second exciton state to defective intermediate states play an important role. In this case, the population of the second exciton state ρ_2 decreases, according to (2*c*), and the second exciton maximum in the luminescence band disappears. Wide intense fluorescence bands are observed in the red region of the spectrum. Physically, this means that, in spite of the fact that the absorption band of the second exciton continues to exist and the level *2* is intensely populated, radiative and nonradiative transitions to intermediate states practically instantaneously occur from it. In this case, the population ρ_3 can be much larger than the population ρ_2 due to the factor G/g in the formula (2*d*). Such a situation was observed experimentally for a sample with 1.8 nm CdSe quantum dots [10-12]. Here we observe four exciton maxima in absorption band and one exciton maximum in luminescence band. If G/g is small then there are two exciton luminescence maxima in the fluorescence spectrum, along with wide intense lines in the red region. Such a situation was observed for 2.3 nm CdSe quantum dots.

The redistribution of the intensity in the luminescence spectra is shown on the right-hand side of Fig. 3. The product GT_1 significantly influences the intensity distribution between two exciton luminescence bands. An increase in the values of the ratio G/g leads to a total decrease in the intensity of the two bands of exciton luminescence (Fig. 3 *d*, *e*), and an increase in the intensity of the fluorescence bands in the red region of the spectrum associated with transitions to the intermediate states *3*.

Thus, it is shown that the redistribution of the intensity in the fluorescence spectra for a system whose quantum states are depicted in Fig. 2, is completely determined by two parameters GT_1 and G/g.

3. Results and discussion

In Fig. 2 and in the system of equations (1), we did not take into account the decays of the exciton state *1* into defect states, since the luminescence corresponding to the 1→0 transition is observed for all temperatures and for two types of samples. Transitions from the exciton level *1* to defect states are less likely than similar transitions from level *2*. They can be easily taken into account in theory and do not affect its basic positions.

In the model we used a level scheme with two exciton states (Fig. 2). It can also be extended to a greater number of exciton states. Four maxima in the absorption band for a sample with 1.8 nm quantum dots are associated with four exciton states. The luminescence lines corresponding to the decay of the three exciton states, in our case, are suppressed by transitions to defect states. The exciton luminescence band consists of a single maximum.

Fig. 3 shows the displacement of the exciton luminescence bands in the red spectral region with temperature increasing. This displacement can be due to the manifestation of exciton-phonon interaction. In the first approximation, phonons can be described by a set of independent linear oscillators. In this case, for high temperatures ($kT \gg h\nu$), the width of the phonon zone located near the ground state of the crystal increases linearly with temperature, i.e.

$$\Delta E_{ph} \propto k\Delta T\,. \tag{9}$$

Transitions from the exciton states *1* and *2* (Fig. 2) are carried out to the upper region of the phonon zone, which is near the Fermi level. Accordingly, the luminescence line is shifted to the red side by an amount proportional to ΔE_{ph}. The effect of the size of the sample under investigation on this shift and crystal lattice defects can be taken into account by introducing the parameter α, which in our approximation does not depend on the temperature T.

Taking into account these remarks we obtain the following temperature dependence of the luminescence energy:

$$E = E_0 - \alpha kT, \tag{10}$$

where E_0 corresponds to the energy at some fixed temperature T_0 (in our case, 70 K).

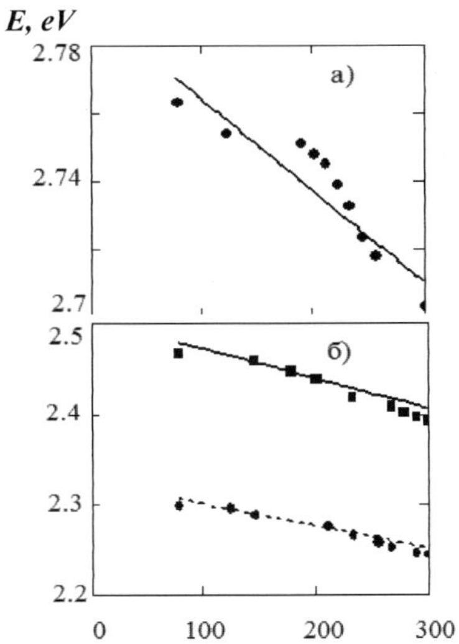

Figure 4. Experimental data [10-12] (points) and result of calculation (solid lines) for the temperature dependence of the position of the luminescence bands: a) luminescence spectrum for a sample with 1.8 nm quantum dots (α = 3.13, E_0 = 2.79 eV), b) two luminescence maxima for a sample with 2.3 nm quantum dots (α = 3.84, E_0 = 2.5 eV, α = 2.82, E_0 = 2.33 eV).

The result of calculation of the position of the luminescence bands maxima according to (10) with the corresponding parameters E_0 and α for samples with 1.8 nm and 2.3 nm quantum dots is shown in Fig. 4 together with the experimental data.

4. Conclusion

It is shown that the simulation of $K^{e,g}$ functions by Lorentz profiles, allows one to describe the shape of inhomogeneously broadened absorption bands and exciton luminescence bands. The introduction of two or more exciton states into the scheme of the levels of the system under study makes it possible, on the basis of the balance equations and the assumption of the Lorentz profile of the exciton band, to describe the redistribution of the radiation intensity in the exciton luminescence spectra with a change in the temperature and size of the quantum dots of the sample. These equations can be generalized to study the effects of blinking fluorescence of single semiconductor nanocrystals [19]. It is shown that the effect of phonons on the dynamics of the system under study leads to a shift of the bands of exciton luminescence to the red spectral region with increasing temperature. The numerical calculation of the temperature dependence of the luminescence spectra in the framework of the proposed model is consistent with the experimental results.

This work was supported by the Russian Science Foundation (project No 14-12-01415 – study of temperature dependencies of photophysical properties of quantum dots). K. Karimullin and K. Magaryan thank for support the grant of the President of the Russian Federation (project No. MK-342.2017.2 – spectroscopy of a new materials based on quantum dots).

References

[1] Bao J and Bawendi M G 2015 A colloidal quantum dot spectrometer *Nature* 2015 **523** 67-70
[2] Shesterikov A S, Gubin M Yu, Gladush M G and Prokhorov A V 2017 Formation of plasmon pulses in the cooperative decay of excitons of quantum dots near a metal surface *J. Exp. Theor. Phys.* **124** 18-31
[3] Aubret A, Pillonnet A, Houel J, Dujardin C and Kulzer F 2016 CdSe/ZnS quantum dots as sensors for the local refractive index *Nanoscale* **8** 2317-25
[4] Zhang F J, Wang B, Pang F F and Wang T Y 2011 A luminescent temperature sensor based on a tapered optical fiber coated with quantum dots *Proc. SPIE* **7990** 79900R
[5] Eremchev I Yu, Osad'ko I S and Naumov A V 2016 Two mechanisms of fluorescence intermittency in single core/shell quantum dot *J. Phys. Chem.* **120** 22646-52
[6] Shchukina A L, Eremchev I Yu and Naumov A V 2015 Looking at a blinking quantum emitter through time slots: the effect of blind times *Phys. Rev. E* **92** 032102
[7] Ortner G, Yakovlev D R, Bayer M, Rudin S, Reinecke T L, Fafard S, Wasilewski Z and Forchel A 2004 Temperature dependence of the zero-phonon linewidth in InAs⁄GaAs quantum dots *Phys. Rev. B* **70** 201301(R)
[8] Valerini D, Creti A, Lomascolo M, Manna L, Cingolani R and Anni M 2005 Temperature dependence of the photoluminescence properties of colloidal CdSe⁄ZnS core/shell quantum dots embedded in a polystyrene matrix *Phys. Rev B* **71** 235409
[9] Favero I, Berthelot A, Cassabois G, Voisin C, Delalande C, Roussignol Ph, Ferreira R and Gerard J M 2007 Temperature dependence of the zero-phonon linewidth in quantum dots: An effect of the fluctuating environment *Phys. Rev. B* **75** 073308
[10] Magaryan K A, Mikhailov M A, Vasilieva I A, Karimullin K R and Klimusheva G V 2014 Temperature dependence of the luminescence spectra of liquid crystal composites with CdSe quantum dots *Bull. Russ. Acad. Sci. Phys.* **78** 1336-40
[11] Magarian K A, Fedyanin V V, Karimullin K R, Vasilieva I A and Klimusheva G V 2013 Luminescence properties of CdSe nanocrystallites in cadmium alkanoate glasses *J. Phys.: Conf. Ser.* **478** 012007
[12] Magaryan K A, Mikhailov M A, Karimullin K R, Knyazev M V, Eremchev I Y, Naumov A V, Vasilieva I A and Klimusheva G V 2016 Spatially-resolved luminescence spectroscopy of CdSe quantum dots synthesized in ionic liquid crystal matrices *J. Luminescence* **169** 799-803
[13] Allen L and Eberly J H 1975 *Optical resonance and two-level atom*s (New York: Wiley)
[14] Osad'ko I S 2002 *Selective spectroscopy of single molecules* (Berlin: Springer)
[15] Högele A, Seidl S, Kroner M, Karrai K, Warburton R J, Gerardot B D and Petroff P M 2004 Voltage-controlled optics of a quantum dot *Phys. Rev. Lett.* **93** 217401
[16] Kroner M, Rémi S, Högele A, Seidl S, Holleitner A W, Warburton R J, Gerardot B D, Petroff P M and Karrai K 2008 Resonant saturation laser spectroscopy of a single self-assembled quantum dot *Physica E* **40** 1994-6
[17] Muller A, Flagg E B, Bianucci P, Wang X Y, Deppe D G, Ma W, Zhang J, Salamo G J, Xiao M and Shih C K 2007 Resonance fluorescence from a coherently driven semiconductor quantum dot in a cavity *Phys. Rev. Lett.* **99** 187402
[18] Tselikov G I, Timoshenko V Yu, Plenge J, Ruhl E, Shatalova A M, Shandryuk G A, Merekalov A S and Tal'roze R V 2013 Photoluminescence properties of cadmium selenide quantum dots embedded in liquid crystal polymer matrix *Semiconductors* **47** 647-9
[19] Osad'ko I S 2016 Blinking fluorescence of single semiconductor nanocrystals: basic experimental facts and the theoretical models of blinking *Physics-Uspekhi* **59** 462-74

IOP Conf. Series: Journal of Physics: Conf. Series **951** (2017) 012012 doi:10.1088/1742-6596/951/1/012012

Formation of microconical structures on Si surface under the action of pulsed periodic Yb-fiber laser radiation

D.S. Polyakov[1], V.P. Veiko[1], N.M. Salnikov[1], A.A. Shimko[2], A.A. Mikhaylova[2]

[1] – ITMO University, Saint-Petersburg, Russia
[2] – Centre for Optical and Laser Materials Research, Saint-Petersburg State University, Saint-Petersburg, Russia

polyakovdmitry1988@gmail.com

Abstract. In this paper the results of microstructuring of monocrystalline silicon by series of nanosecond Yb-fiber laser pulses are presented. It is shown that nanosecond Yb-fiber laser can be successfully applied for creation of microrelief consisting of microcones arrays. Obtained relief provides a reduction of the reflection coefficient below 10% in visible range of the spectrum. It is also established that for creation of microconical relief by nanosecond Yb-fiber laser it is advisable to use "short" nanosecond pulses with pulse durations from several nanoseconds to several tens of nanoseconds.

1. Introduction

Laser structuring of Si surface can be applied for annealing of ion-implanted layers [1, 2], creation of superhidrophobic/superhydrophilic reliefs [4, 5] or antireflective reliefs [5 – 7]. The last application is important in a context of the problem of increasing of efficiency of solar cells, since the antireflective relief allows to reduce the optical losses. Currently, several methods are used to increase the absorptivity of silicon: chemical etching [8], reactive ion etching [9] and surface structuring using a laser radiation [5 – 7]. The laser methods allow to effectively reduce reflectivity and do not require to use any toxical gases or chemicals. Disadvantage of laser methods is that its performance is not high enough. In addition commonly used for creation of antireflective relief femtosecond lasers are difficult to operate, and they still not widely used as a technological tool. In opposite to femtosecond lasers nanosecond Yb-fiber laser is already widely used in manufacturing and has a number of indisputable technological advantages such as high efficiency factor, high reliability and stability of parameters, high processing speed caused by the possibility of operation at high pulse repetition rates (up to 100 kHz and higher). However the opportunities of Yb-fiber laser for creation of antireflective microrelief on silicon surface investigated insufficiently.

So the aim of this work is investigation of geometrical and optical properties of microrelief formed on silicon surface under the action of nanosecond Yb-laser pulses.

2. Description of the experiment

In our experiments we used nanosecond pulses-periodic Yb-fiber laser operated at fundamental wavelength 1.07 µm. Irradiation was carried out by line scanning across the square area 2x2 mm with different pulse durations t_p (4, 30 and 200 ns) and numbers of scan cycles N. The distance between scan lines was 10 µm. The laser beam had a Gaussian profile with a diameter 50 µm at e^{-1} level. The beam movement over the surface was carried out by two-axis galvanometric scanner. Scanning speed was 100 mm/s, pulses repetition rate was 50 kHz. The single pulse intensities were chosen close to the melting threshold and were about $6.7 \cdot 10^8$ W/cm^2, $1.3 \cdot 10^8$ W/cm^2 and $2.3 \cdot 10^7$ W/cm^2 for 4, 30 and 200 ns pulse durations respectively. The scheme of experimental setup and beam motion patterns are shown at Fig. 1.

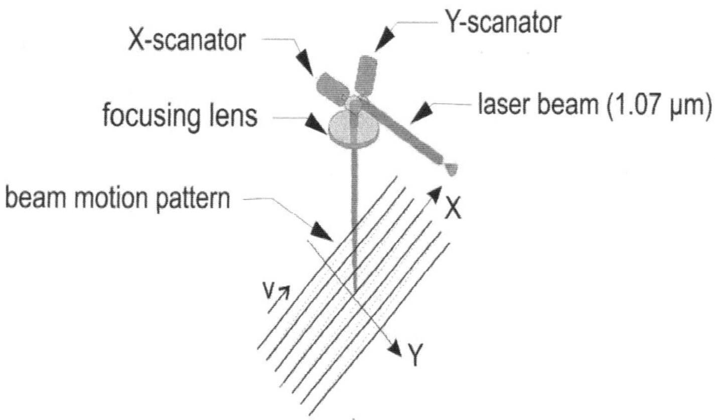

Figure 1. Scheme of experimental setup and beam motion pattern

Figure 2. General view of treated sample

The plates of n-type (100) monocrystalline silicon were used as experimental samples. The experiments were conducted in air under normal conditions. After irradiation the samples were washed in ultrasonic bath. Obtained microreliefs were studied by the methods of optical and atomic-force microscopy (AFM). The reflection spectra were measured by spectrophotometer with an integrating sphere. The structural properties of threated surface were studied by Raman spectra measurements.

3. Results and discussion

The general view of sample treated by pulses with t_p = 30 ns is shown at Fig. 2. It can be seen that antireflective relief is formed. Typical types of microreliefs obtained by treatment with nanosecond pulses with different pulse durations are shown at Fig. 3. It can be seen that in case of irradiation by pulses with durations 4 and 30 ns the relief consisting of arrays of microcones is formed. The typical height of cones is about 4 – 6 μm for t_p = 4 ns pulse duration and about 14 – 16 μm for t_p = 30 ns. The base diameter of microcones is about 10 – 12 μm for t_p = 4 ns and 25 – 30 μm for t_p = 30 ns. So the aspect ratio in both cases is approximately the same. Note also that increasing of numbers of scan cycles N doesn't allow to significantly increase the aspect ratio of obtained microcones. The studies of Raman spectra of the treated samples did not reveal the presence of amorphous areas in the surface layers. It can be seen from Fig. 3 that when irradiation is carried out by "long" nanosecond pulses with t_p = 200 ns the microconical relief is not formed. The obtained in this case relief consist of "microcracks" oriented along the crystallographic directions (100), and some other types of surface periodic structures. That result indicates that the role of the processes of generations and accumulations of dislocations is much important for irradiation by "long" nanosecond pulses. So the processing mode with "long" nanosecond pulses is not suitable for formation of the relief consisting of microcones.

The dependences of reflection coefficient R of obtained samples on wavelength in the range of 300 – 1500 nm are shown at Fig. 4. It can be seen that after laser structuring by the pulses with t_p = 4 ns and t_p = 30 ns significant reduction of reflection coefficient occurs. The reflection coefficient falls approximately 3-4 times in the visible range of the spectrum in comparison with the initial one, and in the near UV there is a decrease by a factor of 5-6. Since the aspect ratio of microconical structures obtained in cases of structuring by the pulses with durations 4 and 30 ns differs insignificantly the reduction of reflection coefficient for this cases in approximately the same. The relief obtained by structuring with "long" nanosecond pulses practically has no antireflective effect.

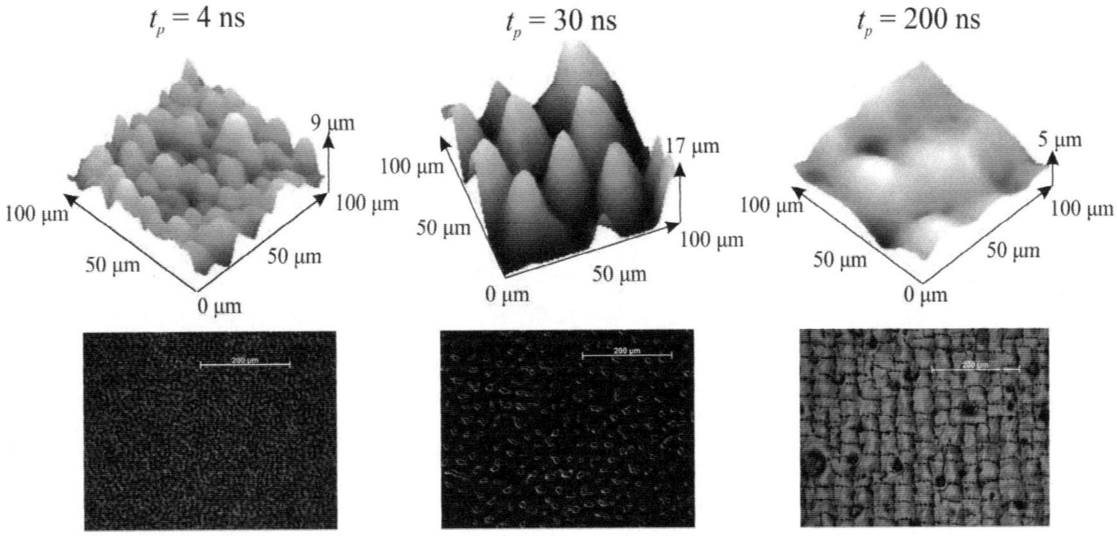

Figure. 3. Optical and AFM images of microrelief obtained by processing with laser pulses with different pulse durations (number of scan cycles $N = 40$ for $t_p = 4$ ns and $t_p = 30$ ns, $N = 30$ for $t_p = 200$ ns)

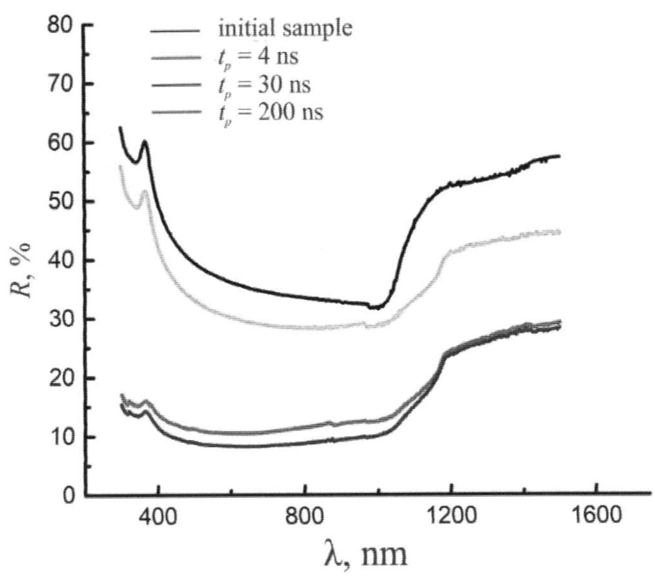

Figure 4. Reflection spectra of obtained samples

The comparison of absorptivity ($A = 1 - R$) of monocrystalline silicon processed by different methods shown at Fig. 5. It can be seen that the absorptivity of the samples structured by nanosecond Yb-fiber laser radiation exceeds the absorptivity of the samples structured by etching in NaOH solution. Absorptivity of femtosecond threated samples is a few percent higher than in case of nanosecond Yb-fiber laser structuring. However the processing speed is several times higher for Yb-fiber laser structuring in comparison with typical femtosecond experiment [7].

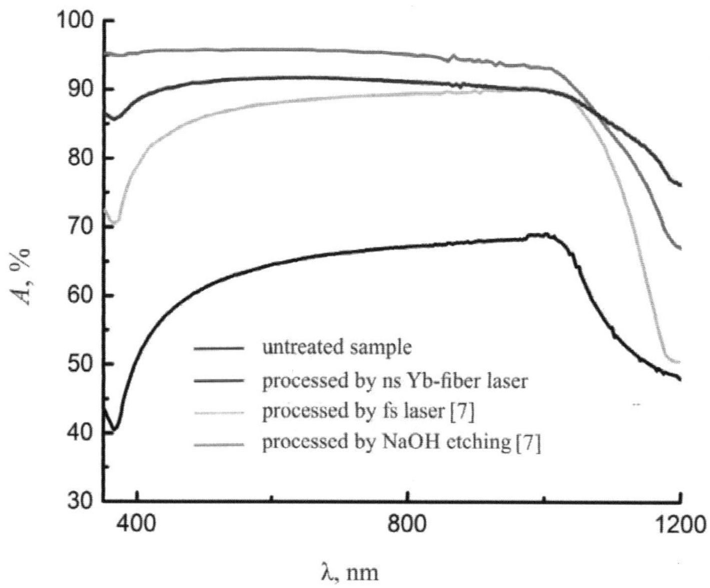

Figure 5. Comparison of absorptivity of silicon samples threated by different methods

4. Conclusions

It is shown that under the action of pulsed periodic radiation of nanosecond Yb-fiber laser it is possible to create microreliefs consisting of microcones array. It is established that for creation of such type of microrelief it is advisable to use “short” nanosecond pulses with durations from several nanosecond to several tens of nanoseconds. Formation of the microconical relief by nanosecond radiation of Yb-fiber laser lead to a decreasing of reflection coefficient in visible range of spectrum in 3 – 4 times. At the same time the using of nanosecond Yb-fiber laser makes it possible to increase the processing speed several times in comparison with structuring by femtosecond lasers.

References

[1] Godbole V P, Chaudhari S M 1988 *Bull. Mater. Sci.* **11** 97
[2] Yang Q et al 2014 *Nano Lett.* **14** 1769
[3] Zorba V, Persano L, Pisignano D, Athanassiou A, Stratakis E, Cingolani R, Tzanetakis P, Fotakis C 2006 *Nanotechnol.* **17** 3234
[4] Vorobyev A Y, Guo C 2010 *Opt. Express* **18** 6455
[5] Vorobyev A Y, Guo C. 2011 *Opt. Express* **19** A1031
[6] Kontermann S, Gimpel T, Baumann A L, Guenther K M, Schade W 2012 *Energy Procedia* **27** 390

IOP Conf. Series: Journal of Physics: Conf. Series **951** (2017) 012012 doi:10.1088/1742-6596/951/1/012012

[7] Kabashin A V et al 2010 *Nanoscale Res. Lett.* **5** 454
[8] Liu S et al 2014 *Solar Energy Matterials & Solar Cells* **127** 21
[9] Kim J, Inns D, Fogel K, Sadana D K 2010 *Solar Energy Matterials & Solar Cells* **94** 2091

IOP Conf. Series: Journal of Physics: Conf. Series **951** (2017) 012013 doi:10.1088/1742-6596/951/1/012013

Bimetallic clustered thin films with variable electro-optical properties

A Antipov, D Bukharov, S Arakelyan, A Osipov, A Lelekova

Vladimir State University named after Alexander and Nikolay Stoletovs, 87 Gorky Street, Vladimir 600000, Russia

E-mail: aantipov@vlsu.ru

Abstract The drop deposition of colloidal nanoparticles was performed from water-based colloidal solutions. The proposed procedure is based on the agglomeration of colloidal particles in laser-assisted evaporation processes. The evaporation process was resulted in the formation of clustered thin films on a glass substrate. In the experiments with bimetallic Au:Ag solutions, the clustered films are grown, the formation of the clustered films with the average height of 100 nm was achieved. Optical properties of the deposited structures were investigated experimentally. It is shown that the obtained films may become transparent and its properties are defined by its morphology.

1. Introduction

Efficient synthesis of thin films with controlled optical properties is a key issue in modern photonics. In particular, possibilities of plasmon-assisted light manipulations by using nanoparticles and nanostructures currently attract considerable attention enabling various applications and opening new research area [1]. Drop deposition of the colloidal particles on glass substrate represents, furthermore, a promising solution to this challenging problem. The optical properties of these films are dominated by a set of plasmonic effects and strongly depend not only on the composition of the film but also on its morphology. In the case of the structures composed of nanoparticles, such parameters as spacing and ordering of nanoparticles define the local field enhancement to be observed either for distinct resonance frequencies or in a much wider spectral ranges [2, 3]. Furthermore, if nanoparticle spacing is comparable with its sizes, optical properties of the random structures can considerably differ from that of the ordered ones. In the case of the thin bimetallic films, one can expect even more complicated optical properties [4].

2. Deposition of bimetallic thin films

Gold and silver particles were formed as a result of CW laser irradiation (power 30 W, laser spot diameter 100 μm) of a target in water [5]. The target was scanned by a laser beam with a velocity of 100 μm/s; the total exposure time was 30 min. After laser irradiation, the solution was exposed to ultrasound for 10 min and then separated on a CM-6M centrifuge. The particle sizes in the colloidal

IOP Conf. Series: Journal of Physics: Conf. Series **951** (2017) 012013 doi:10.1088/1742-6596/951/1/012013

solution was investigated by dynamic light scattering method with a Horiba LB-550 particle-size analyzer. They appear to be in the range from 5 to 10 nm. A mixture of silver and gold colloidal solution with the weight ratio of 1:1 was prepared by vigorous stirring.

We have formed clustered films by method of drop deposition [6]. The basic concept of this method is the deposition of a small-radius (10 □m) drop of colloidal solution combining with the pinning effect. The particles are deposited on the contact area of the droplet with the substrate placed on the thermostabilized table at the temperature of about 80°C. The area of droplet doesn't change during the evaporation processes, because of an additional energy barrier, which impedes the interface displacement. In this case, the surface tension forces, which tend to make a drop spherical, dominate over the gravitational forces, which tend to flatten the drop. The dynamics of drop spreading is determined by the following factors: surface tension, wetting, viscosity, thermal conductivity, and ionic bonds [6]. Thus, when the particle concentration in a drop is relatively low, the particle deposition can be described similar to the behavior of an isotropic medium. As an example, Fig. 1 shows the deposited layers for different concentration of colloidal system.

For metal particles, we can control the cluster morphology at slow deposition process [6]. A more detailed analysis of the film morphology reveals a variety of aggregate shapes. Some of them gain typical fractal shapes which are shown in Figure 1.

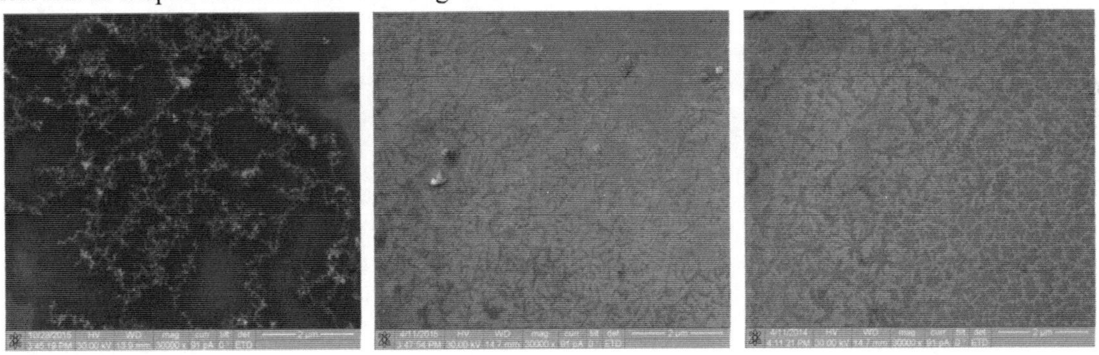

a) b) c)

Figure 1.Three different morphologies of the obtained fractal aggregates: (a) -"dendrites" configuration; (b) -"brunches" configuration; and(c)-"snowflakes" configuration.

3. The optical properties of deposited films

Transmittance spectra strongly depend on the aggregate shape. There is a periodical variation ranges in the transmission spectra of all films at the wavelength of the plasmon resonance of gold (500-540 nm) and silver (390-420 nm). The behavior of bimetallic films transmission spectra is more complicated at the wavelength of the plasmon resonance of silver nanoparticles. Additionally, Figure 2 illustrates this dependence for three typical fractal aggregate configurations shown in Figure 1.

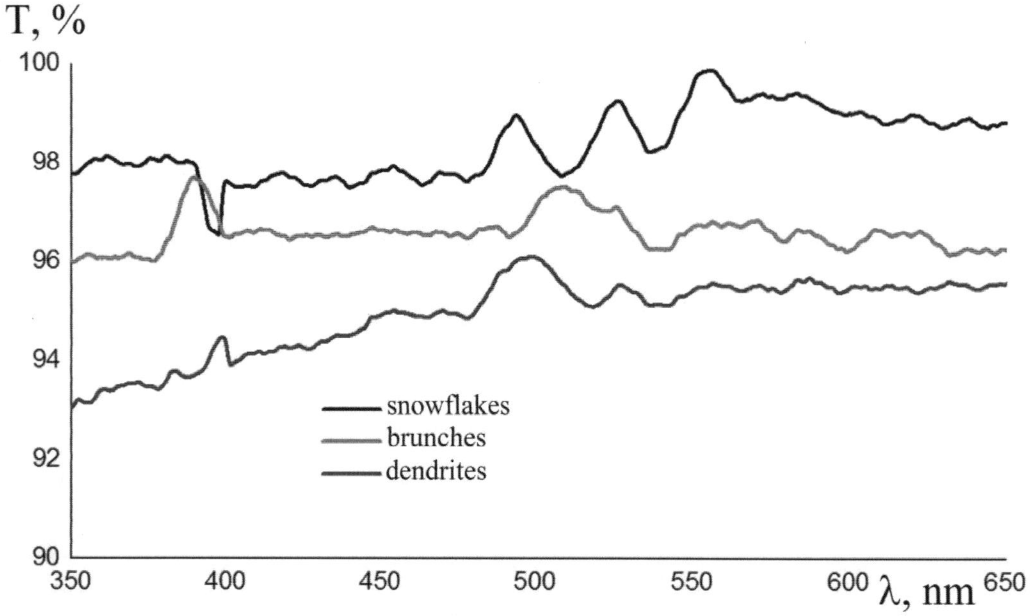

Figure 2. Measured transmittance spectra of fractal aggregates shown in Figure 2.

4. Summary

We have presented an efficient method of the drop deposition of bimetal island films. The possibility of the formation of the clustered films of gold and silver nanoparticles with controlled morphology is demonstrated. The experimental data on optical properties of nanostructured bimetallic clustered films has been obtained. We have demonstrated that the properties of the island films depend on its morphology.

5. Acknowledgments

This study was supported by the Ministry of Education and Science of the Russian Federation (state projects nos. 16.5592.2017/VU and 3.4903.2017/VU), RFBR grant number 16-32-60067 mol_a_dk and the grant of president of Russian Federation by project MK-2842.2017.2 .

References

[1]. Genov D.A., Sarychev A.K., Shalaev V.M. *et al* 2004 *Nano Lett.* **4** 153
[2]. Mayer C., Palkovits R., Bauer G. *et al* 2001 *Journal of Nanoparticle Research* **3** 359
[3]. Bakhti S., Tishchenko A.V., Zambrana-Puyalto X. *et al* 2016 *Scientific Reports* **6** A32061
[4]. Tlahuice-Flores A. 2013 *Journal of Nanoparticle Research*, **15** A1771
[5]. Arakelyan, S.M., Veiko, V.P., Kutrovskaya, S.V. *et al* 2016 *Journal of Nanoparticle Research* **18** A155
[6]. Antipov, A. A., Arakelyan, S. M., Vartanyan, T. A., *et al* 2015 *Optics and Spectroscopy* **119** 119.

IOP Conf. Series: Journal of Physics: Conf. Series **951** (2017) 012014 doi:10.1088/1742-6596/951/1/012014

Mechanisms of graphene exfoliation under the action of femtosecond laser radiation in liquid nitrogen

K S Khorkov, D A Kochuev, V A Ilin, R V Chkalov, V G Prokoshev and S M Arakelian

Vladimir State University named after Alexander and Nikolay Stoletovs, 87 Gorky Street, Vladimir 600000, Russia

E-mail: freeod@mail.ru

Abstract The processes of graphene structures formation under the action of the femtosecond laser radiation on carbon samples in liquid nitrogen are discussed. Mechanisms of graphene sheets exfoliation are proposed depending on the power density of the laser radiation: in the first case, the separation occurs due to the volumetric expansion during heating the region occupied by nitrogen molecules; at a laser radiation energy exceeding the ablation threshold, the surface of graphite begins to breakdown in the region of the action, followed by separation into graphene layers.

1. Introduction

Graphene is a two-dimensional crystal consisting of carbon atoms arranged in a hexagonal lattice. The term "graphene" appeared in 1962, but the discovery of real graphene occurred only in 2004 using the method of mechanical splitting of graphite. According to this method, graphene sheets are separated from crystalline graphite, either as a result of friction of pieces of crystalline graphite against each other, or in "peeling" them off the graphite surface with a special adhesive tape, followed by its dissolution. The micromechanical splitting of graphite became the main way of obtaining graphene in the early studies of its properties, and formed the basis for a whole series of new production methods, initiating active research in the field of graphite layers' separation. To date, their results have been generalized in a number of review papers [1-5].

Now, there is a large abundance of methods for the production of graphene, one of which is the "liquid-phase exfoliation of graphene". This method is based on the intercalation of foreign atoms or molecules in the interlayer space of graphite. Because of the introduction of impurities, the energy of interaction between layers is weakened by increasing the distance between them. Thus, further separation of the graphite into its constituent monolayers becomes possible. As penetration substances are used surfactants having a high energy of interaction with graphene layers such as N, N-dimethylacetamide, g-butyrolactone (GBL) and 1,3-demethyl-2-imidazolidinone (DMEU). For example, in [1] a suspension was prepared from N-polymethylpyrrolidone and a finely dispersed graphite powder, which was subsequently subjected to ultrasonic treatment and centrifugation. As a result of this treatment, a suspension is formed in the surface active liquid containing suspended single-layer graphene sheets, as well as multilayers graphene. Thermodynamically, this possibility is due to the higher energy of interaction of the surfactant with the surface of the graphene sheet than the energy of the interaction of the neighboring layers with each other.

IOP Conf. Series: Journal of Physics: Conf. Series **951** (2017) 012014 doi:10.1088/1742-6596/951/1/012014

To further study the properties of the resulting material, a drop of the treated suspension was applied to a carbon mesh. The obtained samples contained flat and curved graphene scales, as well as flakes consisting of several layers. Another, more complex example of such an approach, is described in [2]. The authors managed to achieve a 90% yield of single-layer graphene films in the following way: graphite powder was pre-annealed for 1 minute at 1000 ° C and finely ground, and then subjected to multistage surfactant interaction together with ultrasonic treatment and centrifugation. Another common method of liquid-phase separation of graphite is the Hammer's method [3], according to which the inner layers of graphite are oxidized by an anhydrous mixture of concentrated sulfuric acid with sodium nitrate and potassium permanganate. After the course of the reaction accompanied by active gas evolution, the mixture is dried and formed an aqueous suspension of graphene oxide (GO) flakes which has enhanced stability by electrostatic repulsion of the layers of GO in water.

2. Experiment

Targets from highly oriented pyrolytic graphite (HOPG) HOPG-1.7-10x10x1-1 and glassy carbon (GC) SU-2000 were used as initial samples for graphene production. HOPG is fundamentally the best material for the elimination of graphene, since it is formed from high-quality graphite layers, almost parallel to each other. The glassy carbon has a rather complex structure. It is a bundle of randomly intertwined carbon ribbons crosslinked by carbon bonds of varying multiplicity. Obviously, glassy carbon cannot be a source of flat graphene sheets. However, for a number of applications of graphene, for example, as a material for electrodes of ionistors (supercapacitors) or for reinforcing nanoceramics, it is necessary that it has a complex extended surface [6].

The surface treatment of the targets was carried out by radiation of ytterbium femtosecond TETA-10 laser. Parameters of the laser radiation: $\lambda = 1029$ nm, radiation pulse duration $\tau = 280$ fs, repetition rate $f = 10$ kHz, pulse energy $\varepsilon = 150$ μJ. The polarization of laser radiation is linear. The exposure scheme is shown in Figure 1. The surface treatment of graphite was performed in a liquid nitrogen medium, which covered it with a layer about 1 cm thick. The diameter of the laser spot on the target surface reached 100 μm [7].

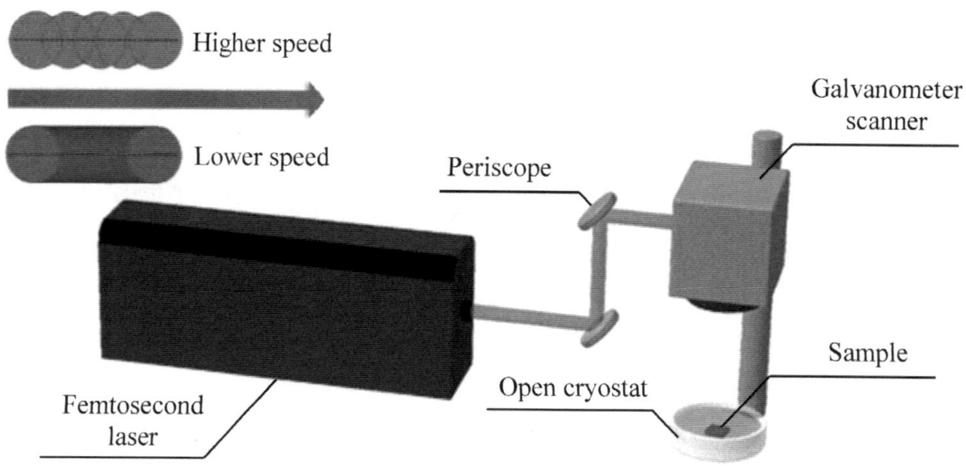

Figure 1. Scheme of the experiment of graphene structures formation and a schematic representation of the overlapping pulses of laser radiation

The speed of movement of the laser beam along the surface of the target was from 10 to 100 cm/s. Accordingly, depending on the scanning speed of the beam, each region of the target surface was exposed from 1 to 10 femtosecond pulses (upper left corner of Figure 1). Graphene splitting occurs at a laser beam velocity of 25 cm/s, when the graphite surface is exposed to four laser pulses (overlapping - 75%). With a larger overlap, a strong destruction of the surface of the target and at the same time an exfoliated carbon material begins.

At the end of the laser treatment, the targets were kept under natural conditions until liquid nitrogen evaporated completely, after which their surface became accessible for investigation by microscopic methods. The results of laser action on graphite targets were studied on the basis of images obtained with the scanning electron microscope Quanta 200 3D (SEM-images) and a probe nanoscale laboratory Integra Spectra (Raman spectra).

3. Processing results

Microscopic studies of the carbon material detached from the surface of HOPG under the influence of femtosecond laser radiation confirmed the production of graphene sheets of various shapes. Graphene tapes with a width of up to 50 μm and a length of more than 150 μm have been identified (Figure 2a), as well as graphene plates of arbitrary shape with a characteristic size of more than 150 μm (Figure 2b).

Graphene tapes obtained during the experiments, although have folds and bends, also have rather large areas (up to 2000 μm^2) of flat areas of the surface.

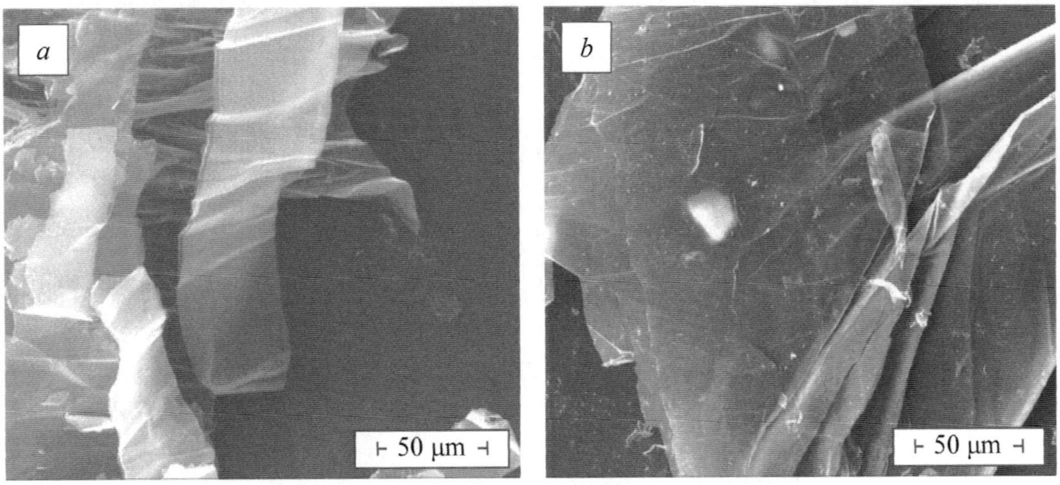

Figure 2. Graphene tapes (a) and plates (b), cleaved from the main sample, under the expose of femtosecond laser radiation at HOPG in liquid nitrogen

The synthesized plates (flakes) also exhibit numerous folds and bends, as well as significant flat areas of the surface. It should be noted that on the surface of large graphene sheets, much smaller (characteristic sizes of the order of 1 μm) and crumpled structures are observed.

On the SEM images (Figure 3) on the graphene structure, can be seen areas (highlighted by an arrow) that have a different number of graphene layers. At the same time, such "steps" are formed at the edges of the sheets, which follow from the process of laser-induced exfoliation of graphene.

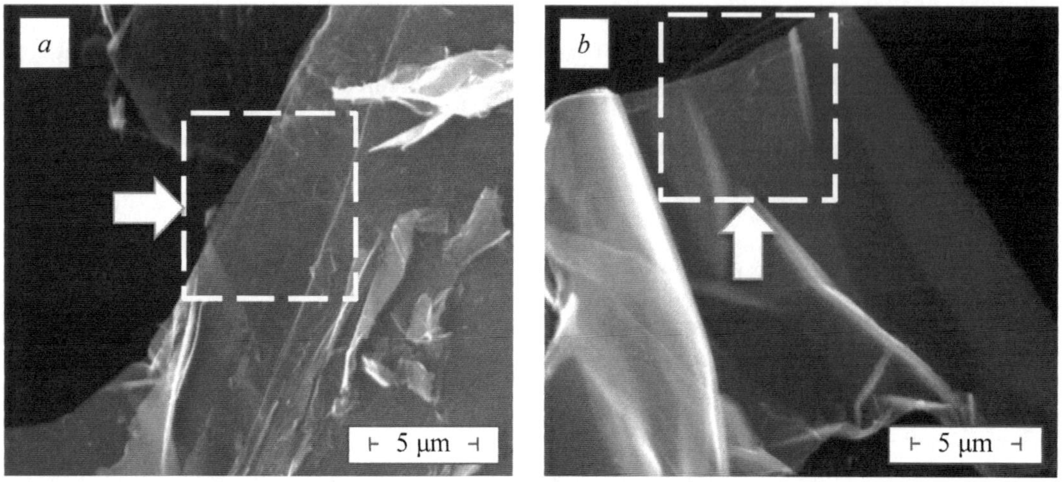

Figure 3. SEM-images of formed graphene structures, the edge region is highlighted and shown by an arrow

Graphene structures obtained by laser treatment of the surface of glassy carbon have a fundamentally different shape (Figure 4). Graphene sheets form lumps with a complex folded structure. Clumps with a characteristic size of the order of 1 μm are recorded

Figure 4. SEM-image of graphene structures on the surface of glassy carbon

4. Discussion of graphene exfoliation mechanisms

A comparison of the results of treatment of HOPG and GC by femtosecond laser radiation in a liquid nitrogen environment showed that the use of the first material is preferable for the purpose of obtaining the flattest and thin graphene structures. Obviously, in this case the primary internal structure of the irradiated materials plays a decisive role.

The thermal expansion of the graphite lattice is anisotropic, as are the values of the thermal and electrical conductivity of this material. The anisotropy for graphite is the stronger, as the higher the ordering of the lattice. The temperature coefficient of linear expansion in the direction of the basal planes is much less than in the perpendicular direction, thereby contributing to exfoliation of the structure under intense heating.

The cooling of graphite to the temperatures of liquid nitrogen reduces the lattice vibrations of the graphite starting material, so that the interlayer distance is reduced to 3.339Å, and when heated, it increases to 3.465 Å (at 1000 °C) [8]. Cooling of samples with liquid nitrogen is often used in view of availability, which satisfies the boiling point and the small size of the nitrogen molecule, which ensures its high fluidity. At low temperatures, the graphite lattice becomes less "plastic", which, probably, will make it possible to destroy interplane bonds by a smaller energy contribution at the site of local heating of nitrogen molecules diffused into thc interplanar distance.

In [9] proposed a mechanism for obtaining carbon "nanoscrolls" – twisted graphene sheets in a spiral. A sample of natural graphite was immersed in liquid nitrogen and irradiated with microwave radiation for 5 seconds. Upon irradiation, the initial sample of graphite was heated to a high temperature, as a result of which it increased in volume. At the same time, the graphene sheets on the graphite surface are cooled by liquid nitrogen and compressed. Thus, a stress state is formed between the cooled surface of the sample and the main volume of the heated graphite. As a result of the stresses created at the boundary, the interplane bonds break, the graphene sheets are peeled off, followed by folding into carbon nanostructures due to a more favorable energy state. Further processing consisted in the action of ultrasound that facilitated the further separation of the resulting structures from the graphite sample. The allocation of the necessary fractional structural composition of the nanoscrolls was carried out by means of centrifugation.

In comparison with experiments on heating a graphite sample with microwave radiation using laser radiation, the peeled graphene structures do not take the form of nanoscrollers, but form flat shapes. This can be due, firstly, to the characteristic heating time of the region exposed to the effect – with short laser pulses there is a rapid local heating, in contrast to uniform heating by microwave radiation. Secondly, in the case of a laser-induced exfoliation, a sufficiently large number of graphene layers is separated, which does not allow the graphene lattice to deform in the general layer and to perform sheet folding in a spiral.c

As it suggested in [8], when a graphite sample is immersed in liquid nitrogen, the latter penetrates into the interlayer distance of the graphite planes as a monolayer of the liquid ("mono-layer 2D-fluid"). Having a smaller size, nitrogen molecules penetrate into the interplanar distance in places that have defects (cracks, chips, layering), and also from the ends of the sample itself (Figure 5).

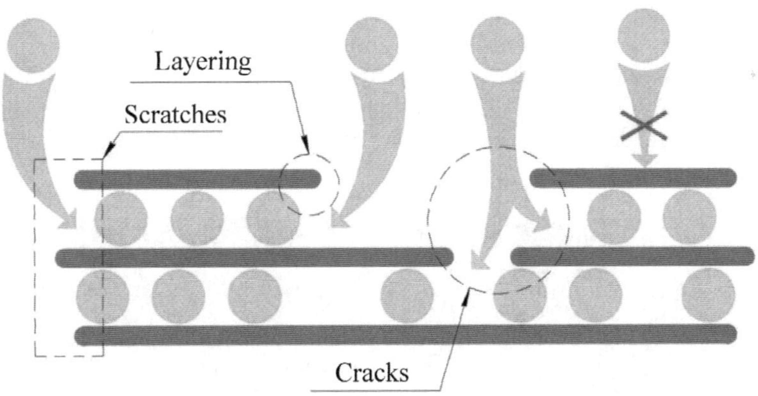

Figure 5. Schematic representation of the process of nitrogen diffusion into the interlayer distance of the graphite lattice

For nanosecond laser pulses, the main process that affects the heating of nitrogen molecules is the transfer of target energy due to thermal conductivity. The laser pulse duration is much longer than the electron-phonon energy transfer time, i.e. during the time of the laser pulse, the electrons transmit energy to the lattice of the substance, and its heating occurs within the laser exposure region, including diffused nitrogen. As a result of laser heating and, as a consequence, an increase in the amplitude of vibrations of nitrogen molecules, a volumetric expansion of the region occupied by nitrogen molecules occurs (a transition from the liquid phase to the gas phase). Thus, interplanar carbon π-bonds (16-40 kJ/mol) are destroyed, which are weaker than σ-bonds (of the order of 450 kJ/mol) [10]. Effective heating of nitrogen is carried out due to the transfer of thermal energy from the graphite lattice in the absence of nitrogen absorption bands at the working wavelength.

In the interaction of ultrashort laser pulses with a graphite sample in a liquid nitrogen medium, several variants of the course of the processes of structural changes in the material can be identified. When the laser radiation power is insufficient for laser ablation processes to occur, the photon energy is absorbed by the carbon lattice and diffused nitrogen. In view of the sharp increase in the lattice vibrations and nitrogen molecules, weak interplane π-bonds collapse (Figure 6). The splitting and splitting of the graphite layers takes place mainly on the defects of the processed structure (cracks, chips, scratches, including heterogeneous lattice defects in view of the gaps facilitating the diffusion of liquid nitrogen molecules.) Accordingly, an increase in the number of nitrogen atoms in the lattice of the material promotes more intensive exfoliation of graphene sheets.

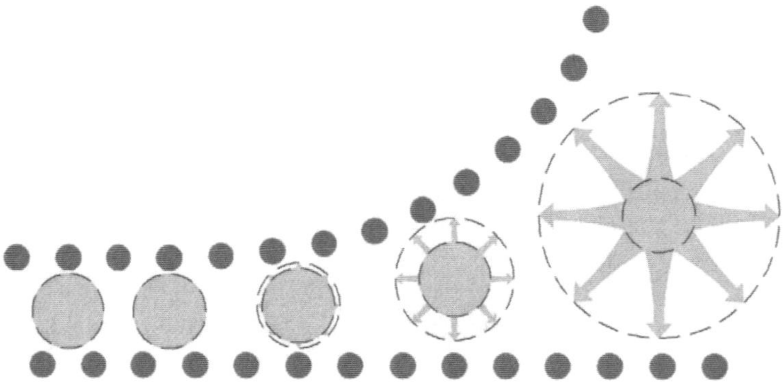

Figure 6. Schematic representation of the process of graphene peeling as a result of thermal expansion of nitrogen

Figure 7 shows the result of the detachment of graphene structures on the surface of the HOPG in the region of the crack. It is clearly seen that the detachment starts from the center of the crack according to the mechanism described above. On a SEM image with a large magnification, it is seen that individual structures exfoliate and bend in the direction from the beginning of the detachment, holding at the base and in some places also exfoliating. The next step in using the obtained graphene structures is to separate from the original sample using ultrasonic treatment, form a solution and deposit on the required substrates.

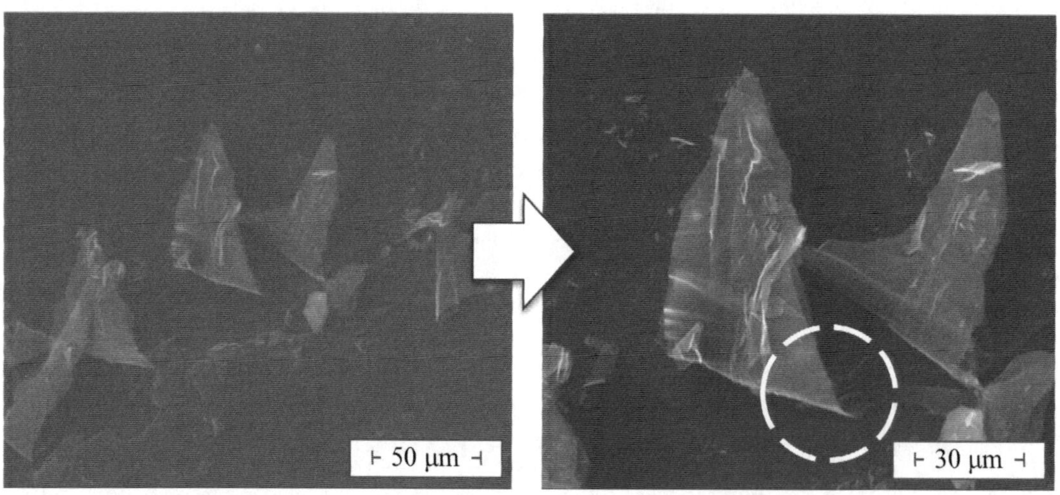

Figure 7. SEM-images of the edges of graphene sheets (with different magnifications)

IOP Conf. Series: Journal of Physics: Conf. Series **951** (2017) 012014 doi:10.1088/1742-6596/951/1/012014

At laser radiation energy above the ablation threshold, the surface of graphite starts to breakdown in the area of the impact. As a result of the processes, occurs during laser ablation, a high-pressure region forms on the walls of the formed cavity. The separation of the graphite structure to multilayer graphene under such mechanism occurs as a result of the reactive action of ablation products. In view of the presence of various defects, the graphene layers are separated from the main sample.

Figure 8 schematically shows the process of graphene structures exfoliation upon reaching the ablation process: black bars denote graphite layers; red hemisphere – laser-induced plasma plume; black arrows – the directions of expansion of the ablation products, directions of propagation of the pressure front. In the case of scanning by laser radiation, the main parameter at the start of the mechanism is the exposure time at a specified value of the density power (Figure 9).

Figure 8. Schematic representation of the process of graphene structures exfoliation

The increase of the energy of laser radiation leads to an increase of temperature and pressure in the impact area due to insignificant heat removal and the impossibility of expanding deep into the material. In this case, nonequilibrium conditions are forming and leads to a change in the phase composition at some depth of the sample and to subsequent formation of bulk crystal structures [11].

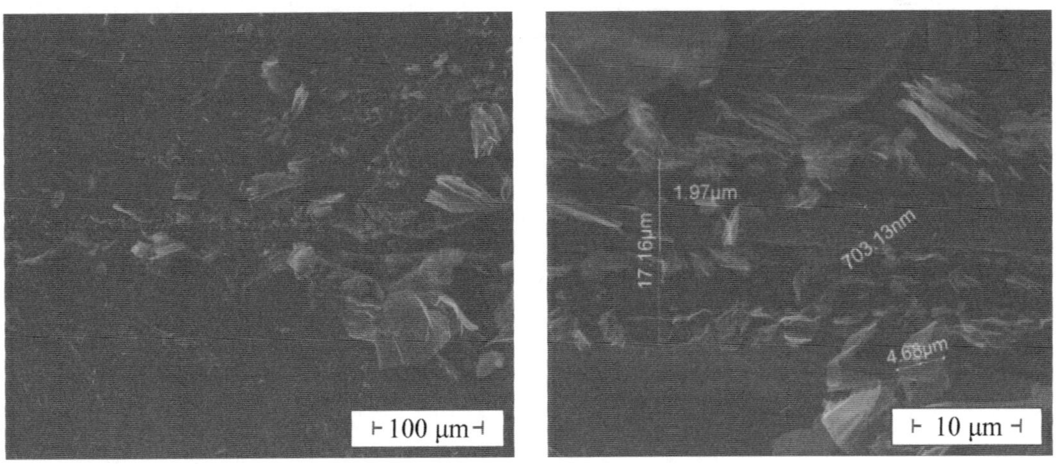

Figure 9. SEM-images of the laser radiation processing area

5. Conclusion

Thus, the general mechanism of graphene exfoliation by femtosecond laser action in liquid nitrogen is the separation of surface layers of the target as a result of disruption of interplanar bonds of carbon samples due to the expansion of the induced transition of diffused nitrogen to the gas phase under laser heating.

At the same time, from the surface of the HOPG are separated flat layers, of which it consists. On the surface of the glassy carbon, a sufficiently thick layer containing both graphene ribbons and amorphous material is pinched. With the using of glassy carbon as a material, this effect becomes more complex, allowing forming crumpled type structures.

This study was performed as part of the state task VlSU 1106/17 and grants of the Russian Foundation for Basic Research 16-42-330651, 16-32-00760.

References

[1] Yeletsky A V, Iskandarova I M, Knijnik A A and Krasikov D N 2011 *Uspekhi fizicheskikh nauk* **181**
[2] Tkachev S V, Buslaeva E Yu and Gubin S P 2011 *Neorganicheskiye materialy* **47**
[3] Graifer E D, Makotchenko V G, Nazarov A S, Kim S D and Fedorov V E 2011 *Uspekhi khimii* **80**
[4] Razumov V F 2010 *Rossiyskiye nanotekhnologii* **5**
[5] Avouris P and Xia F 2012 *Mrs Bulletin* **37**
[6] Xiaolin Li, Guangyu Zh, Xuedong B, Xiaoming S, Xinran W, Enge W and Hongjie D 2008 *Nature nanotechnology* **3**
[7] Abramov D, Arakelian S, Kochuev D, Makov S, Prokoshev V and Khorkov K 2016 *Nanosystems: physics, chemistry, mathematics* **7**
[8] Mortazavi S Z, Parvin P and Reyhani A 2012 *Laser Physics Review Letters* **9**
[9] Zheng J, Liu H, Wu B, Guo Y, Wu T, Yu G, Liu Y and Zhu D 2011 *Advanced Materials* **23**
[10] Knunyants I L 2013 *Kratkaya khimicheskaya entsiklopediya* (Moscow: Ripol Klassik)
[11] Khorkov K S, Zvyagin M Yu, Kochuev D A, Chkalov R V, Arakelian S M and Prokoshev V G 2017 *Bulletin of the Russian Academy of Sciences. Physics.* **81**

Formation of microspheres under the action of femtosecond laser radiation on titanium samples in hydrocarbons

D A Kochuev, K S Khorkov, A V Ivashchenko, V G Prokoshev and S M Arakelian

Vladimir State University named after Alexander and Nikolay Stoletovs, 87 Gorky Street, Vladimir 600000, Russia

E-mail: _b_@mail.ru

Abstract. This work describes the original method of laser synthesis of microspheres which contain titanium carbide. The formation of microspheres is carried out by the action of femtosecond laser radiation on the surface of titanium in the reaction medium – the ultimate hydrocarbon. The resulting microspheres have a high surface smoothness, a narrow particle size distribution, an average size of 1-3 µm. They can be used in applications of additive engineering, powder metallurgy as the main raw material, or as an alloying additive.

1. Introduction

Nowadays, in the technology of materials manufacturing, one of the topical and promising trends is the synthesis of metallic microspheres. The development of technologies in this field led to the searching of efficient methods of obtaining such type of particles. Currently, there are many ways and approaches for obtaining powders of various materials: by mechanical grinding, chemical precipitation, magnetron sputtering, high-temperature sputtering (gas-plasma spraying, electric arc, and rarely powders are produced by laser-induced methods). Each of the proposed methods has advantages and disadvantages, which are determine the field of application of the obtained powder material. Despite the very limited prevalence of laser technological complexes, the method of laser synthesis of powder material has a high interest due to the exceptional properties of laser radiation, unattainable by other sources of concentrated energy.

The method of laser synthesis of powder materials finds its application when it is necessary to get small batches of products. One of the features of laser synthesis of powders is the possibility of getting microgranules from a wide range of materials, the process also has a high spatial localization, which largely determines the low dispersion of particle sizes. The peculiarities of laser action make it possible to achieve a change in the structural and phase states of a material without adding impurities to it (provided that they are absent in the region of laser radiation). This feature makes it possible to get pure powder materials in comparison with other synthesis methods. When introducing the necessary impurities, it is possible to get complex alloys, composites, cold solutions in the required aggregate state. In the field of obtaining powdered materials, the process of material destruction due to the action of laser radiation is of primary interest, in particular, it is required to obtain not a melt, not steam, but directly "splashes" of the processed material, which have a spherical shape and a close granulometric composition. Pulse laser systems which are able to efficiently transmit the energy of laser radiation for the development of the laser ablation process are suitable for such purposes. Using the additional means that can influence the development, dynamics and result of this process, such as the processing medium, or an additional sources of physical impact, will contribute to the achievement of the goal.

In recent years, the processes that occur during laser ablation in a liquid are being intensively studied. Significant progress has been made in this area due to various technological means developed for the study of physical and chemical processes in the laser ablation of solids in liquid media [1-4]. Laser ablation in the liquid has a number of unique properties over the other methods. As for the synthesis of powder materials, the rapid cooling of ablation products promotes the formation of a "highly spherical" surface. The used during the processing liquid medium can have both an inert and reactionary properties. Metallic microspheres obtained by this method can be used in various applications: additive engineering, powder metallurgy, where application is possible both as the main material and as the alloying additive.

2. Equipment and samples

In the experiments was used a Ti:Sapphire-laser system with the following parameters: pulse duration 50 fs, average power 450 mV, wavelength 800 nm, repetition rate 1 kHz. A laser beam with a diameter of 30 μm was focused on the surface of a titanium sample located in the reaction medium (Fig. 1). A titanium plate of grade titanium (VT1-0) was used as a sample. The cuvette represented a container isolated from atmospheric air, having a point of attachment of a sample and a transparent window for introducing laser radiation. The sample was located under a layer of liquid hydrocarbon. Scanning with a laser beam was carried out with the galvanometer scanner with a speed of 0.1-200 mm/s.

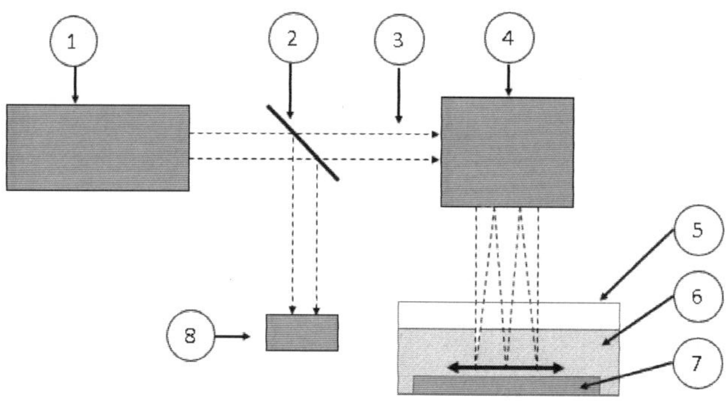

Figure 1. Scheme of the experiment: 1 - femtosecond laser system, 2 - beamsplitter, 3 - laser radiation, 4 - optical head with galvanometer scanner (X-Y), 5 - vessel isolated from the atmosphere with a window for laser radiation input, 6 - reaction medium (liquid hydrocarbon), 7 - titanium sample, 8 - optical powermeter

3. Description of the experiment

The depth of immersion of the sample was selected in such a way that the liquid layer would cover the surface of the sample being treated, but did not make a tangible contribution to the absorption of laser radiation. This results in a dynamic ejection of both the liquid and the sample material from the processing area. During the processing in volume of the cuvette has forming a thick enough "fog". Consider in more detail the process of sorption of carbon by titanium. In [5,6], at the study the interaction of laser radiation with the solid-liquid interface was found that the characteristic boiling time of the liquid occurs at subnanosecond times, since the velocity of the liquid flight is practically independent of the laser radiation power [7,8]. The rate of expansion of the laser-induced plasma is of the order of 10^3 [9, 10] - 10^7 m/s [11, 12]. The speed of motion of the plasma plume strongly depends on the energy of the applied to the impact area and the density of the environment. The rate of expansion of the laser-induced plasma is comparable with the rate of the boiling point of the liquid and

IOP Conf. Series: Journal of Physics: Conf. Series **951** (2017) 012015 doi:10.1088/1742-6596/951/1/012015

also with the rate of liquid expansion. In this way, it is impossible to neglect the interaction of the "reactive flux" directly with hexane molecules, in the liquid or gaseous aggregate state, their dissociation and as a consequence of saturation of the ablation products with carbon.

Consequently, the formation and sorption of carbon occurs before the boiling of the liquid (hexane). During boiling, intense vapor-gas formation occurs, thereby sharply reducing the density of the carbon-containing medium, the probability of sorption of the formed carbon also decreases. During the movement of the ablation products by volume of chamber filled with a "fog" hydrocarbon, the energy of the ejected substance decreases, begins the self-localization into a spatial form with the smallest surface area, which corresponds shape of a sphere. The rapid cooling of the surface of microgranules, in contact with droplets of "fog", reduces the activity of their surface, so they cannot stick together, deform as a result of strikes. The method of synthesis of spherical particles in a liquid medium contributing to rapid cooling is described in [13]. The release of microspheres directly depends on the efficiency of the energy contribution to the formation and ejection of the liquid phase of the material.

4. Results

The studies were carried out using scanning electron microscopy (SEM) and Raman spectroscopy. As a result of treatment, the surface becomes covered with craters formed under the influence of laser radiation. The structure of the surface is not typical for the mode of processing by ultrashort laser pulses (Fig. 2). There are traces of fusion and ejection of the liquid phase of the material on the surface, consequently, the source of the thermal action was the laser-induced plasma, the lifetime of which is much longer than the laser pulse duration and depends from both the applied energy and the pulse repetition rate. During the processing, the lifetime of the laser-induced plasma was longer than the time interval between the laser pulses, which led to a pronounced increase in the plasma energy, as a consequence, the strong thermal action.

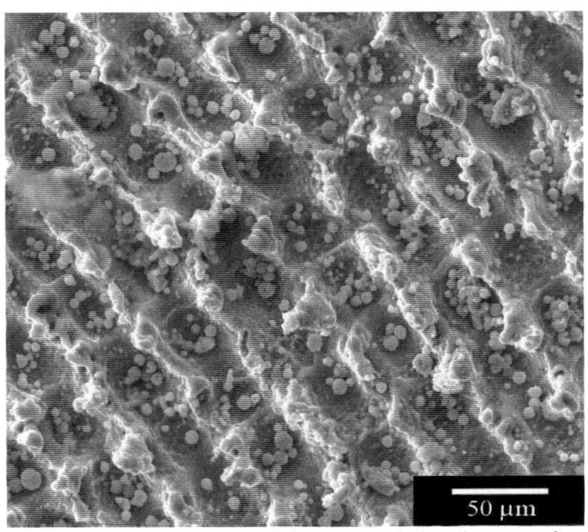

Figure 2. SEM image of the treated titanium surface

Figure 3 illustrates the SEM-image of the sphere surface on a silicon substrate. The microsphere surface has a high smoothness and sphericity. About 80% of the microspheres fit within a range of sizes of 1-3 μm. Granules collected from the surface of the sample have pronounced defects, pores, cracks, irregularities on the surface, probably formed as a result of repeated exposure to a source of laser radiation or as a result of contact with high-energy laser ablation products. The microspheres collected from the bottom of the cuvette have an almost perfect surface without pronounced defects. This feature of the method is expressed in high localization by the action of the energy source.

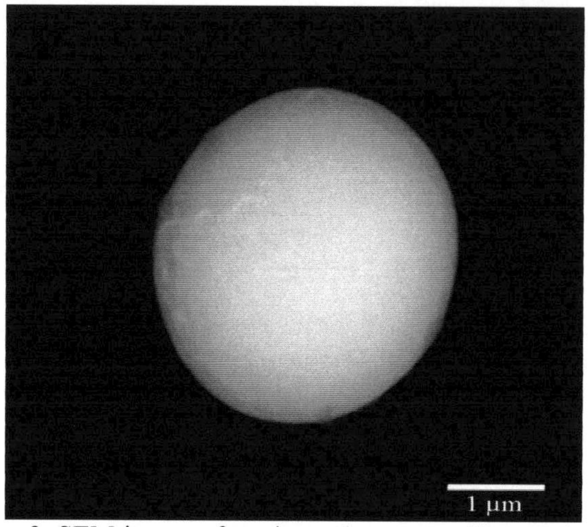

Figure 3. SEM-image of a microsphere on a silicon substrate

The obtained spectra indicate the presence of "polycrystallites" of titanium carbide (wherein, low-intensity spectral lines are quite broad, Fig. 4). To compare the results of the study, were obtained spectra of the surface of a titanium carbide crystal synthesized in an industrial way (135, 249, 431, 601 cm^{-1}), the spectrum is shown in Fig. 5, the positions of the obtained lines coincide. The technology of industrial synthesis of titanium carbide involves the sintering of titanium chips with soot under high temperature conditions and mechanical compaction of this mixture, followed by prolonged homogenization in vacuum at high temperature [14]. In our case, as a result of the release of matter from the region of influence into the volume of the reaction medium, the ejected substance is titanium saturated with carbon atoms of the hexane formed as a result of dissociation. Titanium is a fairly active chemical element, in this system the reaction of formation of titanium carbide is the only possible one. Thus, occurs the formation of a chemical titanium-carbon bond. In view of the sharp drop in temperature, the microsphere material cools, the conditions for homogenization of titanium carbide are not met, which means that decrease in phase inhomogeneity due to the growth of grains of titanium carbide by the drift of nuclei occurring during high-temperature homogenization of the material does not occur.

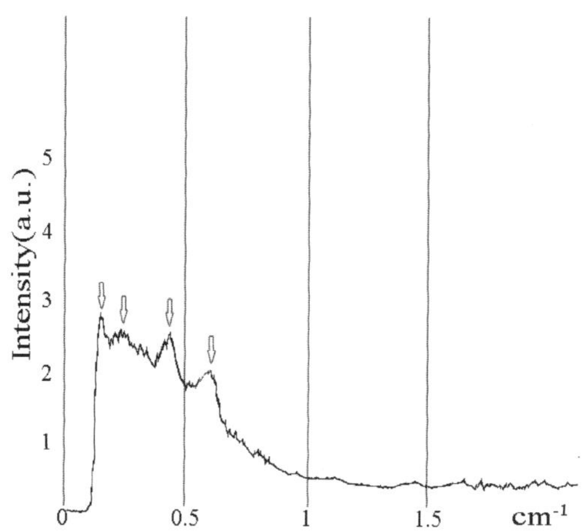

Figure 4. Raman spectra of the microsphere surface, white arrows denote peaks corresponding to titanium carbide

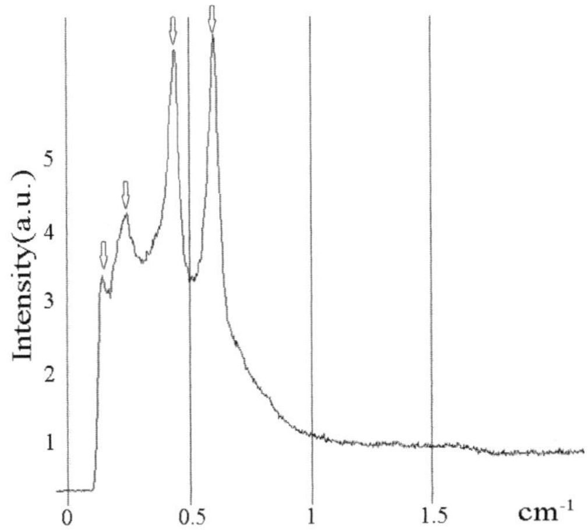

Figure 5. Raman spectra of industrially synthesized titanium carbide, white arrows denote peaks corresponding to titanium carbide

5. Conclusion

The results obtained in this paper can be useful for applications related to additive technologies due to a small range of sizes obtained microspheres: 1-3 μm, high sphericity and smoothness of the formed surface. The carbon deficit, the absence of the expressed grains of titanium carbide (the presence of only their nucleus, crystallization points) will greatly increase the wettability of the components of the composition, for example in NiTi forms [15, 16].

In laser welding there is the problem of burn-out of low-melting alloying additives as a result of laser radiation influence, which is due to the different melting point-evaporation of the components of the composition and different absorption coefficients. The problem of intensive gassing is also widely common; it leads to an increase in porosity of the formed material. The presence of crystallites of titanium carbide at the nuclei stage will significantly reduce the fusion temperature of this composition, which in turn is useful for alloying aluminum, zinc matrices to improve their characteristics (mechanical, strength, operational).

This method is also can be used to introduce carbon into various metallic compositions. The tendency of carbon to the formation of grains of graphite in the structure of metals makes it difficult to distribute it uniformly in the volume of the resulting material. The uses of such alloying add open up new possibilities for the directions of powder metallurgy. Synthesis of powder materials by this approach makes it possible to obtain various powder compositions, because with selecting the processing medium and the intensity of the impact, it is possible to trigger certain processes that contribute structural, phase and chemical changes of processed and synthesized materials [17].

This study was performed as part of the state task VlSU 1106/17 and grants of the Russian Foundation for Basic Research 16-48-330031, 16-08-01226.

References

[1] Barmina E V *et al* 2010 *Quantum Electronics* **40**
[2] Simakin A V, Voronov V V and Shafeev G A 2004 *Proceedings of the General Physics Institute. A.M Prokhorov.* **64**
[3] Kononenko T V *et al* 1999 *Quantum Electronics* **28**
[4] Bozon-Verdura F *et al* 2003 *Quantum Electronics* **33**
[5] Ashitkov S I, Inogamov N A, Zhakhovsky V V *et al* 2012 *Letters in JETP* **95**
[6] Kudryashov S I, Lyon K, Shukla S D *et al* 2006 *Journal of applied physics* **100**
[7] Kudryashov S I and Allen S D 2003 *Journal of applied physics* **93**
[8] Giacomo A De et al 2013 *Physical chemistry* **15**

[9] Burakov V S, Butsen A V and Tarasenko N V 2015 *News of the national academy of sciences of Belarus* **3**
[10] Talala X A, Leyvi A Ya and Yalovets A P 2012 *Bulletin of South Ural State University* **34**
[11] Platonenko V T 1992 *Laser Physics* **2**
[12] Klimentov S M, Garnov S V, Konov V I *et al* 2004 *The russian academy of sciences* **60**
[13] Gordon E B, Karabulin A V, Matyushenko V I *et al* 2014 *High-energy chemistry* **48**
[14] Kiparisov S S 1987 *Titanium carbide: obtaining, properties, application* (Moscow: Metallurgy)
[15] Poletika T M, Chukhlomin A Y, Kulkov S N and Panin V E 1985 *Formation of the structure of TiC-NiTi alloys during liquid-phase sintering* (Tomsk: Preprint)
[16] Starer K R, Wulf 1959 *Trans, of the met. Sat of AIME* **215**
[17] Ionin A A, Kudryashov S I and Samokhin A A 2017 *The successes of physical sciences* **60**

Interaction of pulsed laser radiation with a powder complex based on the Al-Mg-C matrix

A. Voznesenskaya[1], K. Khorkov[1], D. Kochuev[1], A. Zhdanov[1], V. Morozov[1]
[1]Vladimir State University named after A. G. and N. G. Stoletovs, Vladimir, Russia

E-mail: 2obk@bk.ru

Abstract. Experimental work on laser melting of the Al powder composition has been carried out. The influence of the duration of the laser pulse on the result of processing the powder composition has been studied. In this work, the powder material was obtained by the joint mechanical activation of matrix material and filler particles in high-energy ball mills. The research work consisted of analyzing the starting material, the phase composition, the particle size distribution, and the morphology of the powder particles. The obtained samples also studied the phase composition, the presence of pores, cracks, the surface of the formed coating, the average height of the roller. The obtained samples were studied by X-ray diffractometry, Raman spectroscopy, and microsections of the structures obtained by optical microscopy. On the basis of the data obtained, conclusions were drawn about changes in the structural-phase composition, the nature of the distribution, the localization of alloying additives in the course of phase-to-phase transitions, and the change in the phase states of alloying additives

1. Introduction

Currently, additive technologies are actively developing both abroad and in our country. The layered laser sintering of metal powder compositions is one of the directions. The popularity of the application of this method of synthesis of details is growing every day. This can be attributed to the high flexibility of production (from the time of creation of the 3D model to the time of its production, an insignificant time passes), the ability to modulate the properties obtained was made by changing the concentrations of various powders used. A small amount of production waste and minimal subsequent machining are also attractive aspects of layer-by-layer technology.

The essence of the method of selective laser sintering consists in the metered delivery of powder to the region of action of laser radiation. This is done using a special laser head. Further, it moves in space through a system of drives along a given path, then a part of the given geometry is formed, or a coating is applied from the fused powder particles. The accuracy of the manufacturing of parts depends on the thickness of the forming layer (usually it is from 50 to 100 μm). This is due to the size of the spheres in the powder used, the features of the focusing system. The most common is 20 to 60 microns. Modern trends in mechanical engineering require reducing the particle size of the powder to 10 microns, which will increase the accuracy of processing (synthesis of the part or coating), reduce the porosity of the resulting material.

In modern industry, more often than not metal alloys are used, but metal-ceramic compositions. For the production of which various materials (alloying additives) are used, they are introduced into the metal matrix as directly in the process of fusion (as a result of exposure to high temperatures they form

IOP Conf. Series: Journal of Physics: Conf. Series **951** (2017) 012016 doi:10.1088/1742-6596/951/1/012016

a connection with the base metal powder), and it is possible to introduce into the process of powder synthesis directly ceramic elements, carbides, nitrides, oxides. With this approach, the contribution of chemical reactions to the synthesis of the part does not take place, and the effect of laser radiation on such a powder results in the fusion of particles with a minimum transition of the material to the vapor-gas phase from the excess energy produced as a result of the different absorbing capacity of metals and introduced impurities.

The creation of hybrid materials is a new trend in additive engineering applications. The property of this material is laid in the course of its synthesis, i.e. the item is made of materials having different properties (coefficient of thermal expansion, hardness, plasticity, heat resistance). Aluminum matrices filled micro- and nanodispersed carbon components planned to be used as a working material. Carbon nanotubes, fullerenes, onions, nanodiamonds, graphene, and graphite are used as promising reinforcing nanosized filler. Composite materials based on an aluminum matrix filled with carbon "components" [1] are promising for high-tech industries. At present, a significant number of studies have been carried out aimed at obtaining carbon-based composites based on aluminum and its alloys [2-4]. Thus, the problem of the difference melting temperature is serious and requires special attention. In the framework of this paper, it is planned to establish the possibility of using some absorbing materials capable of reducing the necessary energy of laser radiation to form a homogeneous fused area of powder material. It is also planned to investigate the dependence of the influence of the morphology of the powder granules and energy spent on melting the area of the powder material. The application of this approach will allow to provide by reflow due to a change in the absorption coefficient of laser radiation for each material of the powder composition. This will allow to align the energy threshold leading to the reflow of these or other granules. This approach is also possible to use when using powder materials with a large dispersion of the components of the compositions.

Aluminum heat-resistant powders were used as the processed materials. The classic powder for surfacing has a highly spherical shape, a smooth surface of granules, a low dispersion in size. Carbon is planned to be used as an absorbing material. The carbon saturation of the surface of the powder materials was carried out by co-grinding in a planetary mill. As a result of this approach, the surface and shape of the granules has been greatly altered. To establish the phenomenon, which is the dominant role in the absorption process, aluminum was also processed in a ball mill with the addition of corundum. Thus, it was possible to obtain a powder-like material similar in morphology to the surface and form of granules, but without carbon content. In the literature, no data were found concerning the increase in the absorptive capacity of materials with the use of alumina powders for processing with laser radiation with a wavelength of 1.06 μm. In turn, carbon has good absorption in a wide range of wavelengths [5]. The well-developed morphology of the powder granules must absorb laser radiation to a much greater extent than a smooth surface.

When introducing various ceramic components into the composition, there are many difficulties associated with different melting points, phase transitions of the materials used [6].

2. Experimental work

Laser fusion was carried out using a millisecond laser system operating in a pulsed mode of radiation generation. The treatment was carried out in argon: a powder composition was applied to the surface of the metal substrate, the thickness of the powder layer was 150 μm. This thickness of the layer is due to the need to determine the condition for the maximum effective absorption invested in the melting of energy. The efficiency of reflow at the stage of selection of regimes was determined by visual control of the surface of the formed roller. The resulting surface was subjected to pulsed laser action, the radiation wavelength was 1060 nm, the spot size was 2 mm, the overlap when scanning 60%, pulse repetition frequency 25 Hz, the energy in the pulse varied from 1 to 30 joules in the pulse. The diameter of the laser beam on the treated surface was 1.8 mm. The treatment was carried out in 1 layer. The surface treatment of the powder layer was carried out without the use of additional heating. During the action of the laser pulse, a liquid phase was formed, followed by rapid cooling and solidification. The nature of the treatment did not cause a significant heat load on the melted powder-substrate system. The scheme of the experimental setup is shown in Fig 1.

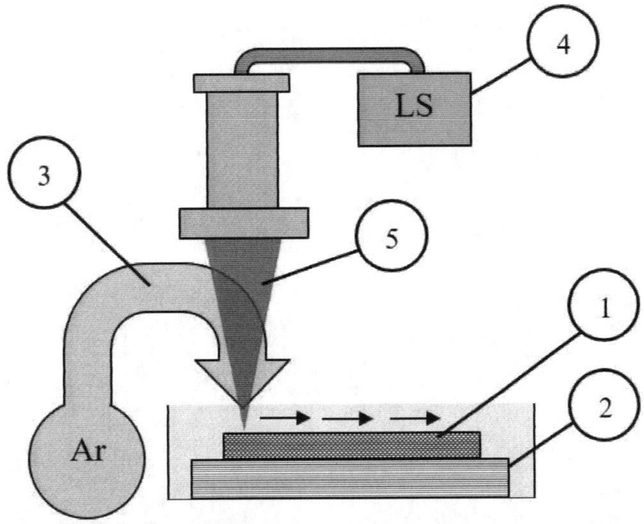

Fig.1. Experimental setup: 1) powder, 2) Metal substrate, 3) The argon supply system, 4) Laser system, 5) Laser radiation

Optimum regimes for uniform reflow of powder materials by the criterion of the minimum sufficient power: 1 - (Al + Mg + 9C) - 4 J, 2 - (Al + Mg + 1C) - 5 J, 3 - (Al + 2Mg + 1ANT) - 5 J, 4 - (Al (heat-resistant)) - 9 J. During processing, materials 1-3 were melted at an energy almost 2 times lower than pure aluminum powder. It is worth noting a fairly narrow window at which a uniform roller is formed (+/- 0.2-0.4 J). When exposed at a lower energy, the powder is collected in droplets, the wettability of the components of the composition is poor; when the energy is exceeded, intense combustion of the material takes place with a characteristic dynamic gassing. In the treatment of pure aluminum, the stable formation of the roller occurs at 9 J, continues to about 12 J, further, cracks are formed, caused by a high temperature load. Probably, if the surface is heated up to temperatures of 250-350 degrees Celsius, according to [7], the formation of cracks can be avoided. At energies above 15 J, active gas formation and combustion of the powder material occur

3. Research

The research work consisted of the analysis of the initial material, the particle size distribution and the morphology of the surface of the powder particles. The morphology of the surface and the granulometric composition of the powders after mechanochemical treatment in a ball mill were investigated by scanning electron microscopy (Fig. 2). The photographs of SEM show that the particle sizes of the powder differ significantly. The particle sizes of Al + Mg + 9C vary from 10 to 100 μm, Al + Mg + 1C is 40-150 μm, Al + 2Mg + 1ANT is 50-120 μm, Al (heat-resistant) is 10-120 μm. Particles of powders have a complex, ambiguous form.

IOP Conf. Series: Journal of Physics: Conf. Series **951** (2017) 012016 doi:10.1088/1742-6596/951/1/012016

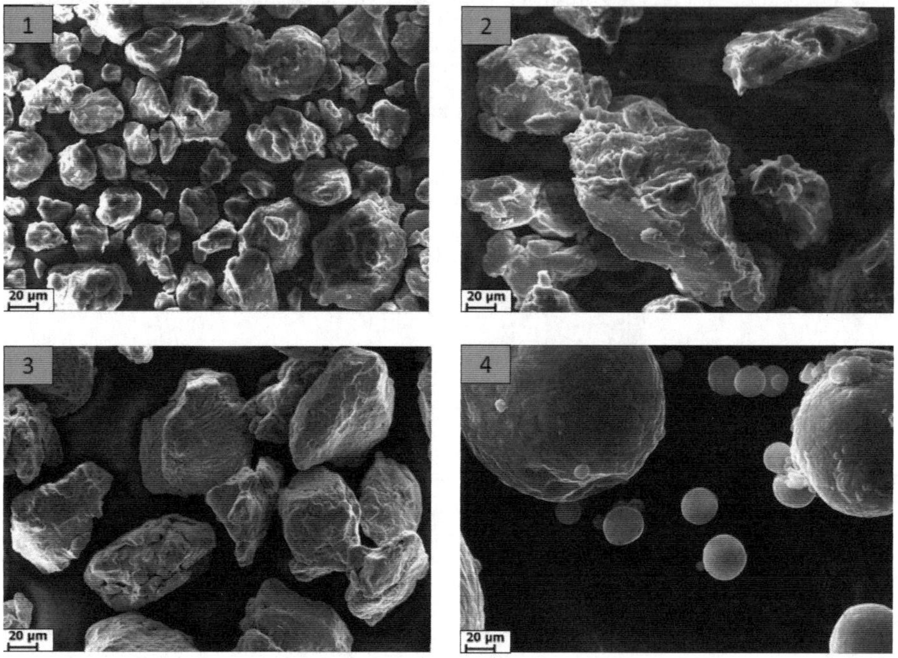

Fig. 2. SEM images of powders: 1) Al + Mg + 9C, 2) Al + Mg + 1C, 3) Al + 2Mg + 1ANT, 4) Al (heat-resistant)

After the work on laser melting of the surface, a study was made of the surface of the Raman scattering (see Fig. 3) and optical microscopy (see Fig. 4), which was fused with spectroscopy. The surface structure, the presence of carbon, aluminum carbide, aluminum oxide were studied. The distribution of these substances on the formed surface was investigated. On the surface of sample 1 there were peaks of aluminum carbide, graphite grains were encountered, islands of fine-grained graphite were encountered on the surface of the sample (Al + Mg + 1C), alumina was recorded on the surface of the sample (Al + 2Mg + 1ANT) at the boundary of the reflow points. For a sample of pure aluminum, aluminum oxide was not recorded.

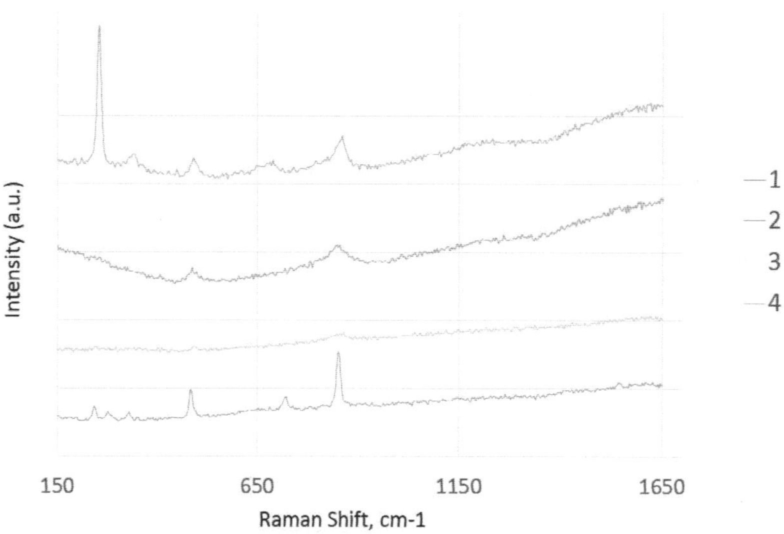

Fig.3. Raman spectroscopy of the obtained samples: 1) Al + Mg + 9C, 2) Al + Mg + 1C, 3) Al + 2Mg + 1ANT, 4) Al (heat-resistant)

The next step was to study the formed roller in the cross section by creating microsections with subsequent etching of the surface. This study was aimed at studying the structure of the formed material in a cross section to detect pores, cracks, foreign inclusions by means of metallographic analysis. The results of the investigation of microsections are shown in Fig.4.

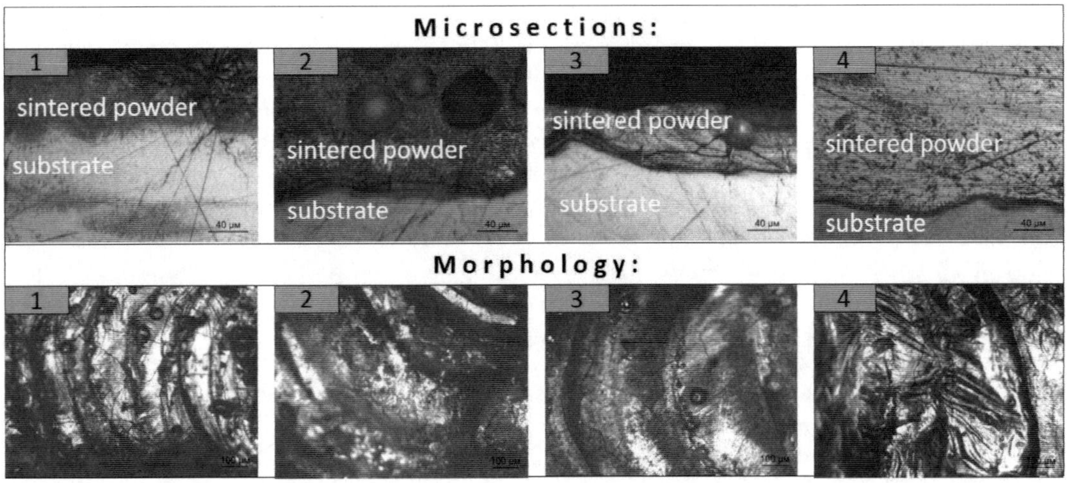

Fig.4. Metallographic examination of sintered aluminum powder by SLS-method: 1) Al+Mg+9C, 2) Al+Mg+1C, 3) Al+2Mg+1ANT, 4) Al (heat-resistant)

Information on the nature of the ongoing sintering processes can be obtained as a result of an analysis of the geometric characteristics of the laser exposure zones - the depth and width of the HAZ. The width and depth of the zone of thermal laser action in the general case depends both on the parameters of laser processing, and on the reflectivity, density, and thermophysical properties of the material being processed.

Fig. 4.1 shown that on the surface of melted zone contain free carbon and the inclusion of aluminum carbide which prevent the wetting between layers. On Fig. 4.2 the observed processes are similar to the first sample, but a smaller content of carbon not lead to formation a expressed carbon-containing layer, preventing to wetting by the melt, there is a possibility of applying several layers. On Fig. 4.3 present needle-like or columnar crystals growing in the direction of intensive cooling. There is a significant number of cracks and low adhesion resistance as the substrate and in the formation of new layers. For Fig. 4.1-4.3 is characterized by a much smaller thickness of the layer formed. This is probably due to the presence of an excessive amount of carbon and corundum that prevents the wetting of the components of the composition between themselves. Also, the developed surface of the powder materials used is a sufficiently gas-saturated medium, which is accompanied by a dynamic gas yield under the action of laser radiation, as a result of the removal of the material from the region of action, a small thickness of the formed layer (about 30% of the powder thickness). On Fig. 4.4 shown the formation of fine-grained structure, the porosity is not observed. However, achieving a stable melting takes 2 times more power than material processed in the case of Fig.1-3.

On the basis of the data obtained, conclusions were drawn about changes in the structural-phase composition, the nature of the distribution, the localization of alloying additives in the course of phase-to-phase transitions, and the change in the phase states of alloying additives. Data are obtained on the pronounced effect of the morphology of the powder granules and on the presence of absorbing materials on the surface of the granules on the absorption of laser radiation energy.

4. Conclusion

In the work done, the effect of surface structuring of powder microgranules was investigated to increase the absorption of laser radiation by laser fusion. Also, an absorbent coating based on finely divided carbon was used. Thus, the use of absorbing materials (carbon) and the change in the morphology of the surface of the granules made it possible to reduce the energy of formation of a uniform roller from the powder by almost 50%, which is a very tangible result. The structure of the formed fusion layer is practically the same as the original material. When using corundum as an abrasive material to change the morphology of the surface of the powder granules, the structure of the formed weld layer has pronounced cracks. This is probably caused by poor wettability of corundum with aluminum. The result of this paper qualitatively shows the possibility of changing the absorption capacity of the granules of the powder composition by modifying the surface of the granules or using highly absorbing materials. Increasing the effectiveness of this approach is possible by improving the technology of changing the morphology of the surface by selecting the optimal phase and quantitative composition of the abrasive material, the material of the absorbing coating, and the use of other physical mechanisms of action.

This study was performed as a part of the state task VlSU 1106/17 and grants of the RFBR number 16-48-330031, 17-20-03084 and 16-08-01226.

References

[1] Aborkin A.V. et al. 2016 *J. Russian nanotechnologies.* The effect of the mechanical activation regime on the morphology and phase composition of the nanostructured composite material Al-2Mg-nC №5-6. P 30-36

[2] Bashkirov E.R,. Kochuev D.A., et al. 2015 *Journal of Magistracy* 3D synthesis of composite materials and parts P 39

[3] Grigoryev S.N., Tarasova T.V. 2015 *J. Metallurgy and heat treatment of metals.* №10 P 5

[4] Yan S.J., Dai S.L., Zhang X.Y. et al. 2014 *J. Materials Science & Engineering* V. 612 P 440

[5] Santos-Beltran A., et al. 2004 *J. Metastable and Nanocrystalline Materials* Microstructural and mechanical characterization of aluminium-graphite composies V. 20–21 P 133–138

[6] Kuznetsov S.I., Nefedov S.A., et al. 2005. *Izvestiya of the Samara Scientific Center of the Russian Academy of Sciences* Influence of composition of ti-al powder composition on the processes of laser sintering and intermetallid synthesis V.7 № 1 P 35-42

[7] Shishkovsky I.V. 2016 *Basics of additive high-resolution technologies* (SPb: Pete.)

[8] Grigoriants A.G., Safonov A.N. 1987 *Laser technology and technology: In 7 books. Book. 3. Methods of surface laser treatment* Ed. A.G. Grigor'yants (Moskow: Higher School) pp 159

[9] Popov M., Medvedev V., et al. 2010 *J. Applied Physics.* Fulleride of aluminum nanoclusters V. 108. № 9 P 094317

[10] Delone N.B. 1989 *Interaction of laser radiation with matter - Course Lectures* (Moskow: Science) pp 280

IOP Conf. Series: Journal of Physics: Conf. Series **951** (2017) 012017 doi:10.1088/1742-6596/951/1/012017

Supramolecular effects as driving force of dipyrrin based functional materials engineering

E Banakova[1], A Bobrov[1], A Kazak[2], Yu Marfin[1], D Merkushev[1], E Molchanov[1], E Rumyantsev[1], M Shipalova[1], S Usoltsev[1], O Vodyanova[1]

[1] Department of Inorganic Chemistry, Ivanovo State University of Chemistry and Technology, Sheremetevsky Str., 10, 153000 Ivanovo, Russia
2 Nanomaterials Research Institute, Ivanovo State University, Ermak Str., 39, 153025 Ivanovo, Russia

E-mail: marfin@isuct.ru

Abstract. Dipyrrin based luminophores are of major interest in different areas of chemistry, material science and molecular biology. Vast variety of the structures with dipyrrin motif were synthesized and investigated up to date. Modern trend in the dipyrrin chemistry is the aimed functionalization of the ligand or complex structure allowing to gain the mechanism based on supramolecular interactions for controlling spectral and photophysical characteristics of compounds for tuning practically valuable properties for specific tasks. Presented paper summarize the results of our research group, working in the field of dipyrrin complexes with *p*-elements: synthesis, spectral characteristics evaluation and possibilities of practical application investigation. Discussion is focused on the opportunities of molecules preorganization for achieving the supramolecular interactions causing the tuning of fluorescence of the compounds in solutions, polymeric matrices and thin films.

1. Introduction

A brief review of existing data and some applications of dipyrrin coordination compounds – chemically stable compounds with intense chromophore and fluorescent properties, would be presented in the following paper. Dipyrrin core – the basic "building" motif of blood heme, chlorophyll, bilirubin and related compounds. The undoubted success of dipyrrhin complexes is primarily caused by the synthetic possibilities of modifying its structure (Figure 1), which allows to "adjust" the spectral-luminescent characteristics, reactivity and physicochemical properties of molecules for specific tasks. To date, more than 30 groups around the world are using this tool for obtainment the new dipyrrin based structures and its using as components of the functional materials of different fields.

IOP Conf. Series: Journal of Physics: Conf. Series **951** (2017) 012017 doi:10.1088/1742-6596/951/1/012017

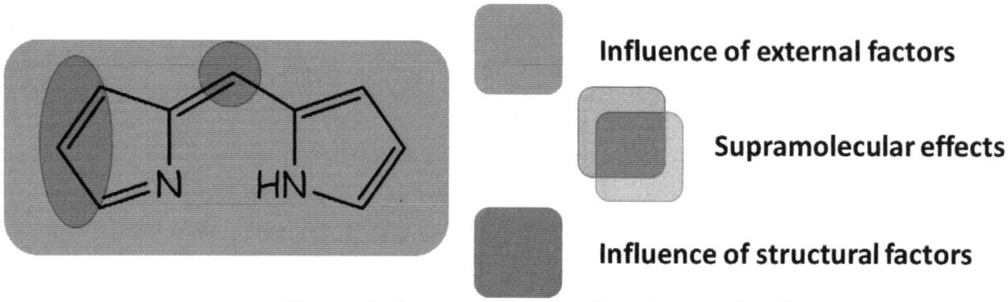

Figure 1: Opportunities for dipyrrin functionalization.

The wide possibilities of directional control of light-gathering and light-transmitting properties due to the variation of "internal", mainly structural factors, and "external" conditions (the nature of the environment) make it possible to predict the high prospects of dipyrrin complexes and materials on their basis as the electron-optical molecular converters for needs molecular sensorics, optical conversion, and modern photovoltaics.

2. Supramolecular effects in the chemistry of Bodipy dyes

Among the variety of complexes with dipyrrin ligands Bodipy are the most interest. Boron dipyrrins (Bodipy) are unique complexes occupying an "intermediate" position between classical coordination and organometallic compounds. The enormous interest of researchers from all over the world to the compounds of this class has by now provided an endless stream of publications that requires generalization.

Bodipy is proved to be a compound of unique kinetic stability in acids, due to covalent binding of Boron atom with Nitrogen heteroatoms and increasing the stability of coordination unit. Bodipy is very resistant to hazardous chemicals including acids and alkali, allowing to expand the range of pH of Bodipy applications including molecular sensorics and hybrid materials synthesis. Stability ranges of different Bodipy dyes could be found in literature as well as the mechanisms of Bodipy dissociation in aqueous and non-aqueous solutions [1]. This results demonstrate the great advantages of Bodipy over other dyes of organic and organometallic nature due to their high chemical resistance to acids and alkalis. The other advantage of Bodipy complexes are they photo- and thermal stability in solutions, which can be further increased via immobilization in the polymeric matrices of organic, or inorganic nature [2]. The listed features of Bodipy dyes determine the breadth of their use as optical converters in laser technologies. However, the use of Bodipy as components of smart materials for molecular sensing requires further modification of the structure of dipyrrins.

For the last several years our team was focused on Bodipy complexes with *meso*-substitution by bulky organic moieties or alkyl chains (Figure 2). This allowed us to investigate the influence of supramolecular organization of Bodipy molecules in solutions, polymeric matrices and thin films via π-π staking causing the formation of aggregates and tuning the photophysical characteristics.

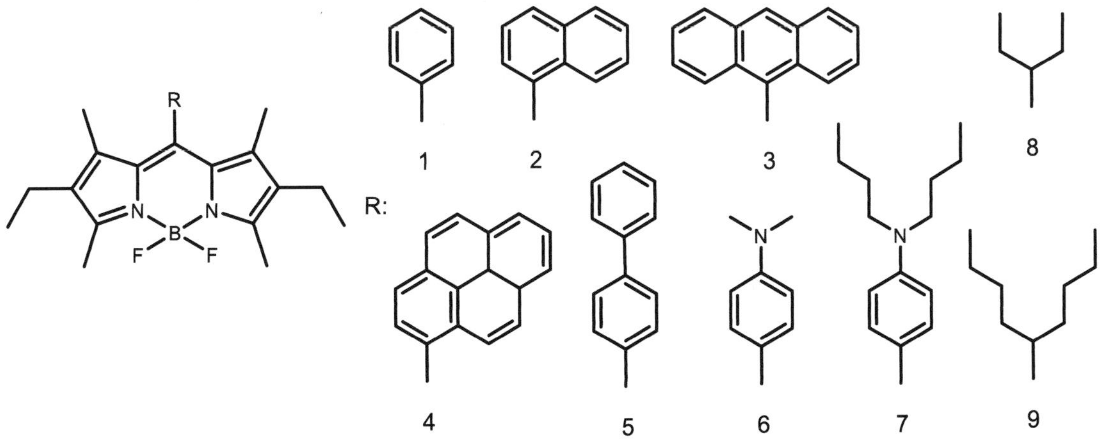

Figure 2: Structures of the Bodipy complexes with bulky substituents.

In dilute solutions the compounds demonstrate intensive fluorescence with the peak at around 540 nm, small Stokes shift and the substituent nature does not affect strongly the spectral properties of the dyes. Spectral and photophysical properties of the discussed Bodipy dyes were evaluated in a range of organic solvents and polymeric matrices. Immobilization does not affect the character of the absorption and fluorescence spectra of the dye. The highest quantum yields changes under the solvent variation was observed for 7, where the changing of the solvent from carbon tetrachloride to ethanol cause fluorescence quenching from 0.633 to 0.028. At the same time, fluorescence intensity is increasing in case of solvent polarity growth for compounds 2-4 with extended aromatic moiety at *meso*-position, it is not typical for Bodipy, where polar solvents are usually regarded as fluorescence quenchers. That fact could be explained by the high hydrophobicity of compounds under investigation. Taking that into account the Bodipy with bulky substituents could be used as the fluorescent media polarity indicators in analytical chemistry and biochemistry [3].

For compound 1 the effect of the substituent interaction with phenyl moiety of the phenyl-modified silica matrix was observed. This effect is caused by π-π interactions of the dye molecule with matrix. This effect could be used for fine control of the matrix formation process and leads to increase of the dye emission in the final material for further practical applications [4].

There is an increased interest in compounds that contain two or three Bodipy domains. Combining two or three dipyrrin moieties with a phenyl fragment through some spacer could lead to an increase in photophysical characteristics with respect to the individual Bodipy. Our assumption is based on the fact that individual domains still have to influence each other, showing either a positive or negative synergistic effect. Several poly-dipyrrin Bodipy bridged by the di- and triphenylamine spacers were synthesized and characterized. Such compounds showed increased absorption characteristics and intense fluorescence. Peaks positions are the same as mono-Bodipy analogies. Obtained compounds demonstrate enhanced aggregation stability, caused by the intramolecular interactions of the separate domains.

8-phenyl and 8-biphenyl substituted Bodipy (compounds 1 and 5) were found to be the fluorescent molecular rotor (Figure 3). The term "molecular rotors" refers to compounds whose molecules consist of two or more fragments that can rotate relative to each other. The major part is called the stator, and the smaller rotator or rotor. The effect of the fluorescent molecular rotor should not be understood purely mechanically, as the rotation of the parts of the molecule relative to each other. The molecule first goes into a twisted state, while the system becomes coplanar and there is a dissipation of energy in the excited state. These molecules exhibit two types of de-excitation pathways once they are in the excited state: a fluorescence radiative pathway and a non-radiative de-excitation pathway. This intermolecular mechanism allowed to use investigated Bodipy as molecular sensors demonstrating the fluorescence changes under the temperature variations unlike the Bodipy with larger substituents [5, 6].

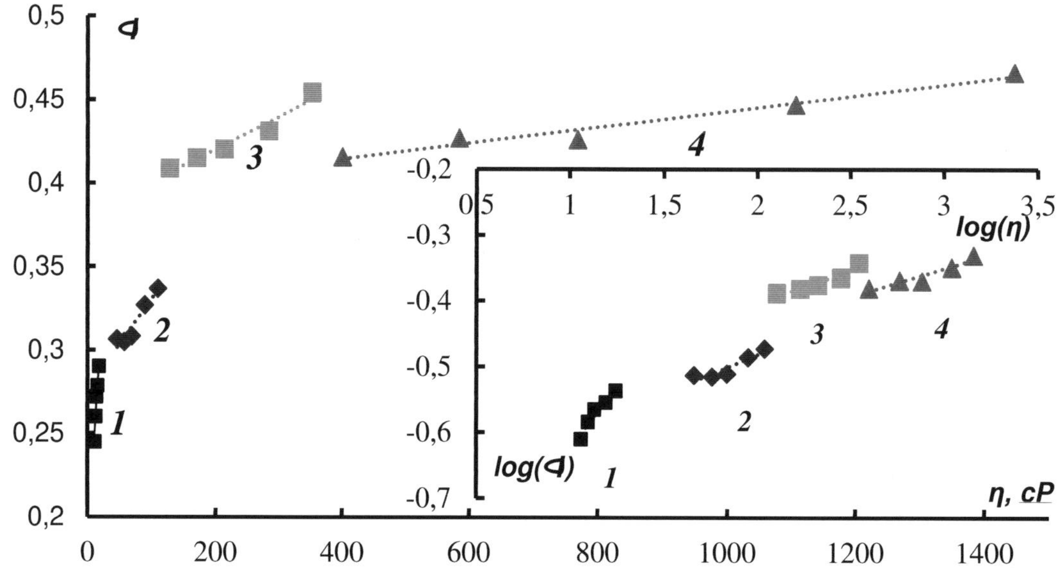

Figure 3: the dependence of the fluorescence quantum yield of 1 on the dynamic viscosity of the binary (2-propanol – glycerol) solvent: 1 – χglycerol = 0; 2 - χglycerol = 0.3; 3 - χglycerol = 0.5; 4 - χglycerol = 0.7. Reprinted form [6].

At medium and high concentrations, the extension of the substituent π-conjugation enhances molecular association caused presumably by π-π interaction between the substituents of the neighboring molecules. Aggregation leads to bathochromic shifts in absorption and fluorescence spectra displaying the ability of the investigated dyes to show aggregation induced emission enhancement. Displacement of the monomer-associate equilibrium in the hybrid materials with polymectyl methacrylate and polyvinylcarbazole was exploited for directional tuning their emission. The fabricated OLED devices were found to demonstrate moderate electroluminescence efficiency. Both self-absorption and aggregation induced emission effects cause desirable emission color changes but decrease efficiency of the active layer (Figure 4) [7].

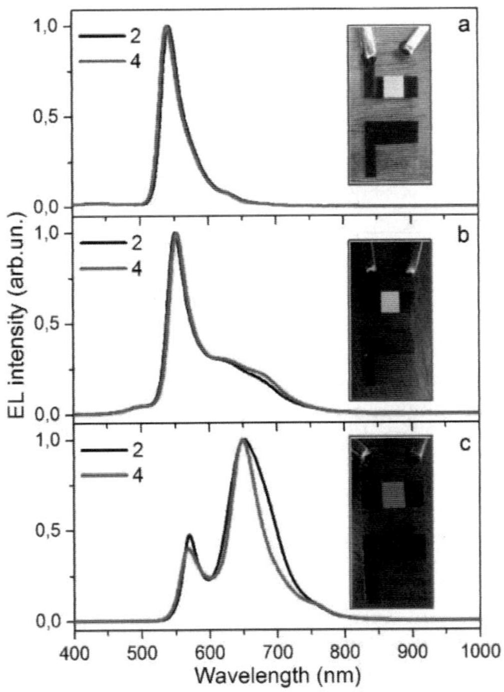

Figure 4: EL spectra of 5% doped (a), 10% doped (b) and neat-layer (c) OLEDs based on the dyes 2 and 4. The insets show EL color tuning for the devices containing 4. Reprinted from [6].

Next we examined investigated Bodipy in Langmuir-Schaefer thin films [8]. It was proved that the substituent's nature greatly affects the photophysical properties of the studied dyes via the shift of a monomer – excimer equilibrium in the obtained films. In particular, compound 7 can form only monomers in the films, while the other dyes demonstrate the different degree of aggregation.

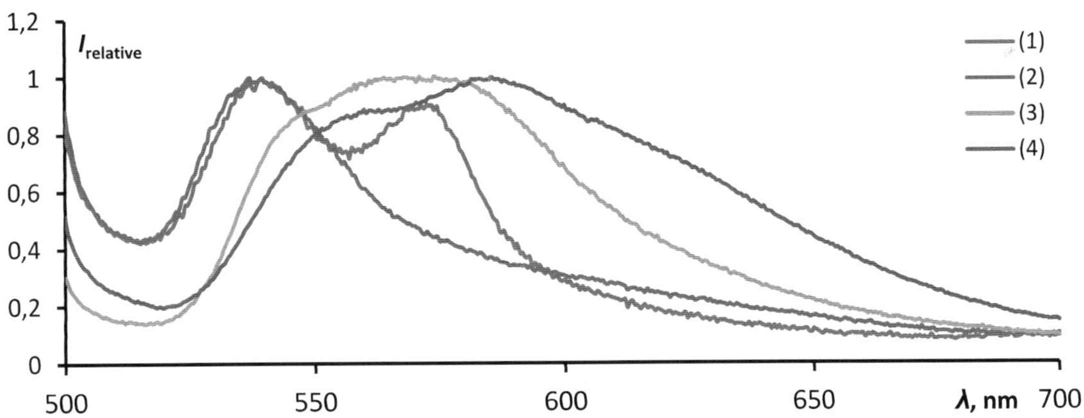

Figure 5: Relative fluorescence spectra of LS-films under the dye variation. The films were transferred onto glass substrate. Line 1 – Compound 6, Line 2 – Compound 7, Line 3 – Compound 8, Line 4 – Compound 9. Reprinted from [8].

It was shown that the film surface reliefs are uniform and without typical for macroheterocyclic compounds aggregation centers. The results of the XRD analysis showed a single-layer arrangement of the BODIPY molecules located parallel to the substrate surface. The microcrystals observed by the fluorescence microscopy were found to form the same structure as thin films but with a higher crystallinity. Intermolecular hydrogen bond formation and π-π staking of the Bodipy cores were found to be the structure forming forces during the films manufacturing. Bodipy 7 is the first example of the solid state monomeric Bodipy with strong fluorescence, making it promising for solid state optoelectronics.

3. Oxophosphoryl complexes of dipyrrins (Podipy) as the evolution of Bodipy ideas to biological applications

Supramolecular mechanisms could be applied for Bodipy usage as fluorescent molecular sensors on biomolecules. Thus, we observed the pronounced changes of dyes fluorescence in presence of the proteins and biofluids. 8-aryl Bodipy derivatives were found to have high fluorescent respond upon the protein addition. The results of fluorescence spectroscopy indicated that the fluorescence of hydrophobic Bodipy dyes substantially increases in the presence of bovine serum albumin (BSA). The prospected reason is binding of the dye to the BSA hydrophobic cavities in the subdomain IIA. π-cation stacking between BODIPY and arginine amino acid residue of the protein was detected. It is also shown that far phenyl moiety of compound 5 can undergo parallel π-π stacking interaction with aromatic fragment of TRP213, being possible way for the energy transfer. Namely, such geometry causes strong influence of the ligand on the protein fluorescence. Aromatic moiety size increase causes shifting of the interaction driving force from mostly non-specific to the π-stacking. Among all probes, the 8-phenyl substituted dye showed the highest sensitivity and the quantitative fluorescent response for BSA in aqueous solution [9].

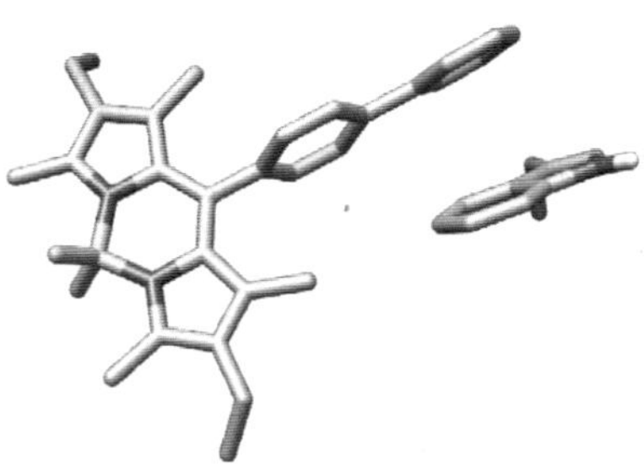

Figure 6: π- π stacking interaction between compound 5 and tryptophan. Reprinted from [9].

Photophysical properties of several Bodipy based fluorescent dyes were investigated in systems containing blood plasma biomolecules. The interaction between the investigated dyes and protein plasma components changes spectral characteristics of the dyes and leads to bathochromic and hypochromic absorption spectra shifts accompanied by changing of fluorescence intensity. It was shown that the static factor of molecular dye-biopolymers complex formation prevails at plasma protein concentration up to 1 g/l, while the higher viscosity range is characterized mainly by nonspecific fluorophore interactions. The increase of fluorescent characteristics of phenyl-substituted Bodipy in the presence of proteins caused by resonance energy transfer and change of physicochemical properties of the molecular environment of the fluorophore was shown for the first time [10].

The development of new methods and the improvement of existing techniques for studying physicochemical processes in living cells is an urgent task of modern biomedicine, the solution of which opens up new opportunities for studying pathologies and controlling the effectiveness of disease therapy. Among the variety of existing solutions in the field of biochemical processes and medical diagnostics, fluorescent technologies - fluorescence spectroscopy and fluorescence microscopy in steady-state and time-resolved modes deserve special attention. These methods require the use of fluorophores as imaging agents, as labels of biomolecules or systems, process and state sensors. In this regard, the development of new luminophores with the required characteristics for specific applications and their approbation is a complex task of synthetic, physical and biological chemistry.

Among the variety of factors reflecting the state of the cell, the local viscosity of the cytosol, the pH gradients, and the polarity of the biopreparation sites are among the most significant. The difficulty of accurately mapping these parameters in the cell in norm and pathology is a key problem, the solution of which is directed at this project. The use of existing organic luminophores for determining the indicated parameters is largely limited by the solubility of the sensor molecules in water. The use of co-solvents for the solubilization of fluorophores makes interpretation of the results difficult, due to the influence of the solvent on the characteristics of the investigated systems. In this connection, the most promising solution will be the use of water-soluble fluorophores with the necessary spectral properties and possibilities for spectral response to changes in the parameters under study.

The proposed fundamentally new solution of the preparation of fluorophores for cell diagnostics based on oxiphosphoryl dipyrrin complexes has a number of competitive advantages, the most important of which are the presence in the structure of the compound of the oxophosphoryl groups, which determines compound's water solubility without additional modifications, and the resistance to quenching of fluorescence in the aquatic environment. From the one hand, such compounds could be used with no additional cosolvent simplifying the interpretation of experimental results. From the other hand, combination in one molecule of the polar oxophosphoryl group with high hydrophilicity with hydrophobic *meso*-phenyl substituent could farther increase the discussed supramolecular effects for such compounds rising the fluorescent response.

At the time of writing, only three articles on the synthesis of PODIPY [11], quantum-chemical study of properties [12], as well as the use of mercury (II) cations as sensors [13] can be found in the literature. It should be noted that examples of the use of Podipy complexes as markers or sensors of biological molecules are currently not available in the scientific literature. At the same time, that the synthesis of PODIPY is currently not worked out, the general approaches to obtaining compounds of this class coincide with those for the widely studied BODIPY complexes.

We used standard synthetic routes for obtaining dipyrrin ligands, then the Podipy was obtained according to [11] (Figure 7). Synthesized dyes were tested in the range of solvents under the variation of solvent polarity, acidity and abilities for the specific interactions with the dye (Figure 8). Spectral properties of Podipy generally the same as for Bodipy dyes: narrow absorption and fluorescence peaks at about 510-520 nm and 530-550 nm respectively. We should mention hypsohromic shifts of the peaks comparing to Bodipy in the same solvents, indicating the lower polarization of ligand electronic system and, thus, lower stability of the complexes. General spectral changes were observed under the solvent variation, reversible character of the band shifts indicate the solvatochromic or aggregation induced mechanism of the process.

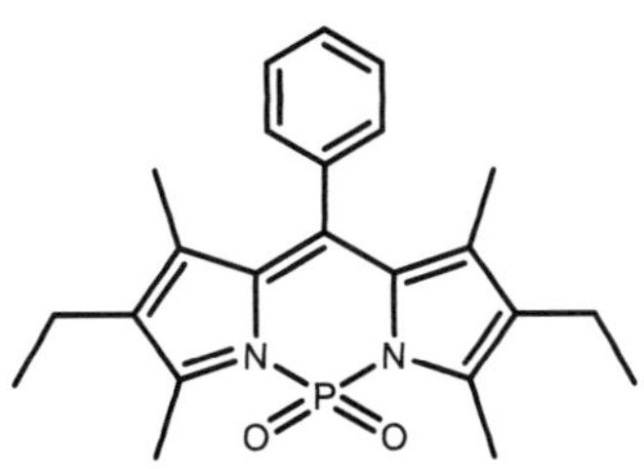

Figure 7: Structure of the Podipy complex.

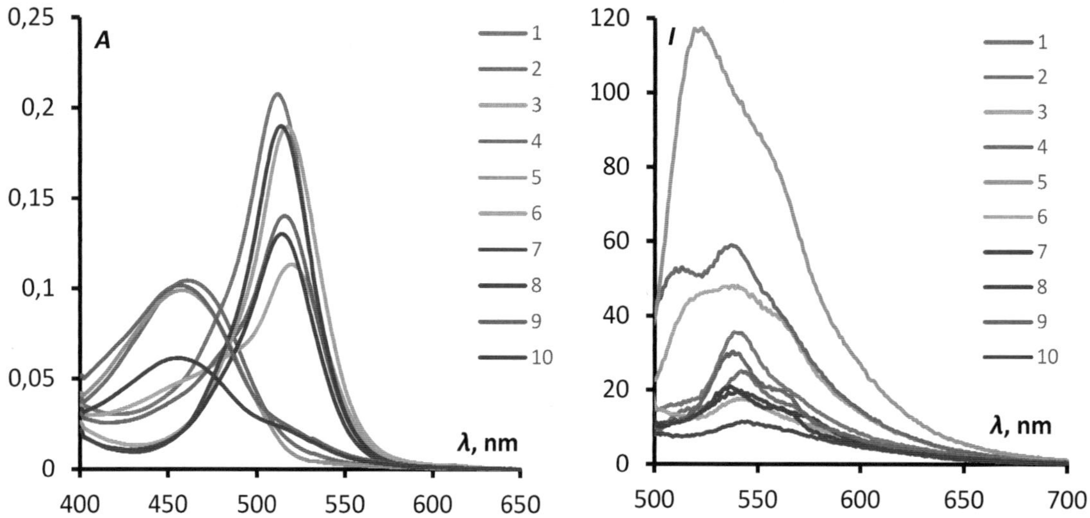

Figure 8: Electron absorption and fluorescence spectra of Podipy complex in the range of solvents: 1 – CH_2Cl_2; 2 – C_6H_6; 3 – C_2H_6OS; 4 – C_2H_5OH; 5 – C_3H_7NO; 6 – CCl_4; 7 – $CHCl_3$; 8 – H_2O; 9 – C_6H_{12}; 10 – C_6H_{14}.

It should be noted, that investigated Podipy demonstrate the moderate water solubility preserving the intensive fluorescent respond unlike Bodipy 1 with the same structure of the ligand. The presence of hydrophilic oxophosphoryl site and hydrophobic phenyl moiety lead to the interesting supramolecular effect, thus, even at small concentrations aggregates formation were detected with the batochromic shifts of fluorescence band. Further research is required for synthesis of the Podipy structures with enhanced amphiphilic properties for address adjustment of spectral characteristics by controlling the supramolecular organization.

Acknowledgments
The results of section 2 were obtained within the state assignment of Grant of the President of the Russian Federation (Grant No MK-8835.2016.3) and the Program of the Ministry of Education and Science of the Russian Federation (Grant No 16.1037.2017/4.6). The results of section 3 were obtained within the RSF grant (project No. 17-73-10408).

References

[1] Evgeniy V. Rumyantsev, Sergey N. Alyoshin Yuriy S Marfin 2013 Kinetic study of Bodipy resistance to acids and alkalis: Stability ranges in aqueous and non-aqueous solutions *Inorganica Chim. Acta* DOI 10.1016/j.ica.2013.08.015.

[2] Marfin S Y, Rumyantsev E V., Yutanova S L and Antina E V. 2013 Spectral and photophysical properties, photo and heat resistance of dipyrrolylmethene borofluoride complex and its hybrid material with polymethylmethacrylate *Russ. J. Gen. Chem.* DOI 10.1134/S1070363213020242.

[3] Marfin Y S, Vodyanova O S, Merkushev D A, Usoltsev S D, Kurzin V O and Rumyantsev E V. 2016 Effect of Pi-Extended Substituents on Photophysical Properties of BODIPY Dyes in Solutions *J. Fluoresc.* DOI 10.1007/s10895-016-1891-3.

[4] Bobrov A V., Usoltsev S D, Marfin Y S and Rumyantsev E V. 2017 Spectral Properties and Possibilities of meso-Substituted BODIPY Usage in Sol–Gel Process and Materials *J. Fluoresc.* DOI 10.1007/s10895-017-2190-3.

[5] Marfin Y S, Merkushev D a., Levshanov G a. and Rumyantsev E V. 2014 Fluorescent Properties of 8-phenylBODIPY in Ethanol – Ethylene Glycol Mixed Solutions *J. Fluoresc.* DOI 10.1007/s10895-014-1447-3

[6] Marfin Y S, Merkushev D A, Usoltsev S D, Shipalova M V. and Rumyantsev E V. 2015 Fluorescent properties of 8-substituted BODIPY dyes: Influence of solvent effects *J. Fluoresc.* DOI 10.1007/s10895-015-1643-9.

[7] Merkushev D A, Usoltsev S D, Marfin Y S, Pushkarev A P, Volyniuk D, Grazulevicius J V and Rumyantsev E V BODIPY associates in organic matrices: Spectral properties, photostability and evaluation as OLED emitters *Mater. Chem. Phys.*, 2016, DOI 10.1016/j.matchemphys.2016.11.053.

[8] Marfin Y S, Usoltsev S D, Kazak A V., Smirnova A I, Rumyantsev Y V., Molchanov E E, Kuznetsov V V., Chumakov A S and Glukhovskoy Y G 2016 Synthesis and spectral properties of preorganized BODIPYs in solutions and Langmuir-Schaefer films *Appl. Surf. Sci.* DOI 10.1016/j.apsusc.2017.04.014.

[9] Vodyanova O S, Kochergin B A, Usoltsev S D, Marfin Y S, Rumyantsev E V., Aleksakhina E L and Tomilova I K 2017 BODIPY dyes in bio environment: Spectral characteristics and possibilities for practical application *J. Photochem. Photobiol. A Chem.* DOI 10.1016/j.jphotochem.2017.09.049.

[10] Marfin Y S, Aleksakhina E L, Merkushev D A, Rumyantsev E V. and Tomilova I K 2016 Interaction of BODIPY Dyes with the Blood Plasma Proteins *J. Fluoresc.* DOI 10.1007/s10895-015-1707-x.

[11] Jiang X D, Zhao J, Xi D, Yu H, Guan J, Li S, Sun C L and Xiao L J 2015 A new water-soluble phosphorus-dipyrromethene and phosphorus-azadipyrromethene dye: PODIPY/aza-PODIPY *Chemistry* DOI 10.1002/chem.201406535.

[12] Fihey A, Favennec A, Guennic B Le and Jacquemin D 2016 Investigating the properties of PODIPYs (phosphorus-dipyrromethene) with ab initio tools *Phys. Chem. Chem. Phys.* DOI 10.1039/c5cp05653e.

[13] Jiang X D, Yu H F, Zhao J L, Sun C L, Xie Y and Xiao L J 2015 A colorimetric chemosensor based on new water-soluble PODIPY dye for Hg2+ detection *Chinese Chem. Lett.* DOI 10.1016/j.cclet.2015.07.002.

Verification of the quantum dimension effects in electricsl condactivity with different topology of laser-induced thin-film structures

S Arakelian, A Kucherik, S Kutrovskaya, A Osipov, A Istratov, I Skryabin

Vladimir State University named after Alexander and Nikolay Stoletovs, 87 Gorky Street, Vladimir 600000, Russia

E-mail: kucherik@vlsu.ru

Abstract A clear physical model for the quantum states verification in nanocluster structures with jump/tunneling electroconductivity are under study in both theory and experiment. The accent is made on consideration of low-dimensional structures when the structural phase transitions occur and the tendency to high enhancement electroconductivity obtained. The results give us an opportunity to establish a basis for new physical principles to create the functional elements for the optoelectronics and photonics in hybrid set-up (optics + electrophysics) by the nanocluster technology approach.

1. Introduction

We focus our attention on the fact that physical properties of nanocluster/granular systems are very sensitive to the form, size and distance between their composing elements. The phenomena are very well known for any material in general, but to change these parameters and to carry out the stable conditions for ordinary solid state object we need both to put the object under extremal high pressure ($\gtrsim$ 106 atm) and to work at low (liquid He) temperature range ($\lesssim$30K) [1]. In contrast, the nanocluster structures can be easily modified both in necessary direction and by controlled way in femto-nanophotonics experiments.

In our experiment we use the method of laser-assisted deposition metal particles from colloidal system [2]. For the deposition, we have used ytterbium fiber laser (λ = 1.06 μ) with a pulse duration of 100 ns, pulse repetition rate of 20 kHz, and a pulse energy up to 1mJ. The diameter of the laser beam in the focal plane was 50 microns. Nanoparticle array formation on the substrate surface was performed by the scanning laser beam along the same direction.

The Volt-Amper characteristics (VAC) of the thin film were measured using a four-probe scheme. The electrical transport properties in analogy with tunnel and quantum correlated states are considered; namely the tunneling/thermal activation effects observed.

2. Features of the electrical conductivity in granular films: experiments and preliminary discussion

2.1. Noble metal structures

The experimental results for the thin films of gold, are shown in

IOP Conf. Series: Journal of Physics: Conf. Series **951** (2017) 012018 doi:10.1088/1742-6596/951/1/012018

Figs. 1,2. Presented data are by averaging over 10 measurements; the film thickness (over Z-axis) was about 50 nm; nanoparticle size – about several nm.

We can see the jumps (Fig. 1) on the obtained I(U) dependences for different thickness. Moreover, we have a six orders difference in values of R (Fig. 2) by topology variation only (different fractal dimension D: from D=1.39 to D=1.93 measured by a special procedure [3]. The tendency is, to decrease R when the nanostructure sample becomes more homogeneous as a 2D-structure.

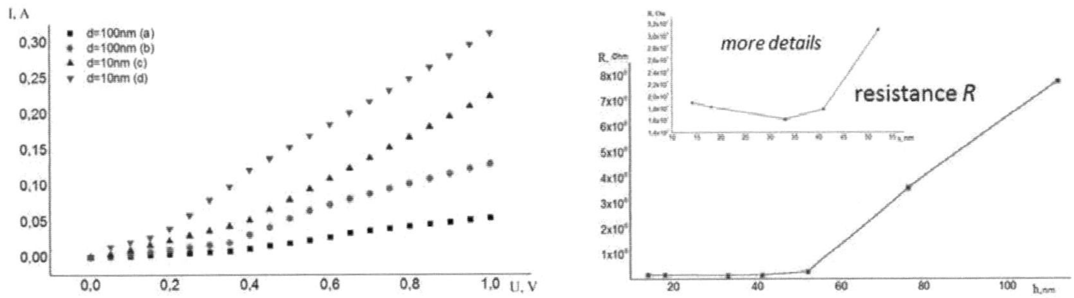

a) b)

Figure 1. (a) Experimental data for voltage-current characteristics (VAC) of the deposited layer for clustered Au-films with different thickness h;

(b) The dependences for R vs h; more detailed picture for initial part is shown in insertion, and can be presented as an analogue of the Kondo effect R (T) but in our case in version R (h).

To recognize a namely mechanism of conductivity for the cases, we carried out different experimental conditions by changing the topology of the film structure so as to realize the most optimal conditions for a given type of conductivity.

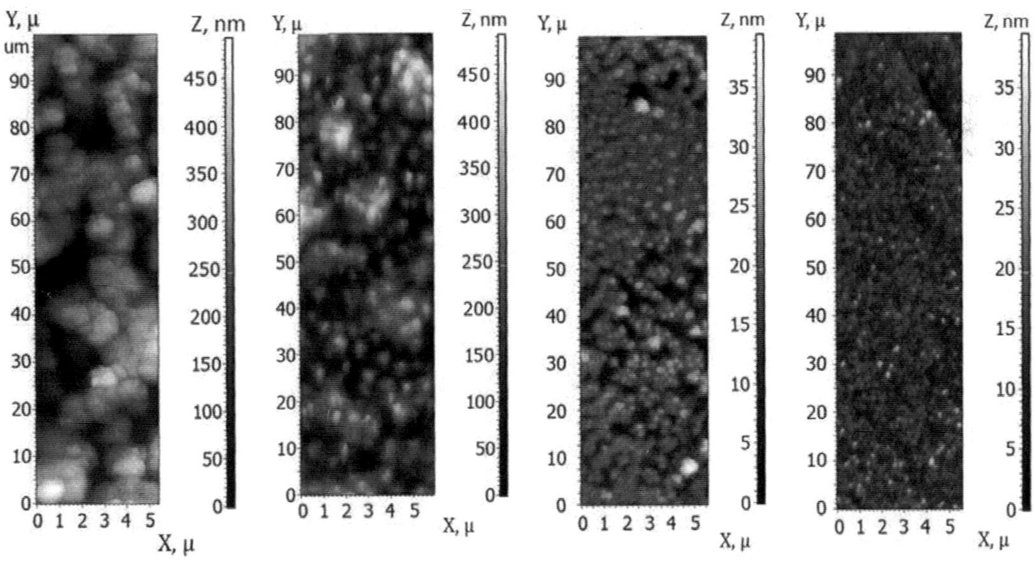

a) b) c) d)

Figure 2. Experimental and theoretical (by fractal approach) comparison of values for electroresistance R for Au-nanostructure: a) Rmeas=5.4•108 Ohm, D=1.39, Rcalc=9.3•108 Ohm;b) R meas = 3.6•107 Ohm, D=1.57, Rcalc=5.5•107 Ohm; c) R meas = 5.4•106 Ohm, D=1.84, Rcalc=9.3•106 Ohm; d) R meas = 3.6•102 Ohm, D=1.93, Rcalc=5.5•102 Ohm, where Rmeas – measurements; Rcalc – calculations; D – fractal dimension.

2.2. Bimetallic structures

To observe such effects is reasonable to use the bimetallic gold and silver nanostructurised films, in which the dielectric barrier (silver oxide) will be formed between the metal (Au and Ag) particles. For bimetallic cluster conductivity of films, deposited on the glass surface, a nonmonotonic dependence of the electroresistance of the film thickness from the amount of deposited materials is obtained (Fig. 3).In fact, we observed a competition between opening of new channels and area of the grains: for the (Fig. 3a)-case than more thickness weaker resistance in contrast with the (Fig. 3b)-case than more thickness more resistance.

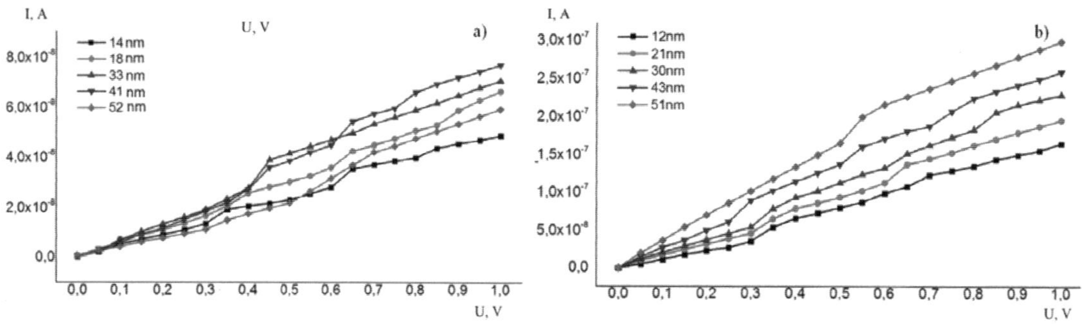

Figure 3. The VAC of bimetallic (Au+Ag) deposited films with an average thickness from Hav~10 nm to Hav~50 nm, and characteristic size (i.e. cluster size) of a~4 nm to a~15nm for two structures of different topology.

First view of last obtained dependences can be explain by mechanism like Wannier-Stark ladder in super-lattices when e-conductivity has two contributions (cf. [4]): metallic conductivity for delocalized states and jump/tunneling mechanism for localized states (insulator).

At the same time the thermal activationeffect impacts also on current value in the formed island films structure (Fig. 4). The activation energy measured by the slope of presented Arrenius like curves gives the value in 0.98 – 1.1 eV range.

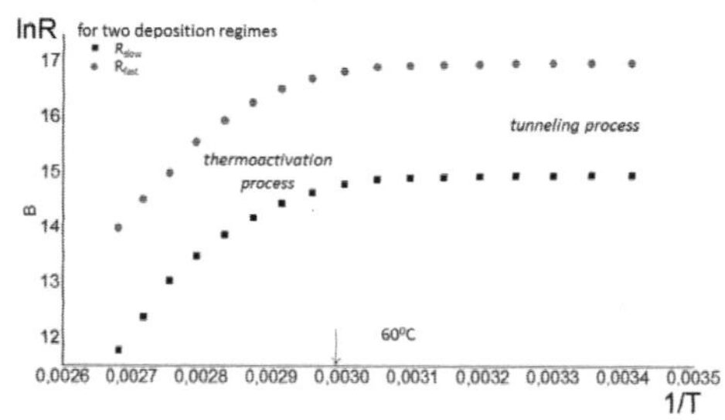

Figure 4. Temperature dependence of resistance R for the Au+Ag samples of thickness 50 nm, obtained under different deposition conditions (slow and fast deposition) in laser experiment.

2.3. TheVolt-Amper measurements for PbTe-semiconductor; tunnel effect recognition

The observed VAC for different experimental conditions is shown in Fig. 5. Quantum tunnel conductivity was detected for specific conditions of the experiment (see Fig. 5a,b). We present the results of detecting the tunnel conductivity for different morphology of NPs on the substrate. Moreover, the occurrence of jump conductivity is universal—there have been shown free types of the electrical resistance measurements, the fact depending only on the topology of the conducting layer (fixed for a given sample). So, the implementation of the electrons transition from the bound state to a free state can be taken into account in the frames of shell model clusters.

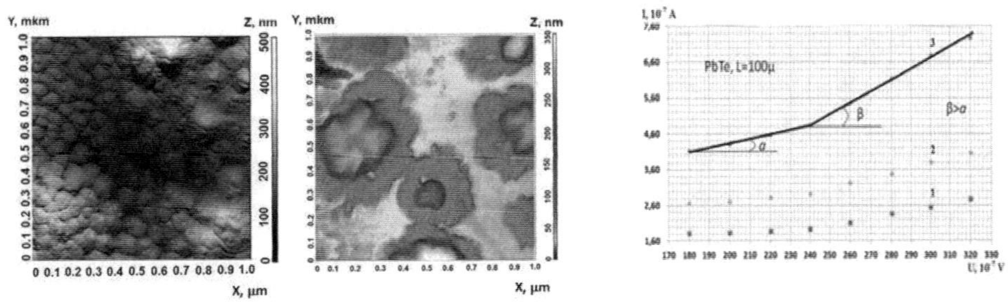

a) b)

Figure 5. VAC for the PbTe cluster film (thickness L=100 µm):(a) AFM-image of deposited PbTe-clusters: two pictures with different magnification; (b) the Volt-Amper experimental data view in the frames of the tunneling mechanism conductivity for the PbTe deposited layers: variation parameters – the average cluster sizes (a) and the average distance between neighbors clusters (d)are marked, correspondly: (1) 120 nm and 20 nm; (2) 210 nm and 8 nm;(3) 120 nm and 8 nm. The last case ($\beta>\alpha$) demonstrates optimal conditions for the tunneling effect.

2.4. Fabrication of the carbon linear structure

The stabilizing of long linear carbon chains [5] (carbyne chains) has been achieved by standard technique of laser ablation in different liquids in the presence of noble/gold nanoparticles which leads to the consolidation of the ends of linear chains on the surface of gold particles and prevents them from instabilities/destruction, i.e. twisting further in tangles. In addition, the symmetry environment of isotropic liquid from any side assisted as well to the LLLC stabilization [6].

Because of the linear carbon system has been stretched/fixed by noble metal atoms like linear system between two metallic contacts (we may practically have a large dipole moment that should result in high electroconductivity in created elements. In fact, the dependence presented in Fig. 6 strongly support the idea.

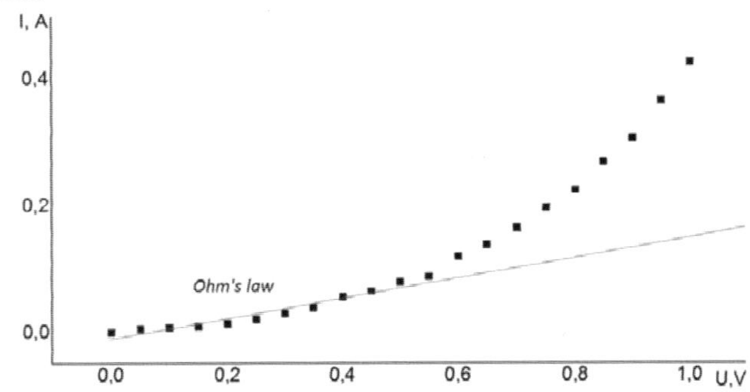

Figure 6. The VAC for Au-carbyne thin films with thickness 30nm: the enhancement of electroconductivity in comparison with linear Ohm's dependence due to 1D-structure.

Conclusion

We studied the cluster structures taking into account in both theory and experiment the correlations in nanoparticle-ensemble with the topological quantum states. The problem of high temperature superconductivity (due to topological surface structures with localized states resulting in coupled states on new dimensional principles) may be considered.

Future activity in the field should be focused on the more detailed comparison of obtained experimental results in dramatic increase of electroconductivity in induced (by controlling way) nanocluster systems and concrete mechanism of its development in different conditions. The results should give us the directions for possible applications of the phenomena in photonic devices of new generation.

IOP Conf. Series: Journal of Physics: Conf. Series **951** (2017) 012018 doi:10.1088/1742-6596/951/1/012018

The reported study was also supported by the Ministry of Education and Science of the Russian Federation (state project no.16.1123.2017/PCh), RFBR grant 16-42-330461 r_a, President grant for young scientist MK-2842.2017.2.

References

[1]. Landau L.D., Lifshitz E.M. *Theoretical physics electrodynamics of continuous media* V, VIII, Nauka Moscow 1982
[2]. Arakelyan, S.M., Veiko, V.P., Kutrovskaya, S.V. *et al* 2016 *Journal of Nanoparticle Research* **18** A155
[3]. Antipov, A.A., Arakelyan, S.M., Kutrovskaya, S.V. *et al* 2014 *Technical Physics Letters* **40**, 529.
[4]. Dragunov, V.P., Neizvestnyi, I.G., Gridchin, V.A. *Fundamentals of Nanoelectronics* Logos Publ. Moscow 2006
[5]. Kucherik A.O., Arakelian S.M., Garnov, S.V. *et al* 2016 *Quantum Electronics* **46** 627
[6]. Kucherik A.O., Arakelyan S.M., Vartanyan T.A. *et al* 2016 *Optics and Spectroscopy* **121** 263

IOP Conf. Series: Journal of Physics: Conf. Series **951** (2017) 012019 doi:10.1088/1742-6596/951/1/012019

Study of laser cooling in deep optical lattice: two-level quantum model

O.N.Prudnikov[1,2], **R. Ya. Il'enkov**[1], **A.V. Taichenachev**[1,2], **V.I. Yudin**[1,2], and **E.M. Rasel**[3]

[1] Institute of Laser Physics, 630090, Novosibirsk, Russia
[2] Novosibirsk State University, 630090, Novosibirsk, Russia
[3] Institut für Quantenoptik, Universität Hannover, Welfengarten 1, D-30167 Hannover, Germany

E-mail: oleg.nsu@gmail.com

Abstract. We study a possibility of laser cooling of ^{24}Mg atoms in deep optical lattice formed by intense off-resonant laser field in a presence of cooling field resonant to narrow $(3s3s)\,^1S_0 \to (3s3p)\,^3P_1$ ($\lambda = 457$ nm) optical transition. For description of laser cooling with taking into account quantum recoil effects we consider two quantum models. The first one is based on direct numerical solution of quantum kinetic equation for atom density matrix and the second one is simplified model based on decomposition of atom density matrix over vibration states in the lattice wells. We search cooling field intensity and detuning for minimum cooling energy and fast laser cooling.

Pacs 32.80.Pj, 42.50.Vk, 37.10.Jk,37.10.De

1. Introduction

Nowadays deep laser cooling of neutral atoms is routinely used for broad range of modern quantum physics researches including metrology, atom optics, and quantum degeneracy studies. The well-known techniques for laser cooling below the Doppler limit, like sub-Doppler polarization gradient cooling [1], velocity selective coherent population trapping [2, 3] or Raman cooling [4, 5] are restricted to atoms with degenerated over angular momentum energy levels or hyperfine structure. However, for atoms with single ground state ^{24}Mg, ^{40}Ca, ^{88}Sr, ^{174}Yb are of interest for developing optical time standard these techniques can not be applied directly. For example, for ^{24}Mg atoms with the ground state $^1S_0$ the Doppler cooling temperature ($k_B T_D \approx \hbar\gamma/2$) can be reached on closed singlet transition $^1S_0 \to {}^1P_1$ ($\lambda = 285.3$ nm). For lower temperature additional cooling on $^3P_2 \to {}^3D_3$ optical transition with degenerated over angular momentum energy levels can be applied [6, 7]. However, the experimental realization of laser cooling on $^3P_2 \to {}^3D_3$ optical transition does not result significant progress. The atoms were cooled to temperature $T \approx 1\,mK$ is about Doppler limit only [7]. The quantum simulation of laser cooling are also shows the limitation of cooling temperature to about Doppler limit in conventional MOT, formed by laser waves with circular polarization [8].

An alternative way of deep laser cooling of these elements is to use narrow lines and "quenching" techniques of narrow-line laser cooling [9, 10, 11] successfully applied for ^{40}Ca atoms, the recent the recent progress on cooling 88Sr on narrow line in dipole trap was also reported in [12], but, to our knowledge, still do not show significant progress for ^{24}Mg atoms.

Recently, laser-driven Sisyphus-cooling scheme was proposed for cooling atoms in optical dipole trap [13]. This scheme utilize the difference in trap-induced ac Stark shift for ground and exited levels of atom coupled by resonant laser light. The laser cooling scheme has clear semiclassical interpretation: been excited by resonant laser light on the bottom of shallow optical potential related to the ground state an atom moves further in steepest potential related to excited state. Spontaneous emission returns it back from the steepest potential in the excited state to the shallow potential in the ground state. The loosing a portion of energy in each act of these processes leads to atom cooling after several cycles due to "Sisyphus effect" [13]. This semiclassical model was applied for description of laser cooling of Yb and Sr in optical dipole trap.

In the following paper we study application of this cooling scheme to ^{24}Mg atom on narrow $(3s3s)\,^1S_0 \to (3s3p)\,^3P_1$ ($\lambda_1 = 457$ nm, $\gamma_1 = 196$ s^{-1}) optical transition for the atoms trapped in optical lattice at $\lambda_L = 1064$ nm. Additional light field resonant to $(3s3p)\,^3P_1 \to (3s4s)\,^1S_0$ optical transition ($\lambda_2 = 462$ nm, $\gamma_{21} = 109$

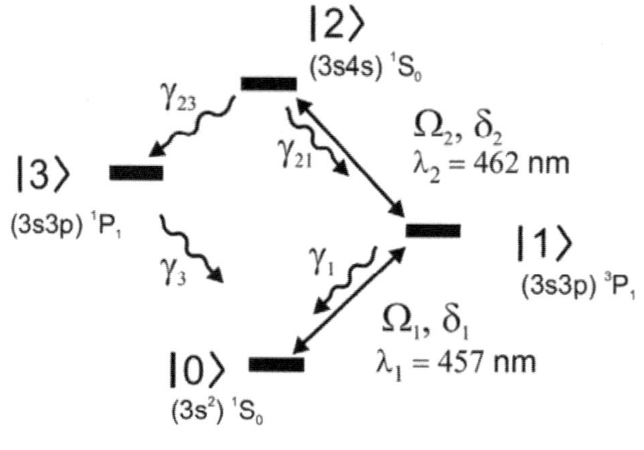

Figure 1. *Relevant energy levels for optical quenching and cooling of* ^{24}Mg.

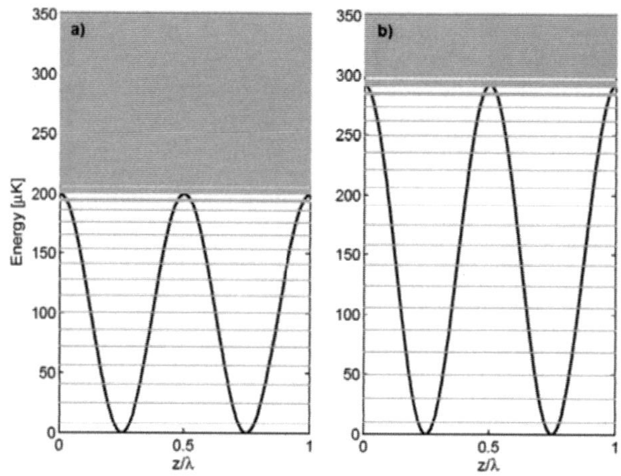

Figure 2. *Ac Stark shift and vibration energy levels for ground* $^1S_0$ *(a) and excited* $^3P_1$ *(b) states of* ^{24}Mg *atoms in optical lattice potential with* $\lambda_L = 1064nm$.

s^{-1} and $\gamma_{23} = 2.1 \cdot 10^7\ s^{-1}$) is applied for optical quenching (see Fig.1), i.e. increasing the effective linewidth of optical transition [11]. We find the semiclassical description of laser cooling of Mg atom with narrow optical transition can't be used here. The simulation of force on atom shows it is strongly varied on a scale of velocity $\Delta v \ll \hbar k/M$, i.e. on a range that corresponds recoil momentum obtained by atom in the processes of absorption and emission of light fields photons. For description of laser cooling we use quantum approaches that allow to take into account optical pumping and photon recoil effects in laser cooling process. In the paper we point our attention to minimum laser cooling temperature for described scheme and cooling time as well.

2. Description of the models

We consider motion of ^{24}Mg atom in the potential of optical lattice with $\lambda_L = 1064\,nm$ that provide higher polarizability of atom in the excited state $(3s3p)\,^3P_1$ than in the ground state $(3s3s)\,^1S_0$. In the following paper we restrict our consideration by two-level model assuming the quench field results to increasing effective linewidth of optical transition to γ_{eff} [11]:

$$\gamma_{eff} = \gamma_1 + \gamma_2 \frac{\Omega_2^2}{\gamma_2^2 + 4\,\delta_2^2}, \quad (1)$$

where Ω_2 is Rabi and δ_2 is detuning of quench field. Thus, for example, to get $\gamma_{eff} = 100\gamma_1$ at $\delta_2 = 0$ one have to apply the quench field intensity $I_q \approx 1.6\,W/cm^2$. The simulated polarizability difference for the considered wavelength λ_L is about $\alpha_e/\alpha_g = 1.46$. Here (fig.2) the quantum nature of atomic motion becomes essential. For considered optical lattice potential depth of the ground state $U_g = 200\,\mu K$ with vibration energy separation of the lowest states are $\hbar\omega_g \approx 16.6\,\mu K$. The excited state optical lattice potential depth $U_e = 292\,\mu K$ and the lowest states separation are $\hbar\omega_e \approx 20.1\,\mu K$. The large energy levels separation require quantum model for description of laser cooling dynamics of atoms in the potential of optical lattice. Really, the semiclassical models can't be applied here because of the simulation of force on atom shows it is strongly varied on a velocity range of $\Delta v \ll \hbar k/M$. As well the semiclassical parameter $\varepsilon_R = \hbar k^2/(2M\gamma) \gg 1$ is not small, that also contradicts requirements for semiclassical approach [1, 14, 15].

For description of laser cooling of Mg in the potential of optical lattice we consider two quantum approaches. The first one is based on decomposition of atom density matrix on optical potential vibration level states. Restricting by limited number of lowest vibration states we simulate the stationary distribution over the vibration levels in the optical lattice potential, as well as the laser cooling dynamics to steady state distribution. This approach is similar to method was described in [16]. However, in our model we also take into account the optical coherence of different vibration states.

The second method we consider is based on direct numerical solution of quantum equation for atom density matrix that allows to take into account not only the fixed number of the lowest vibration level states but density matrix of atoms entirely, which also include tunneling effects and above barrier motion. However, in this method, due to the high complicity of the problem we omit the recoil effects from the pumping field resonant to $(3s3s)\,^1S_0 \to (3s3p)\,^3P_1$ optical transition that is equivalent to orthogonal orientation of wave vectors of pumping and optical lattice light waves in one dimensional model.

2.1. Two-level model: exact numerical solution of quantum density matrix equation

We consider the motion of Mg atom in the potential of optical lattice is a standing light wave propagating along z direction with linear polarization along x. The cooling light field also linear polarized along x with wavevector along z or y. The quantum equation for atom density matrix describes evolution of internal and external states of atoms

$$\frac{\partial}{\partial t}\hat{\rho} = -\frac{i}{\hbar}\left[\hat{H}_0 + \hat{V}_{ed}, \hat{\rho}\right] + \hat{\Gamma}\left\{\hat{\rho}\right\} \tag{2}$$

with $\hat{H}_0$ is Hamiltonian, $\hat{V}_{ed}$ describes interaction with cooling field and $\hat{\Gamma}\{\hat{\rho}\}$ describes relaxation of density matrix due to spontaneous decay.

As was mentioned above, we restrict our consideration by effective two-level model with $(3s3s)\,^1S_0$ is the ground (g) and $(3s3p)\,^3P_1$ is excited state (e), assuming the influence of the quench field Ω_2 results to adjustable linewidth by modification of decay rate from γ_1 to γ_{eff} only, as described in [11]. Further in the paper we omit parameters indexes Ω_1, δ_1 and γ_1 by writing Ω, δ and γ instead. The Hamiltonian of atom in the potential of optical lattice has the form:

$$\hat{H}_0 = \frac{\hat{p}^2}{2M} + \hbar\omega_g(z)|g\rangle\langle g| + \hbar[\omega_e(z) + \omega_0]|e\rangle\langle e| \tag{3}$$

with optical potentials in the ground $\hbar\omega_g(z) = U_g\cos^2(kz)$ and $\hbar\omega_e(z) = U_e\cos^2(kz)$ in excited states ($\hbar\omega_0 = E_e - E_g$ is energy difference of unperturbed ground and excited states). The wavevector $k = 2\pi/\lambda_L$ is defined by the optical lattice field. Applying rotating wave approximation the equation for atom density matrix components in coordinate representation $\hat{\rho}(z_1, z_2)$ takes the followign form:

$$\begin{aligned}
\left(\frac{\partial}{\partial t} - \frac{i\hbar}{M}\frac{\partial}{\partial q}\frac{\partial}{\partial z}\right)\rho^{ee} &= -\gamma_{eff}\,\rho^{ee} + i\frac{U_e}{\hbar}\sin(2kz)\sin(kq)\rho^{ee} - \frac{i}{\hbar}\left[\hat{V}\rho^{ge} - \rho^{eg}\hat{V}^\dagger\right] \\
\left(\frac{\partial}{\partial t} - \frac{i\hbar}{M}\frac{\partial}{\partial q}\frac{\partial}{\partial z}\right)\rho^{gg} &= \hat{\gamma}\{\rho^{ee}\} + i\frac{U_g}{\hbar}\sin(2kz)\sin(kq)\rho^{gg} - \frac{i}{\hbar}\left[\hat{V}^\dagger\rho^{eg} - \rho^{ge}\hat{V}\right] \\
\left(\frac{\partial}{\partial t} - \frac{i\hbar}{M}\frac{\partial}{\partial q}\frac{\partial}{\partial z}\right)\rho^{eg} &+ \left(\frac{\gamma_{eff}}{2} - i\tilde{\delta}(z,q)\right)\rho^{eg} = -\frac{i}{\hbar}\left[\hat{V}\rho^{gg} - \rho^{ee}\hat{V}\right] \\
\left(\frac{\partial}{\partial t} - \frac{i\hbar}{M}\frac{\partial}{\partial q}\frac{\partial}{\partial z}\right)\rho^{ee} &+ \left(\frac{\gamma_{eff}}{2} + i\tilde{\delta}(z,-q)\right)\rho^{ge} = -\frac{i}{\hbar}\left[\hat{V}^\dagger\rho^{ee} - \rho^{gg}\hat{V}^\dagger\right],
\end{aligned} \tag{4}$$

with $z = (z_1 + z_2)/2$, $q = z_1 - z_2$, and the function $\tilde{\delta}(z,q)$:

$$\tilde{\delta}(z,q) = \delta - (U_e - U_g)\frac{1 + \cos(2kz)\cos(kq)}{2\hbar} + (U_e + U_g)\frac{\sin(2kz)\sin(kq)}{2\hbar}.$$

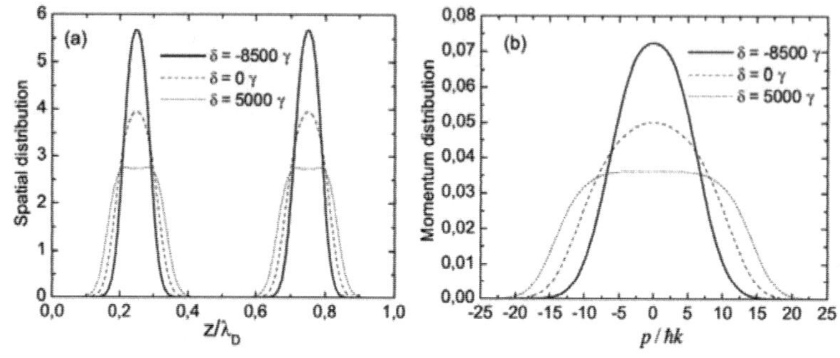

Figure 3. *Spatial (a) and momentum (b) distribution of* ^{24}Mg *atoms in optical lattice* ($U_g = 200\,\mu K$, $U_e = 292\,\mu K$) *for orthogonal orientation of cooling wave and optical lattice trap* ($\theta = \pi/2$) *and cooling field intensity* $I \simeq 34\ W/cm^2$ ($\Omega/\gamma = 20000$) *and different detunings.*

The spontaneous income part to the ground state $\hat{\gamma}$ in coordinate representation for two-level model has simple form :

$$\begin{aligned}\hat{\gamma}\{\rho^{ee}\} &= \tilde{\gamma}(q)\rho^{ee} \\ \tilde{\gamma}(q) &= 3\gamma_{eff}\left(\frac{\sin(k_1q)}{(k_1q)^3} - \frac{\cos(k_1q)}{(k_1q)^2}\right)\end{aligned} \tag{5}$$

with $k_1 = 2\pi/\lambda$ is wavevector of emitted photon ($\lambda = 457$ nm). The cooling field induces transitions between the ground and excited states. This part is described in (4) by operator

$$\hat{V} = \Omega/2\, \exp(ik_1 z\cos(\theta)) \tag{6}$$

with Ω is Rabi frequency of cooling field and θ is angle between the axis z and cooling wave propagation direction. For the case of orthogonal orientation of the cooling wave propagation to the optical lattice trap $\theta = \pi/2$ the equation for density matrix (4) can be solved numerically by the method suggested in [17, 18]. It should be noted the considered method allows to get steady state solution for density matrix with taking into account quantum recoil effects as for atoms in the optical lattice trap as for nontrapped atoms.

The figure 3 shows spatial and momentum distribution of Mg atoms in the optical lattice for orthogonal orientation of cooling wave ($\theta = \pi/2$), for field intensity $I \simeq 34\ W/cm^2$ ($\Omega/\gamma = 20000$) and different detunings.

The obtained numerical solution for steady state density matrix $\hat{\rho}(z_1, z_2)$ contains whole information on internal and translational states of atoms in the potential of optical lattice. In particular one can extract the population of vibration levels in the ground and excited states that in coordinate representation for atom density matrix takes a form:

$$\begin{aligned}\rho_n^{ee} &= \int \psi_n^{*(e)}(z_1)\rho^{ee}(z_1, z_2)\psi_n^{(e)}(z_2)dz_1dz_2\,, \\ \rho_n^{gg} &= \int \psi_n^{*(g)}(z_1)\rho^{gg}(z_1, z_2)\psi_n^{(g)}(z_2)dz_1dz_2\end{aligned} \tag{7}$$

where $\psi_n^{(e,g)}(z)$ are n-th vibration level eigenfunctions.

The energy of cooled atoms can be found by different way. First of all one can use the relation for the temperature of cooled atoms in the well known form

$$k_BT = < p^2 > /M\,, \tag{8}$$

with $< p^2 >= Tr\{\hat{p}^2\hat{\rho}\}$. This relation neglects the atom localization effects in the optical potential. The most accurate relation for energy is expressed by the following averaging:

$$E = Tr\left\{\left(\frac{\hat{p}^2}{2M} + \hbar\omega_e(z)\right)\hat{\rho}^{(ee)}\right\} + Tr\left\{\left(\frac{\hat{p}^2}{2M} + \hbar\omega_g(z)\right)\hat{\rho}^{(gg)}\right\}\,. \tag{9}$$

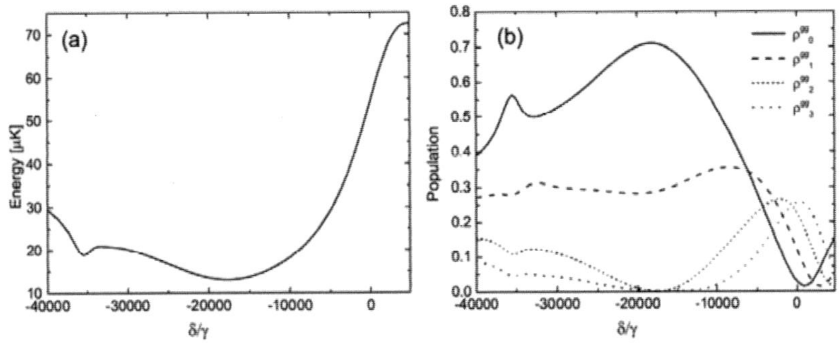

Figure 4. *The energy (a) of cooled ^{24}Mg atoms in the optical lattice and population of the lowest vibration levels (b) as function of cooling field detuning for $I \simeq 0.34\ W/cm^2$ ($\Omega/\gamma = 2000$) obtained by direct numerical solution of eq.(4). ($\theta = \pi/2$)*

As an alternative way one can find the average energy over the vibration states

$$E = \sum_n E_n^{(e)}\,\rho_n^{(ee)} + E_n^{(g)}\,\rho_n^{(gg)}. \tag{10}$$

For considered parameters all above definitions give very close values that denotes the main contribution to energy are given by atoms on the lowest vibration energy levels in the region where the optical potential has close to parabola shape. The energy of Mg atoms for different detuning is shown on figure 4. The total population of excited vibration level states here do not exceed 2% and are not shown on figure 4(b). In the region of detuning $\delta > -6000\gamma$ we find inversion of the lowest vibration levels population resulting to energy growth. For the higher intensity of cooling field this effect is also exists and moves to larger detuning area.

2.2. The model based on vibration states decomposition

As we see from the simulations above based on numerical solution of basic equation for atomic density matrix (4) for the considered parameters Mg atoms can be cooled and well localized in the potential of optical lattice. Thus for description of laser cooling and laser cooling time we can also apply an alternative approach based on decomposition of atom density matrix over vibration level states.

$$\begin{array}{ll} \rho^{ee}_{nm} = \langle e,n|\rho^{ee}|e,m\rangle & \rho^{eg}_{nm} = \langle e,n|\rho^{eg}|g,m\rangle \\ \rho^{ge}_{nm} = \langle g,n|\rho^{ge}|e,m\rangle & \rho^{gg}_{nm} = \langle g,n|\rho^{gg}|g,m\rangle \end{array} \tag{11}$$

with elements

$$\rho^{ab}_{nm} = \langle a,n|\rho^{ab}|b,m\rangle = \int \psi_n^{*(a)}(z)\rho^{ab}(z_1,z_2)\psi_m^{(b)}(z)dz_1dz_2\,. \tag{12}$$

and indexes (a), (b) are corresponds to indexes of exited and ground states (e) and (g). The equation for these components has similar to (2) form:

$$\begin{aligned} \tfrac{\partial}{\partial t}\rho^{ee}_{nm} &= -\gamma_{eff}\,\rho^{ee}_{nm} - i\left(\omega_n^{(e)} - \omega_m^{(e)}\right)\rho^{ee}_{nm} - i\textstyle\sum_k\left(\Omega_{nk}\rho^{ge}_{km} - \rho^{eg}_{nk}\Omega^{\dagger}_{km}\right) \\ \tfrac{\partial}{\partial t}\rho^{eg}_{nm} + \left[\tfrac{\gamma_{eff}}{2} - i\delta\right]\rho^{eg}_{nm} &= -i\left(\omega_n^{(e)} - \omega_m^{(g)}\right)\rho^{eg}_{nm} - i\textstyle\sum_k\left(\Omega_{nk}\rho^{gg}_{km} - \rho^{ee}_{nk}\Omega_{km}\right) \\ \tfrac{\partial}{\partial t}\rho^{ge}_{nm} + \left[\tfrac{\gamma_{eff}}{2} + i\delta\right]\rho^{ge}_{nm} &= -i\left(\omega_n^{(g)} - \omega_m^{(e)}\right)\rho^{ge}_{nm} - i\textstyle\sum_k\left(\Omega^{\dagger}_{nk}\rho^{gg}_{km} - \rho^{ee}_{nk}\Omega^{\dagger}_{km}\right) \\ \tfrac{\partial}{\partial t}\rho^{gg}_{nm} &= \hat{\gamma}\left\{\rho^{ee}\right\}_{nm} - i\left(\omega_n^{(g)} - \omega_m^{(g)}\right)\rho^{ee}_{nm} - i\textstyle\sum_k\left(\Omega^{\dagger}_{nk}\rho^{eg}_{km} - \rho^{ge}_{nk}\Omega_{km}\right) \end{aligned} \tag{13}$$

where $\hbar\omega_n^{(e)}$ and $\hbar\omega_n^{(g)}$ are the vibration levels energy of the excited and the ground states. Matrix elements Ω_{nm} define coupling of the ground and excited vibration states:

$$\Omega_{nm} = \langle e,n|\hat{V}|g,m\rangle = \int \psi_n^{*(e)}(z)\hat{V}(z)\psi_m^{(g)}(z)dz\,. \tag{14}$$

with $\hat{V}$ is defined in (6). The relaxation term $\hat{\gamma}\{\rho^{ee}\}_{nm}$ describes income to ground vibration states due to emission of spontaneous photons:

$$\hat{\gamma}\{\hat{\rho}\}_{nm} = \gamma_{eff}\sum_{\nu\mu}\Gamma^{\nu\mu}_{nm}\rho^{ee}_{\nu\mu}$$
$$\Gamma^{\nu\mu}_{nm} = \int \psi^{*(g)}_n(z_1)\psi^{(e)}_\nu(z_1)\tilde{\gamma}(z_1-z_2)\psi^{*(e)}_n(z_2)\psi^{(g)}_\nu(z_2)dz_1dz_2 \quad (15)$$

and $\tilde{\gamma}(q)$ is defined in (5). The equation (13) describe evolution of diagonal elements (vibration levels population) and nondiagonal elements as well. To make better insight into these complex equation we consider first low intensity limit for cooling field $S_{nm} \ll 1$

$$S_{nm} = \frac{|\Omega_{nm}|^2}{\gamma^2_{eff}/4 + \left[\delta^2 - (\omega^{(e)}-\omega^{(g)})^2\right]} \quad (16)$$

and consider evolution of the population of the ground and excited states that described by balance equation in this limit:

$$\frac{\partial}{\partial t}\rho^{ee}_{nn} = -\gamma_{eff}\left(1+\sum_{m=1}^{Ng}S_{nm}\right)\rho^{ee}_{nn} + \gamma_{eff}\sum_{m=1}^{Ng}S_{nm}\,\rho^{gg}_{mm}$$
$$\frac{\partial}{\partial t}\rho^{gg}_{nn} = \gamma_{eff}\sum_{n=1}^{Ne}\left(\Gamma^{nn}_{mm}+S_{nm}\right)\rho^{ee}_{nn} - \gamma_{eff}\left(\sum_{n=1}^{Ne}S_{nm}\right)\rho^{gg}_{mm} \quad (17)$$

for fixed number of the excited (Ne) and the ground (Ng) vibration states. The equation (17) can be solved analytically. Restricting by two vibration levels in the ground and excited states we get

$$\rho^{gg}_{00} = \left[1+\frac{\Gamma^{00}_{11}S_{00}+\Gamma^{11}_{11}S_{10}}{(1-\Gamma^{00}_{11})S_{01}+(1-\Gamma^{11}_{11})S_{11}}\right]^{-1}$$
$$\rho^{gg}_{11} = \left[1+\frac{(1-\Gamma^{00}_{11})S_{01}+(1-\Gamma^{11}_{11})S_{11}}{\Gamma^{00}_{11}S_{00}+\Gamma^{11}_{11}S_{10}}\right]^{-1} \quad (18)$$

desribes for ρ^{gg}_{00} smooth curve on detuning with minimums determined by $\omega^{(e)}_0 \to \omega^{(g)}_0$ and $\omega^{(e)}_1 \to \omega^{(g)}_0$ resonances (i.e. for detuning $\delta = \omega^{(e)}_0 - \omega^{(g)}_0$ and $\delta = \omega^{(e)}_1 - \omega^{(g)}_0$) and maximum for detuning close to $\delta = \omega^{(e)}_0 - \omega^{(g)}_1$ and $\delta = \omega^{(e)}_1 - \omega^{(g)}_1$.

3. Results

The fugure 5 shows the energy of cooled Mg atoms in the optical lattice and population of the ground state vibration levels obtained by both described above methods (section 2.1 and 2.2). Both methods demonstrates close results with the differences far from the minimum cooling energy when the populations of the top vibration levels are not negligible and tunneling effects and above barrier motion of atoms can not be neglected. Nevertheless, the simplified model (13) well describe the laser cooling near the minimum of energy and allows to estimate the cooling time.

As well the populations and energy of ^{24}Mg atoms in dipole trap and coplanar geometry of pumping field is shown on figure 6. We see the balance equations (17) well describe atom behavior in the trap for low intensity of pumping field. The modulations of population and total energy on detuning corresponds to resonances $\omega^{(e)}_0 \to \omega^{(g)}_1$, $\omega^{(e)}_0 \to \omega^{(g)}_2$, $\omega^{(e)}_0 \to \omega^{(g)}_3$, and $\omega^{(e)}_0 \to \omega^{(g)}_4$. With increasing the pumping field intensity the balance equation approach (13) violates. Additionally, we see the field shift of the resonances (Fig. 6(b) and 6(d)).

Additionally the model based on decomposition on vibration state allows to solve dynamical problem and estimate the cooling time. To find the cooling time we assume the atoms populate the highest vibrational energy level of the ground state optical potential at $t = 0$. The time evolution of vibration levels population has a complex dependence. We fit $\rho^{gg}_{00}(t)$ by exponential function of the form $\rho^{gg}_{00}(t) = a - b\exp(-t/\tau)$ with τ describes the cooling time. Additionally we note, the energy of cooled atoms does not depend on parameter γ_{eff} (i.e. on quench field intensity) in the range of our simulations $\gamma < \gamma_{ef} < 100\gamma$, while the cooling time τ is inversely proportional to γ_{eff} in considered model. This allows us to represent the cooling time τ in the more general form through dimensionless value $\tilde{\tau}$ figure 7(b)

$$\tau = \tilde{\tau}/\gamma_{eff}. \quad (19)$$

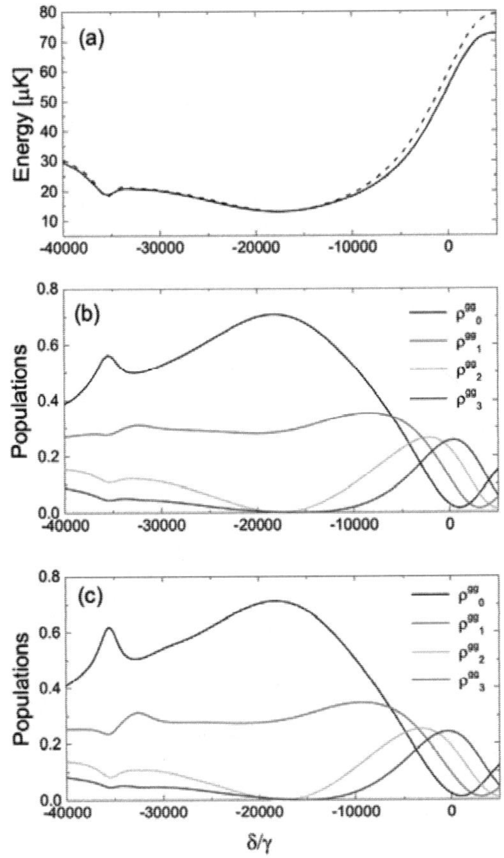

Figure 5. *The energy of cooled ^{24}Mg atoms in the optical lattice (a) as function of cooling filed detuning obtained by exact numerical solution of eq.(2) (solid line) and by decomposition of vibration state method eq.(13) (dashed line). Populations of the ground state vibration levels (g) as function of detuning for exact numerical solution (b) and decomposition on vibration state model (c). Pumping field $\Omega/\gamma = 2000$ ($I = 0.34W/cm^2$), $\gamma_{eff} = 100\gamma$. Orthogonal orientation of pumping field and the trap $\theta = \pi/2$.*

4. Conclusion

We study laser cooling of ^{24}Mg atoms in the optical lattice trap with additional pumping field resonant to narrow $(3s3s)\,^1S_0 \to (3s3p)\,^3P_1$ ($\lambda = 457$ nm) optical transition and quench field resonant to $(3s3p)\,^3P_1 \to (3s4s)\,^1S_0$. The effect of quenching we consider in simplified form as widening of optical transition to γ_{eff} only. We have considered two quantum models. The first one is based on the direct numerical solution of quantum kinetic equation for atom density matrix and the second one is simplified model is based on decomposition of atom density matrix over vibration states in the trap of optical lattice. Additionally to vibration state model we consider simplified balance equations that allow clarify modulations on detuning the atom energy and population of vibration levels.

Both considered models describe cooling of atoms on the lowest vibration levels with difference appears for high intensity of pumping field (above $50W/cm^2$, or Rabi above 25000γ) when the populations of the top vibration levels are not negligible and tunneling effects and above barrier motion of atoms can not be neglected. Nevertheless, the vibration state model well describe the laser cooling to minimum of cooling energy (about 10 μK). Additionally this model allows to estimate the cooling time. The parameters of pumping field for cooling to minimum energy do not coincide with conditions for fast cooling. We find parameters that allow cooling the atoms for reasonable cooling time $\tau \approx 10/\gamma_{eff}$ ($\tau \approx 0.5$ ms for $\gamma_{eff} = 100\gamma$) to energy $E \approx 12\,\mu K$.

Finally we note the important assumption we used for accounting the quench field (1) might not be enough

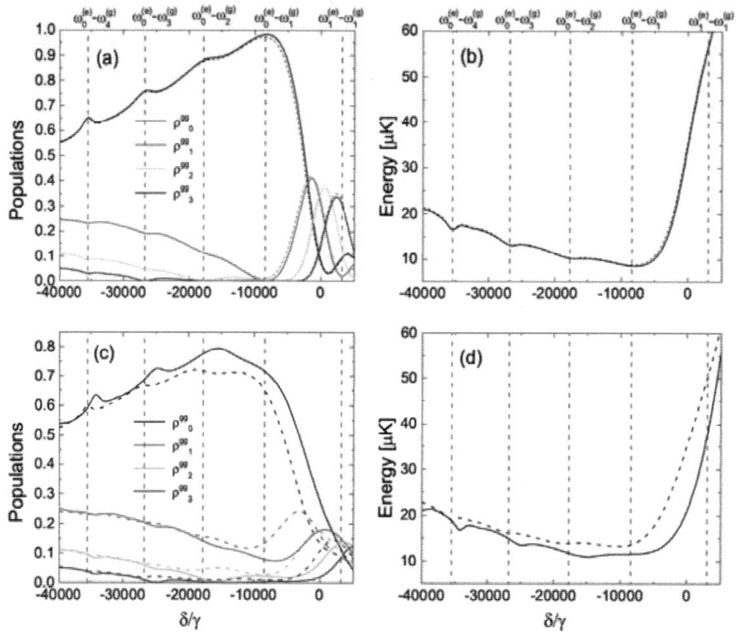

Figure 6. *Populations and energy of cooled* ^{24}Mg *atoms in the optical lattice for different intensity of cooling field ((a) and (b) for* $\Omega/\gamma = 2000$ *and (c) and (d) for* $\Omega/\gamma = 10000$*) as function of pumping filed detuning obtained by exact numerical solution of eq.(13) - solid lines and by balance eq.(17) - dashed lines. Coplanar geometry of pumping field and optical lattice trap* $\theta = 0$.

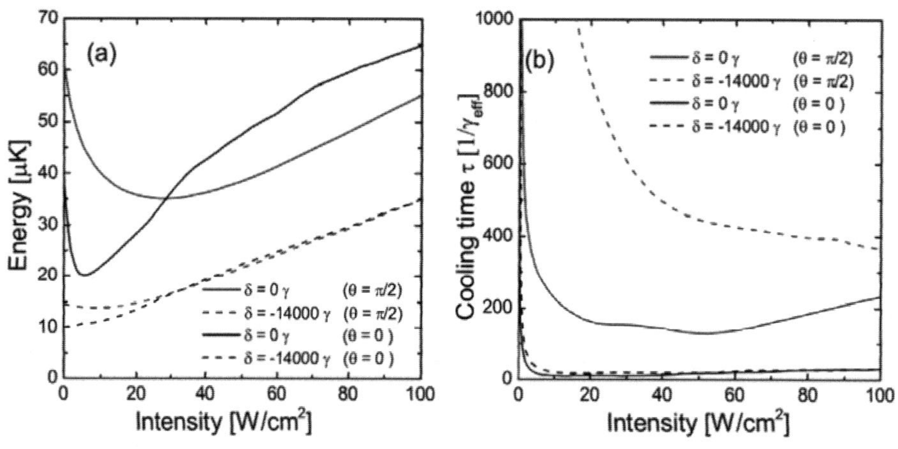

Figure 7. *Energy of cooled* ^{24}Mg *atoms in the optical lattice (a) and cooling time* $\bar{\tau}$ *(b) for different orientation of cooling wave* θ *and different detunings obtained by solution based on decomposition on vibration state model as function of cooling field intensity.*

adequate the real situation and demands more detailed consideration on a base of three level scheme in dipole trap. We hope to finish this consideration in the near future. The work was supported by Russian Science Foundation (project N16-12-00054). V.I.Yudin acknowledges the support of the Ministry of Education and Science (3.1326.2017/4.6) and RFBR (grant 17-02-00570). The work of R. Ya. Il'enkov was personally supported by Russian Science Foundation (project N17-72-10139).

5. References

[1] Dalibard J and Cohen-Tannoudji C 1989 *J. Opt. Soc. Am.* B **6** 2023

[2] Lawall J, Bardou F, Saubamea B, Shimizu K, Leduc M, Aspect A, and Cohen-Tannoudji C 1994 *Phys. Rev. Lett.* **73** 1915
[3] Adams C S, Lee H J, Davidson N, Kasevich M, and Chu S 1995 *Phys. Rev. Lett.* **74,** 3577
[4] Kasevich M and Chu S, 1992 *Phys. Rev. Lett.* **69** 1741
[5] Reichel J, Bardou F, Dahan M Ben, Peik E, Rand S, Salomon C, and Cohen-Tannoudji C 1995 *Phys. Rev. Lett.* **75** 4575
[6] Kulosa A P, Fim D, Zipfel K H, Ruhmann S, Sauer S, Jha N, Gibble K, Ertmer W, Rasel E M, Safronova M S, Safronova U I, Porsev S G 2015 *arXiv: 1508.01118v1, physics.atom-ph*
[7] Riedmann M, Kelkar H, Wubbena T, Pape A, Kulosa A, Zipfel K, Fim D, Ruhmann S, Friebe J, Ertmer W, and Rasel E 2012 *Phys. Rev.* A **86,** 043416
[8] Prudnikov O N, Brazhnikov D V, Taichenachev A V, Yudin V I, Goncharov A N, 2016 *Quantum Electronics* **46** 661
[9] Curtis E A, Oates C W, and Holberg L 2001 *Phys. Rev.* A **64** 031403
[10] Binnewies T, Wilpers G, Sterr U, Helmcke J, Mehlstäbler T E, Rasel E M, and Ertmer W 2001 *Phys. Rev. Lett.* **87** 123002
[11] Mehlstäbler T E, Keupp J, Douillet A, Rehbein N, Rasel E M 2003 *J.Opt. B: Quantum Semiclass. Opt.* **5** S183
[12] Chalony M, Kastberg A, Klappauf B and Wilkowski D 2011 *Phys. Rev. Lett.* **107** 243002
[13] Ivanov V V, Gupta S 2011 *Phys.Rev.* A **84** 063417
[14] Javanainen J 1990 *Phys. Rev.* A **44** 5857
[15] Prudnikov O N, Taichenachev A V, Tumaikin A M and Yudin V I 2004 *JETP* **98** 438
[16] Castin Y and Dalibard J 1991 *Europhys. Lett.* **14** 761
[17] Prudnikov O N, Taichenachev A V, Tumaikin A M and Yudin V I 2007 *Phys. Rev.* A **75** 023413
[18] Prudnikov O N, Taichenachev A V, Tumaikin A M and Yudin V I, 2011 *JETP* **112** 939

Metal-carbyne clusters for SERS realization

A Kucherik[1], A Antipov[1], S Kutrovskaya[1], A Osipov[1], A Povolotckaia[2], S Arakelian[1]

[1] A.G and N.G. Stoletov Vladimir State University (VSU), 87 Gorki st., Vladimir, Russia, 600000
[2] Saint-Petersburg State University (SPbSU), Universitetskaya 7/9, Saint-Petersburg, Russia, 199034

E-mail: kucherik@vlsu.ru

Abstract. Metal–carbyne materials, synthesized by laser irradiation of colloidal systems consisting of carbon and noble metal nanoparticles, are promising objects to realize surface-enhanced Raman scattering. The dependence of the Raman scattering intensity of the material composition has been investigated. The possibility of detecting of the organic dye Rhodamine 6G Raman spectrum was demonstrated.

1. Introduction

Hybrid carbon-based nano- and composite materials have a unique place in modern nanoscience owing to their exceptional electrical, thermal, chemical and mechanical properties. This type of materials have found application in areas diverse as nanoscale electronic components, energy storage and conversion, sensors, drug delivery and *etc*. Carbon quantum dots are a newly developed class of carbon nano-materials that have attracted much interest and attention thanks to demonstrate a promising optical properties depending on the degree of crystallization, size and shape. The amplification of the observed phenomena can be realized by the addition of the noble metal nanoparticles, which have a plasmon resonant peaks in the visible region. The linear carbon chain (carbyne) is the promising carbon material for the creation of the hybrid and composite materials.

Carbyne is an allotrope of carbon composed of *sp*-hybridized carbon atoms. It is an n-type semiconductor with an energy gap vary from as little as 0.2 eV to more than 4 eV, in which electronic transport properties are structurally sensitive. Carbyne has a chain of covalently bonded carbon atoms with sp-hybridization of orbitals of two possible types: alternation of n single and triple bonds (–C≡C–)n, or alternation of n double bonds (=C=C=)n. Optical and electrical properties of carbyne depend on its morphology [1-3]. Thus, the synthesis of carbyne with a variable structure of bonds has a considerable interest not only from a fundamental point of view, but also this material is applicable in the development of new integrated circuits of advanced, next-generation hybrid electronic devices.

The strategy to incorporate metal nanoparticles in a carbon matrix is a great method to integrate the different properties to various materials, enabling to realize multifunctional composites. Such materials may be used in advanced applications like nanobiotecnology, optoelectronics, etc. In particular, a widely used approach is given by the combination of linear carbon chains and metals. In effect, linear carbon chain is the one of the most promising materials because of its unique physical and chemical properties and wide potential applications in nanophotonics. The possibility to obtain composites with large visible photoluminescence spectra for optoelectronic devices becomes especially interesting.

In this paper, we report the results of studying the formation of metal–carbyne clusters under laser irradiation of colloidal systems, consisting of shungite carbon and noble-metal nanoparticles. It was shown that this irradiation leads to the formation of clusters, in which metal nanoparticles are interrelated by carbyne chains. The Raman spectra of these structures exhibit SERS and depend significantly on the variation of metal particles concentration in colloidal system.

2. Laser synthesis of metal-carbyne material

In our experiments, the proposed method is based on the laser irradiation of colloidal systems consisted of shungite carbon [4] and noble metal (Au and Ag) nanoparticles [5].Such colloidal system was placed under Yb-fiber laser (1.06 μm) with a pulse duration of 100 ns and a pulse energy of up to 1mJ. This process resulted in the fabrication of long linear carbon chains (LLCC) in liquid. Such mechanism proceeds because of the homogeneous symmetry interaction in liquid for the carbon system that stabilizes the linear structure [6]. In this study, we have analyzed the possibility of control over the properties of fabricated composites by changing the irradiation conditions and the particle concentration in colloid. The laser irradiation of the colloidal system was performed by scanning a focused laser beam (focal spot diameter 50μm) in a cuvette volume with a scanning speed of 50 μm/s; the total irradiation time was about 15 min.

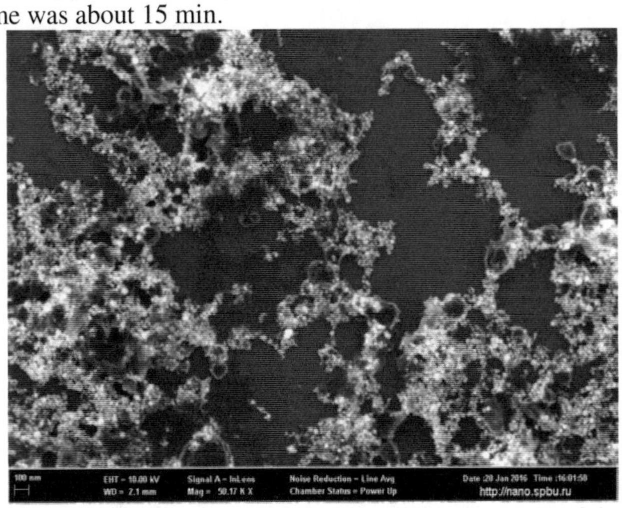

Figure 1. SEM-image of formed metal-carbyne materials.

In the processes of laser irradiation the metal nanoparticles decorate the carbyne fiber. The variation of a laser radiation and concentration on metal nanoparticles in colloidal system allow to change the morphology of metal-carbyne clusters syntheses [5]. The UV-Vis absorption spectra of the colloidal systems presented on the figure 2. The absorption peaks around 400 and 530 nm are the plasmon bands due to pure Ag and Au nanoparticles, respectively. The UV-Vis spectra for the C-Au-Ag systems showed a blue shift of the absorption peak around 500 nm. The absorption band near 290 nm corresponds to the carbon phase.

IOP Conf. Series: Journal of Physics: Conf. Series **951** (2017) 012020 doi:10.1088/1742-6596/951/1/012020

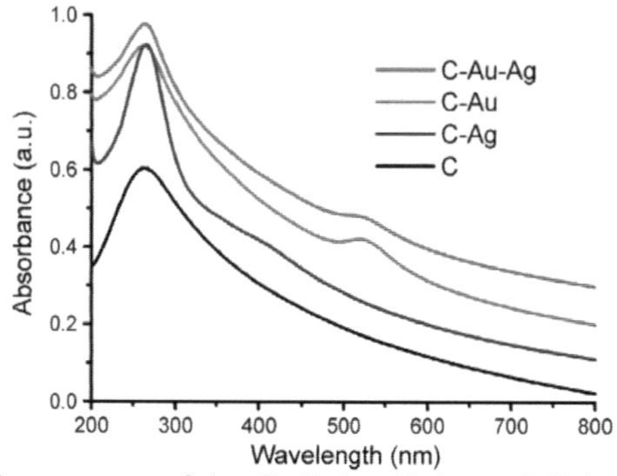

Figure 2. The absorption spectra of the C, C:Au, C:Ag and C:Au:Ag colloidal systems.

The more detailed study of the carbon material structure reconstruction under laser irradiation was performed by using Raman spectroscopy.

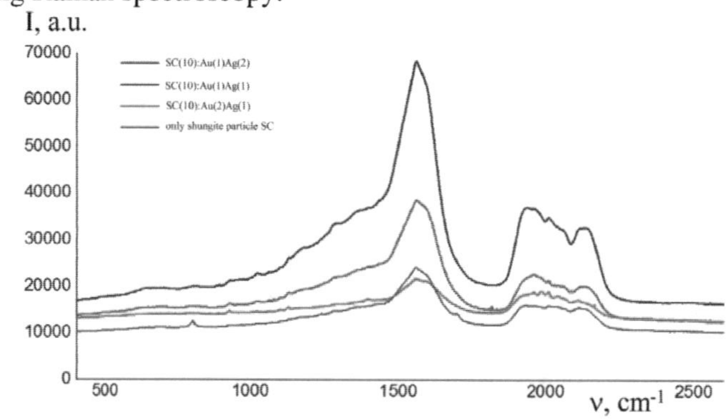

Figure 3. The Raman spectra of C:Au:Ag colloidal systems with different component ratio.

After the irradiation we have detected the peaks in the area of 1900-2200 cm^{-1}. This area corresponds to carbyne spectra [6, 7]. The D-peak for all samples was reduced and forming of the LLCC-peak [8] was observed. Cumulene structure (=C=C=C=) was found by the absorption lines 1950 $cm^{-1}$and 1070 cm^{-1}. Polyyne structure of carbon atomic chains was determined by the lines of the triple bond stretching vibrations in area of 2100-2300 cm^{-1}. The change of chain length resulted in the frequency shift towards the shortwave area during the irradiation of colloidal system. This kind of structures can be also modified under the laser radiation. In this case, the irradiation of the metal-carbon colloidal system can result in the obtaining of the more complicated C:Au:Ag systems (as in analogy [5, 6]). The variation of spectra in area of 1900-2200 cm^{-1} corresponded to the formation of the carbon chains with different length.

3. Deposition of metal-carbon films

For the deposition, we have used ytterbium fiber laser (λ = 1.06 μ) with a pulse duration of 100 ns, pulse repetition rate of 100 kHz, and a pulse energy up to 0.1mJ. The diameter of the laser beam in the focal plane was 30 μm. Periodic array formation on the surface substrate was performed by scanning the laser beam with 60μm step (in XY dimension) with irradiation time about 3s for each step.

Depending on the conditions of laser exposure for a single pulse, the value of the absorbed energy was 10^{-9}J. Under the conditions of multi-pulse exposure the local heating may occur, resulting in the stability loss of the colloidal system [9]. In addition, agglomerates can be formed due to particles heating. This happens when particles absorb a larger volume of laser energy and move thermally in the colloid. The local heating of the liquid results in the formation of convective currents, providing particle motion to the substrate surface (fig 3).

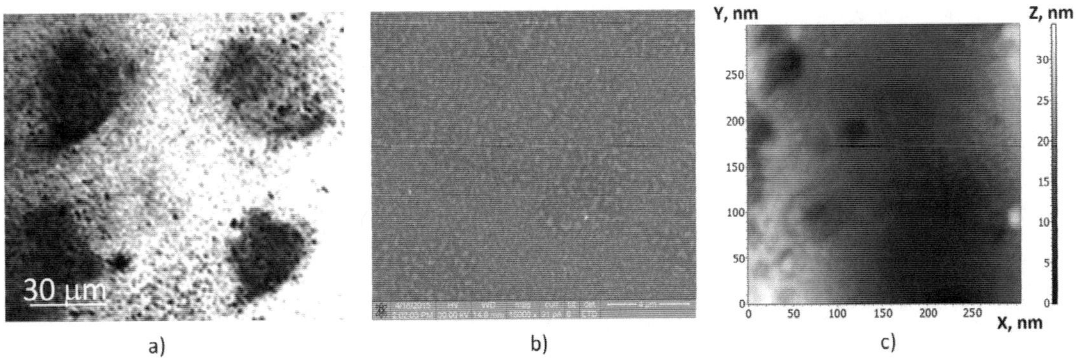

Figure 4. Deposited metal-carbyne thin films: a) optical image; b) SEM-image; c) AFM-image of the deposited films.

4. SERS research

The SERS research using deposited films was performed at Senterra spectrometer (Bruker), with the pump laser wavelength of 532 nm, the power of 0.1 mW and the focal spot diameter of 2 μm.

In experiments we have determined a limit of sensitivity of metal-carbon films. To obtain this value, we deposited the solution of dye Rhodamine 6G with different concentration on the surface of metal-carbon films. The SERS investigations (see figure 5) had demonstrated, that the sensitivity limit was about 10^{-7}M.

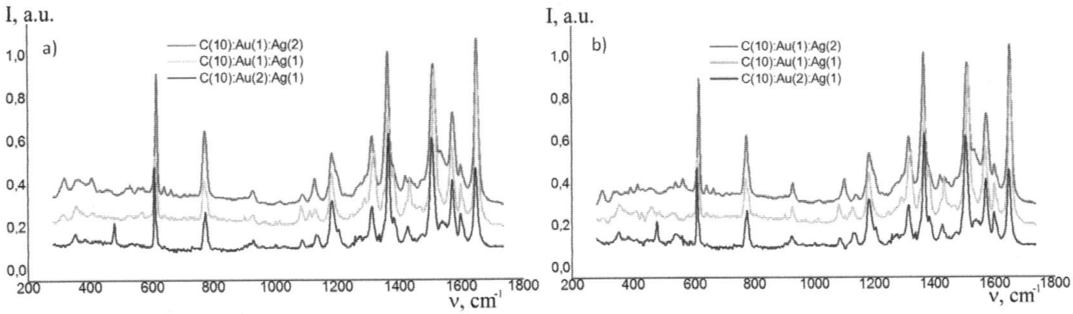

Figure 5. Raman spectra of deposited dye Rhodamine 6G with concentration: a) 10^{-6}M; a) 10^{-7}M.

The use of films with variable composition as a substrate for metal-carbon nanostructures under the same measurement conditions allowed to detect and identify the dye, its Raman spectra can be clearly seen in Figure 4. The changing of the gold and silver nanoparticles ratio results in changes of the amplification degree of different bands in the dye spectrum. The peaks of plasmon resonances for silver and gold are in the range of 410 and 540 nm, thus, different vibration modes are amplified by gold and silver particles, and with different gain.

5. Conclusion

The results of laser synthesis of a metal-carbon clusters and complexes are shown in this work. The laser with nanosecond pulse duration allows to modify the structure of carbon bonds and creates best conditions for obtaining the linear carbon chains with different hybridization. The obtained structures can be used for the SERS effect registration with the possibility of the sensitivity control in different areas of the spectra because of the changing of the initial component concentration and morphology.

Raman and UV/VIS absorption spectra were measured at Center for Optical and Laser Materials Research, St. Petersburg State University.

The reported study was also supported by the Ministry of Education and Science of the Russian Federation (state project no. 16.5592.2017/VU), RFBR grants # 16-42-330531r_a, #16-32-00759mol_a and by the grant of president of Russian Federation by project MK-3053.2017.2.

6. References

[1] Agarwal N *et al* 2016 *J. Phys. Chem.* C **120** 11131
[2] Eisler S *et al* 2005 *J. Am. Chem. Soc.* **127** 2666
[3] Cataldo F 1997 *Polymer Int.* **44** 191
[4] Kucherik A *et al* 2016 *Quant. El.* **46** 627
[5] Kucherik A *et al* 2016 *Opt. and Spectr.* **121** 263
[6] Pan B *et al* 2015 *Sci. Adv.* **1** e1500857
[7] Ravagnan L *et al* 2002 *Phys. Rev. Lett.* **89** 285506
[8] Shi L *et al* 2016 *Nat. Mat.* **15** 634
[9] Arakelyan S *et al* 2016 *J. Nanopart. Res.* **18** 155

IOP Conf. Series: Journal of Physics: Conf. Series **951** (2017) 012021 doi:10.1088/1742-6596/951/1/012021

AFM lithography for TAMM plasmons observation

A Shagurina, S Kutrovskaya, I Skryabin, A Kel'

Vladimir State University named after Alexander and Nikolay Stoletovs, 87 Gorky Street, Vladimir 600000, Russia

E-mail: 11stella@mail.ru

Abstract The results of the formation of planar nanostructures with relief, repeating the trajectory of movement of an AFM probe were presented. The parameters affecting their geometric dimensions: height, width, uniformity in the layer, etc. were investigated. The method of induced deposition of silver/golden clusters on the silicon wafer surface p-type in the presence of an external electric field was developed. The possibility of using such structures as hybrid circuits using photoelectronic transducers was discussed.

1. Introduction

Noble metals nanostructures such as silver and golden nanoparticles are well known due to their ability of supporting surface plasmon localization at optical frequencies [1-2]. Such optical nonlinear gain effectsare widely used in nanoparticle biological and chemical sensing, surface enhanced Raman scattering or in photoelectric applications. Their spectral response strongly depends on the nature of the metal, the size, the shape and the spatial arrangement of nanoparticles. The precise control of these parameters during fabrication process is the main goal for adjusting the optical properties of such structures in many applications [3].

Among the amount of different notable phenomena appeared in plasmonic structures the effect of the directional propagation of light observed in the silicon-metal interfaces deserves special attention. Such structures could potentially find a promising applications in all-optical information processing devices since it can transmit optical signals. A great number of various methods of fabrication of nanostructures supporting for plasmon resonances on metal-silicon interfaces were recently proposed. Most of them are based on complicated multi-step lithographymethods which can be combined by perforce with different chemical etching techniques, electron beam evaporation and so on. All this procedures demand rather expansive equipment and well optimized technology. In this paper we propose an alternative easy and relatively cheap method of fabrication of metal nanostructures grownon a siliconsubstrate. The main peculiarity of the proposed technique is the ability to control the shape and morphology of the designed structure. The proposed method is based on coating thin layers assembled of silver or golden clusters with a prominent precision accessible by modern possibilities of scanning probe microscopy.

2. Method

A new method for electro-induced deposition of metal clusters was based on a method local electro-induced lithography by atomic-force microscope (AFM). The substrate is a p-type silicon with

IOP Conf. Series: Journal of Physics: Conf. Series **951** (2017) 012021 doi:10.1088/1742-6596/951/1/012021

surface roughness about 0.2 nm. The local relief was formed along the trajectory of a conductive AFM probe by local electric-induced dissociation of a water solution of a metal salt in the saturated layer on the silicon. A bias voltage between the AFM tip and the sample was applied. Deposition was carried out in air at room temperature with an atomic-forces microscope platform based nanolab NTEGRA Aura probe in contact mode (see fig.1).We choose the silica as a material for deposition substrate because of the great importance of silica-based devices in a context of fabrication of all-optical photonic circuits.

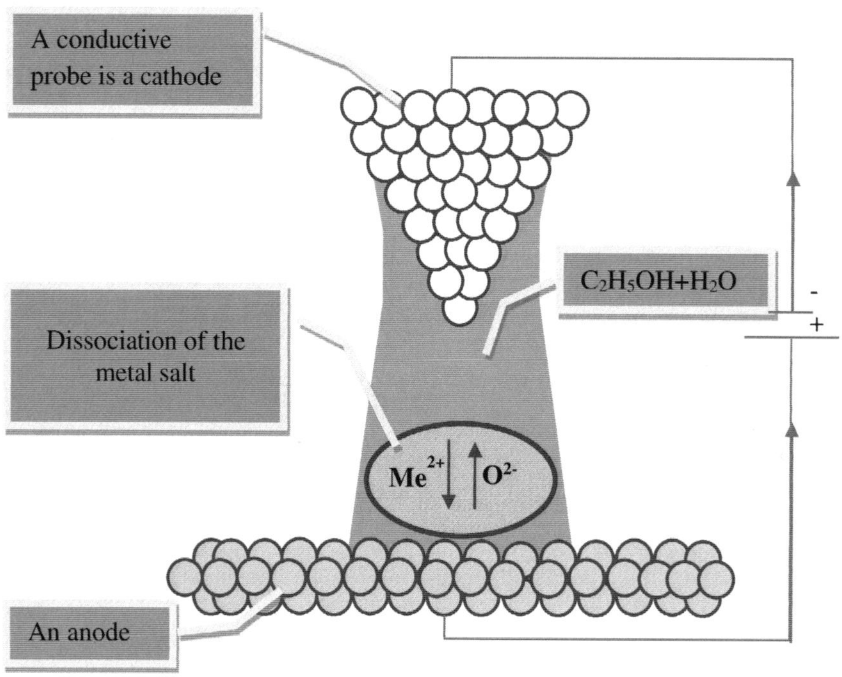

Fig. 1 The scheme process of AFM litograthy.

The average thickness (varied from 0.7 to 4 nm) of the obtained structures is well controlled with bias voltage applied between the tip and the sample. The average value of the track width determined by the diameter of the water meniscus between the conductive probe and the silicon. It was not more than 100 nm (see Fig.2) which is much less then optical wavelength. It indicates that single metal tracks fabricated with proposed method allows for a tight concentration of an optical energy in a plasmon mode.

The ability to use thin and narrow metal stripes placed on dielectric substrate as plasmon waveguide was repeatedly demonstrated both experimentally and theoretically [4]. Although a lot of different schemes of metal plasmon waveguide fabrication was proposed recently, the goal of engineering of integrated all-optical plasmonic circuits still represents an open challenge.

Figure 2 demonstrates how complicated schemes of plasmon-supported waveguides can be fabricated using proposed technique.The structure shown on Fig. 2 is an array of the net of square cells split by thin metal stripes. Since the period between the stripes can be kept in order of optical wavelength (400 nm in Fig. 2b), such a structure represents promising platform for fabrication of two-dimensional metallic photonic crystal for a plasmon excitations propagating along the stripes – cf. with [5]. Besides the obtainedgrating could be coupledto an optical waveguide mode propagating in the substrate. Such systems are known for their narrow-band optical response [6] and are rather sensitive to a change of environmental refractive index which induces spectral shifts of supported plasmon resonances of the structure.

IOP Conf. Series: Journal of Physics: Conf. Series **951** (2017) 012021 doi:10.1088/1742-6596/951/1/012021

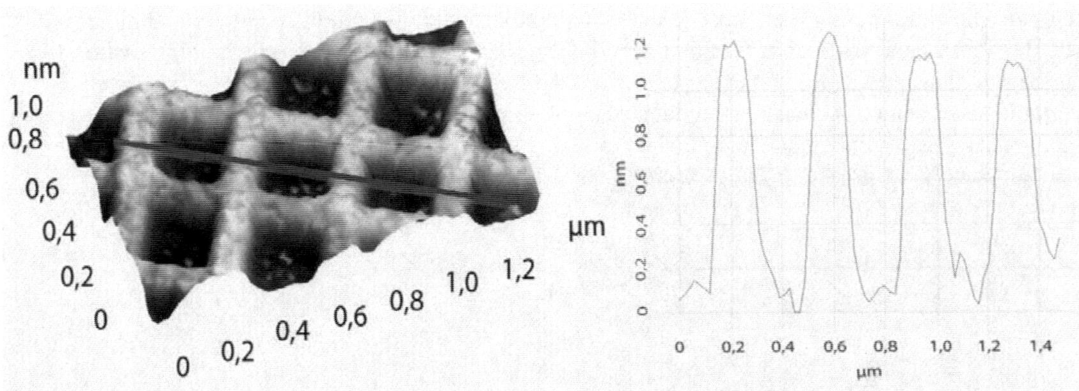

Fig. 2The AFM-image of silicon substrate after lithography at a voltage 8 B (a)and the profile of silver deposited clusters along the line (b).

Also, width and uniformity of the filling pattern of silver can be increased by repeated passage of the probe from a given area (Fig.3). These structures can be used as a possible hybrid circuits for photoelectron transducers.

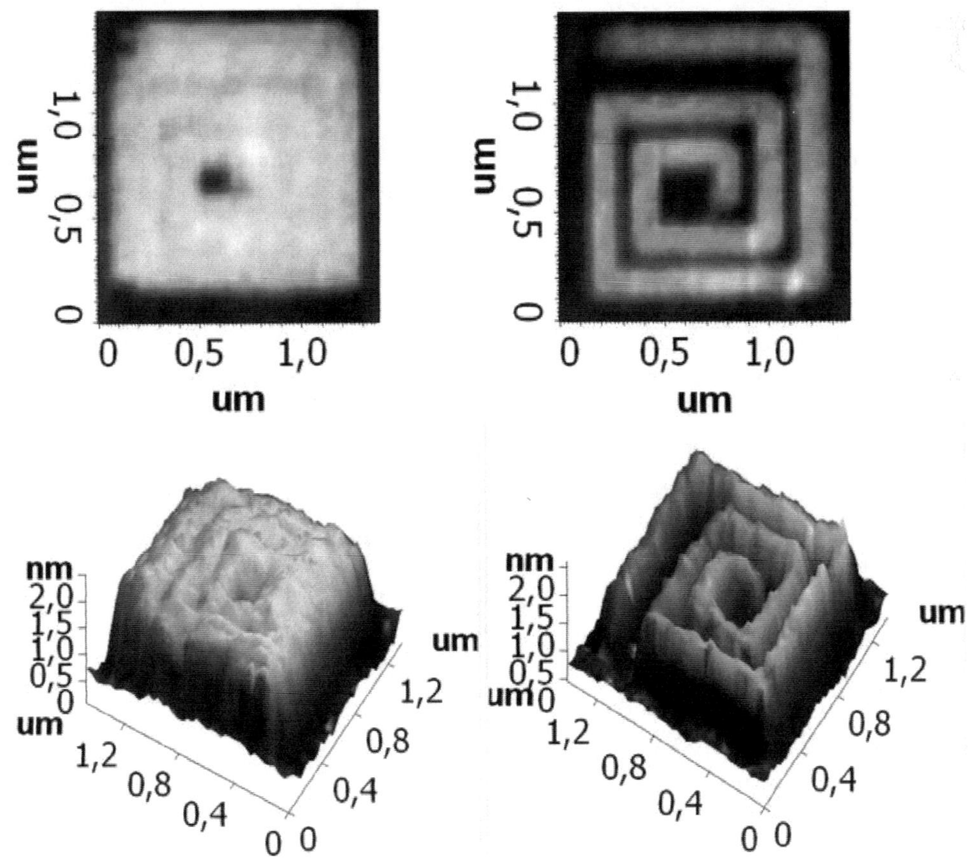

a) b)

Fig.3. AFM-image and 3D-image clusters: 3 passes (a) 1 pass (b)

Because the relief height isn't more than ten nanometers, such structures in the vertical direction should be transparent to the optical radiation. While in the plane of the nonlinear light interaction occurs due to the coincidence of the lattice period, the frequency of the plasmon resonance of silver/gold and a waves excitation on the interface of silicon-metal. Such structures can be applied as optical microcavity and logical element (see fig.4)

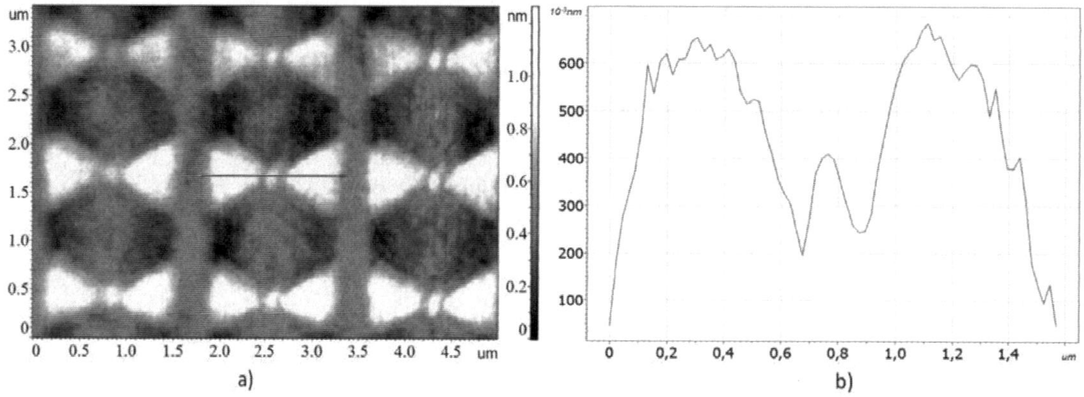

Fig. 4 The AFM-image ofgoldenmassive on silicon substrate after AFM lithography at bias voltage about 8 B (a) and and its profile (b)

The structure shown on Fig. 4 represents the arranged ensemble of bowtie gold nanoantennas. Such structures are known for their tight concentration of electromagnetic field which enable for an enhancement of emission rate of a structure embedded between the tips of a single bowtie. Due to dense concentration of a bowties in a massive the neighboring structures become coupled in the sense of tunnel coupling between adjacent bowties. This coupling modifies the optical response of fabricated structure.

3. Conclusion

The method of electro-induced deposition of metallic layers on the silicon substrate is offered. The formation of planar structures for photonics is discussed.

The reported study was also supported by the Ministry of Education and Science of the Russian Federation (state project no. 16.1123.2017/PCh), RFBR grants 16-32-60067 mol_a_dk, and by the grants of president of Russian Federation byprojectsMK-2842.2017.2 and MK-2988.2017.2.

References

[1]Antipov A A, Arakelyan S M , Vartanyan T A et al 2015 *Optics and Spectroscopy* V 119 **119**
[2]A B Evlyukhin and S I Bozhevolnyi 2015 *Physical Review B* V 92 **245419**
[3]Arakelyan, S M, Veiko V P, Kutrovskaya S V 2016 *Journal of Nanoparticle Research* V 18 **155**
[4]L Gao, L Tang, F Hu, R Guo, X Wang, Z Zhou 2012 *Optics express* V 20 **11487**
[5]Dionne J A, Sweatlock L A, Atwater H A, & Polman A 2005 *Physical Review B* V 72 **075405**
[6]X. Zhang et al. 2006 *Nano Letters*, V 6 **651**
[7] S F Feng et al 2009 *Opt. Express* V 17 **426**

Colloidal quasicrystal for photonics

I Skryabin, S Kutrovskaya, S Arakelian, A Pikalova, A Fedotova

Vladimir State University named after Alexander and Nikolay Stoletovs, 87 Gorky Street, Vladimir 600000, Russia

E-mail: sk-ig@mail.ru

Abstract In this work a method of the laser synthesis of colloidal nanoparticles and the formation of liquid photonic crystals are discussed. Optical properties of the systems depending on the particle concentration, its sizes and shapes were determined.

1. Introduction

At present there is the great deal of research related to the application of semiconductor quantum dots having the optical transitions in the visible range of the spectrum. Nowadays more and more attention is paid to nanocrystals of lead chalcogenides (PbS, PbSe, PbTe) – narrow-gap semiconductor compounds with a band gap of 0.41, 0.278 and 0.31 eV at 300 K with transitions, occupying a wide range of near-infrared region of the spectrum (0.8-5 microns) [1-2].

Colloidal quantum dots are the best alternative to conventional various organic and inorganic fluorophores which are good at not only in photostability, but also on the molar extinction coefficient, and having a number of unique photophysical properties. Semiconductor quantum dots located inside complexes with the nanoparticles (NPs) of noble metals or carbon chains can be used in a varied applications such as sensor systems and optoelectronic devices [3-4].

2. Laser synthesis of colloidal systems

The semiconductor nanoparticles were prepared by using laser ablation in a liquid under the radiation of a continuous laser beam (wavelength is 1.06 µm; power density is about 105W/sm2) [5]. In this case the photon energy of laser quantum is much greater than the band gap of used semiconductors. The radiuses of the produced nanoparticles (d, nm) were determined by using a particle size analyzer 550 Horiba LB-550, the functioning principle of which is based on the effect of dynamic light scattering. Physics of quantum dots creation for lead chalcogenides materials associate with the value of the Bohr radius of the exciton aB, which is defined as:

$$a_B = \hbar^2 \varepsilon / e^2 m^* , \qquad (1)$$

where $\hbar$ – is the Planck's constant, ε – the high-dielectric constant of material,

e – is the electronic charge, a m^* – the effective mass of the exciton, which is calculated through the effective masses of electrons m_e^* and holes m_h^* :

$$1/m^* = 1/m_e^* + 1/m_h^* . \qquad (2)$$

IOP Conf. Series: Journal of Physics: Conf. Series **951** (2017) 012022 doi:10.1088/1742-6596/951/1/012022

Because of the fact that the lead chalcogenides have large values of high-frequency dielectric constant and small values of the effective masses of electrons and holes, the exciton Bohr radius is large value calculated for the isotropic case, for example for PbTe it is about 50 nm.

The comparison of the NPs size with the exciton Bohr radius shows that for nanocrystals of lead telluride, the strict condition is right (see figure 1). So, for PbTe anoparticles the size quantization conditions are realized. The result is there are energy gaps in the electronic spectra of the quantum dots, which can significantly exceed the value of the band gap of the bulk lead chalcogenides.

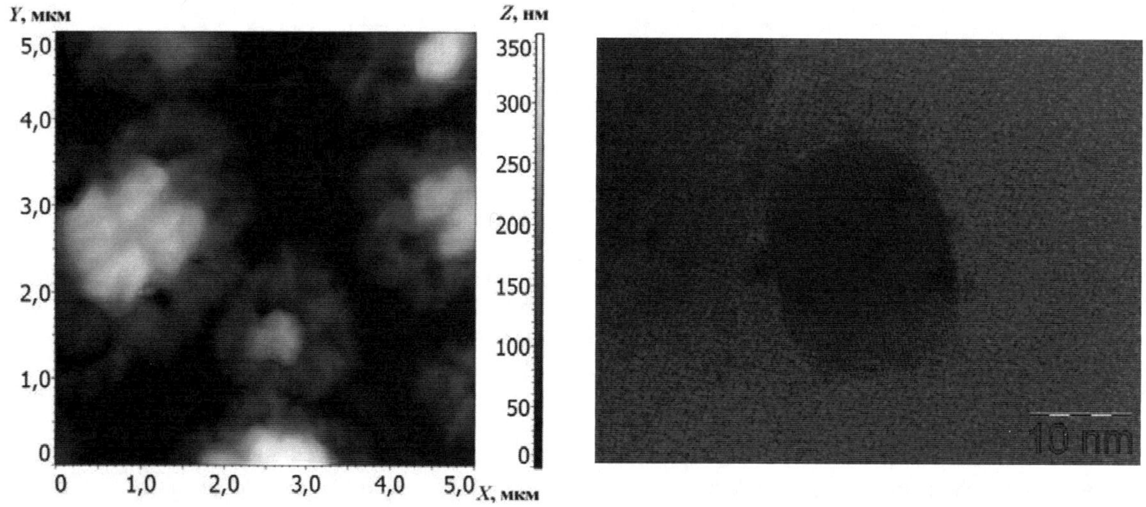

a) b)

Figure 1. Images of PbTe NPs as a regular polygons: a) clusters of nanoparticles were fixed by atomic force microscope Ntegra Aura. The concentration of a colloidal solution was 1μg/ml, b) a single NP and its diffraction pattern is shown in the top right corner. The estimated period of the lattice constant ≈ 0.64 nm.

The quantum dots of PbTe/PbSe were quasiperiodically distributed in a volume of a viscous liquid medium by the effect of self-organization: the compensation between the Coulomb repulsion and van der Waals attraction. The scheme of the diffraction pattern monitoring at the action of laser illumination of the liquid quasicrystal by visible range λ = 520 nm(see fig. 2). To reduce the thermal noise, a cell with a liquid was cooled down to 0 ° C and was placed on a heat stabilized table. The temperature was measured by a contact less method using the IR thermometer Testo 830-T2. Fourier transform of the diffraction pattern confirms the presence of sensible and latent period in distribution of the semiconductor NPs in photonic liquid crystal.

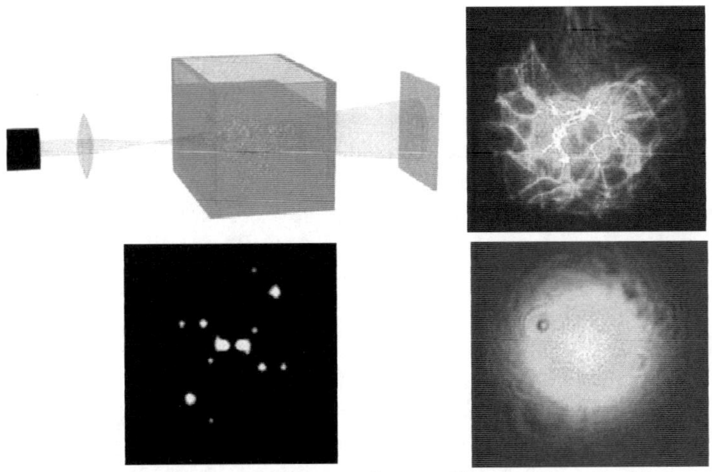

Figure 2.. a) The scheme of diffraction observation at illumination liquid quasicrystal by laser radiation; b) the diffraction pattern of colloidal glycerol solution with a NPs concentration is 1 μg/ml; c)Fourier image of 2b confirms the presence of sensible and latent period in the liquid system; g) the diffraction pattern of glycerol in cuvette at the same conditions. There is scattering of radiation transmitted through the medium.

3. Properties of liquidsystems

The experiments to study the effects of the interaction of an electromagnetic radiation with intensity range from 102 to 104 W/cm2 with a liquid photonic crystal were performed. The systems may be applied in IR fiber optics. However, in our case the fluorescence intensity is insufficient for stable manifestations of optical effects due to the low concentrations of NP semiconductor in colloidal systems. A presented decision is the using Purcell effect by creating complexes including metal and semiconductor particles. We have used PbTe system in combination with NPs Au (the average size was about 10 nm, see fig. 3a) because the effect of increasing the absorption of the semiconductor quantum dots in the presence of plasmonic nanoparticles was well reproduced. FTIR-spectroscopy was used to analyze changes in properties of colloidal systems by the absorption spectra (see fig. 3b).The stable increasing of an absorption for the complex systems are fixed.

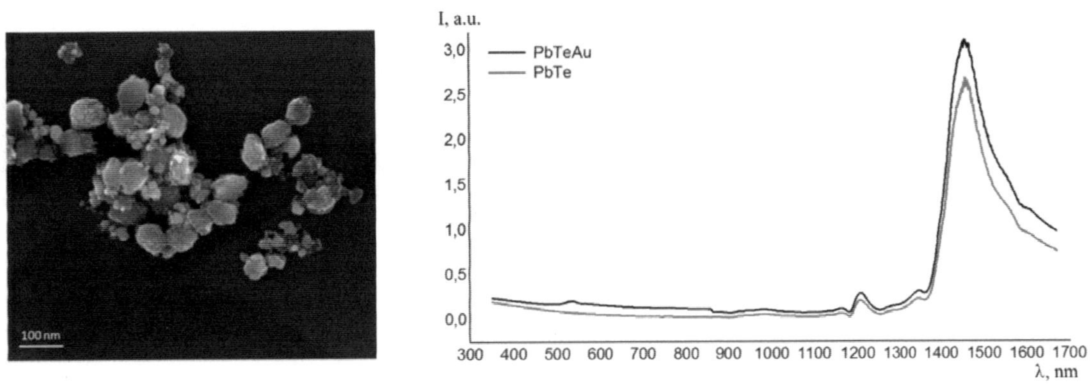

a) b)

Figure 3. a) TEM image complexes including small metal Au particles around polygon semiconductor particles PbTe; b) the absorption spectra of initial liquid system on PbTe and on the complexes PbTe-Au

The distance between the semiconductor quantum dots and metal nanoparticles is a key factor that affects the efficiency of exciton-plasmon interactions between them. When the quantum dot lays closely to the particles, the mechanism of non-radiative relaxation of the exciton quantum dot to the surface of the metal nanoparticles can be realized. There is a decrease in the fluorescence quantum yield of the quantum dot. On the other hand, when the quantum dot is located a far from the nanoparticles, the fluorescence quantum yield is also reduced. As the induced optical field of plasmon nanoparticles decays rapidly with distance from its surface. Thus, only at distances comparable to the particles size the efficiency the exciton-plasmon interactions between quantum dots and metal nanoparticles would be optimal [6].

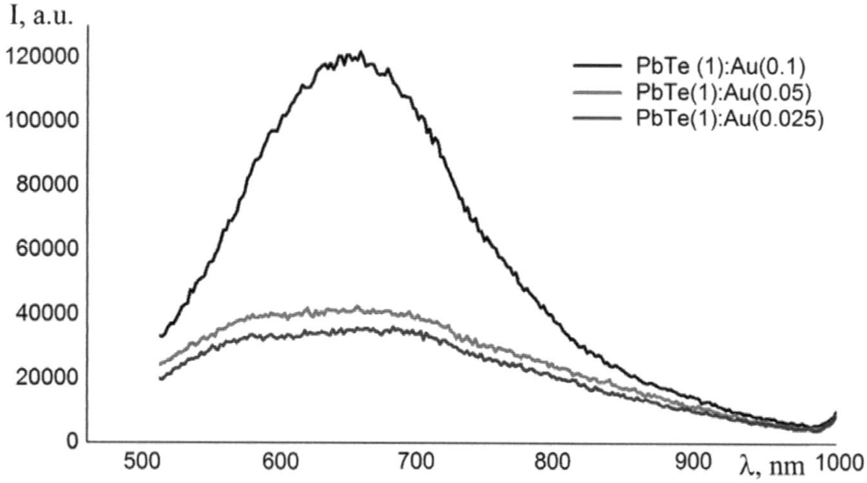

Figure 4. The fluorescence spectrum of semiconductor-metal complexes in different mass proportions, distributed in the volume of glycerol. The wavelength pumping was 488 nm.

The luminescence time of complex PbTe-Au decrease up to t = 2,3 *10 -9 s in comparison to initial PbTe liquid system t = 1,4 *10 -8 s. It is connected the develop of the Purcell effect in media without strong separation boundary metal-semiconductor and quasiperiodically distribution of elements.

4. Conclusion

The results of the quasiperiodical forming of liquid systems on semiconductor quantum dots are shown in this work. Its optical properties depending on the parameters of the colloidal system were investigated. The fluorescence of the colloidal system of metal- semiconductor systems PbTe - Au in the visible spectrum was detected. The concentration of gold nanoparticles in a colloidal system affects to the manifestation of this effect.

Raman and FTIR-absorption spectra were measured at Center for Optical and Laser Materials Research, St. Petersburg State University.

The reported study was also supported by the Ministry of Education and Science of the Russian Federation (state project no.16.1123.2017/PCh), RFBR grants16-32-60067 mol_a_dk, 16-42-330461_r_a.

References

[1] Zhang C-Y, Yeh H-C, Kuroki M T, Wang T-H 2005 *Nature Materials* V 4 **826**
[2] Tian J, Cao G, 2003 *Nano Reviews & Experiments* V 4 **22578**.
[3] Wood V, Bulovic V, 2010 *Nano Reviews & Experiments* Vol 1 **5202**
[4] Kwak J, Bae W K, Lee D, at al 2012.*Nano Letters* V 12 **366**

IOP Conf. Series: Journal of Physics: Conf. Series **951** (2017) 012022 doi:10.1088/1742-6596/951/1/012022

[5] Arakelyan S M, Veiko V P, Kutrovskaya S V 2016. *Journal of Nanoparticle Research* V 18 **155**
[6] Kulakovich O, Strekal N D, Yaroshevich A at al 2002 *Nano Letters* V 2 **1449**.

IOP Conf. Series: Journal of Physics: Conf. Series **951** (2017) 012023 doi:10.1088/1742-6596/951/1/012023

Laser synthesis of hybrids Si-Au complex

A Osipov, S Kutrovskaya, V Simonov, A Kucherik

Vladimir State University named after Alexander and Nikolay Stoletovs, 87 Gorky Street, Vladimir 600000, Russia

E-mail: osipov@vlsu.ru

Abstract The paper demonstrates that the laser ablation approach can be used for synthesis (with average nanoparticle size control) of silicon nanoparticles supporting strong resonant optical responses in the visible spectral range. Porous and monolithic crystalline silicon targets are considered. The generated colloidal solutions have been used for the formation of one-layered covering composed of silicon nanoparticles. The obtained results may be used for the realization of many functional metasurfaces consisting of randomly distributed resonant nanoparticles.

1. Introduction

A huge spectra of photonic applications require silicon nanoparticles [1, 2]. Its optical properties, including resonant responses, are considerably dependent on the size, shape, and crystallization degree [3]. The development of fabrication techniques with controlled nanoparticle parameters is very important. The methods of laser ablation in liquid [3, 4] allow to control the average size and shape of the obtained particles choosing suitable irradiation conditions (pulse duration, energy density etc.). The order deposition of such particles in thin films allows to create metasurfaces for controllable manipulation of the reflection and transmission properties [6,7].

The paper presents results of the silicon nanoparticle synthesis by the continuous laser irradiation of the thin silicon target placed in ethanol. Porous and monolithic silicon targets are considered.

2. The obtaining of colloidal particles

For obtaining of silicon nanoparticles the method of CW-laser ablation was used [8, 9]. Application of the CW-laser radiation source with moderate intensity provided the possibility to obtain nanoparticle with a small deviation of the average size [4, 10]. In our experiments the targets of crystalline monolithic and porous silicon were used. To avoid the oxidation of nanoparticles we realized the laser ablation experiment in the 99% ethanol [3].

The laser beam (diameter 100 μm) was focused on the surface of the target (using the long focal-length lens with the focal length of 100 mm). The power variation was in the diapason of 10 – 100 W, which was resulted in the radiation intensity of $10^5 - 10^6$ W/cm^2. As the conventional time of laser irradiation may be calculated by $\tau = \frac{d}{v}$ (v – scanning speed), the laser beam diameter and scanning speed variations resulted in the irradiation time in the interval from 1 s up to 10 s (the mechanized table moving speed 10 μm/s – 100 μm/s). The whole irradiation time was 60 minutes.

The laser irradiation of the porous target resulted in a wider distribution function of nanoparticle sizes as compared with crystalline target (fig.1). In the both cases spherical nanoparticles with average diameter of 100 nm were obtained (fig. 2).

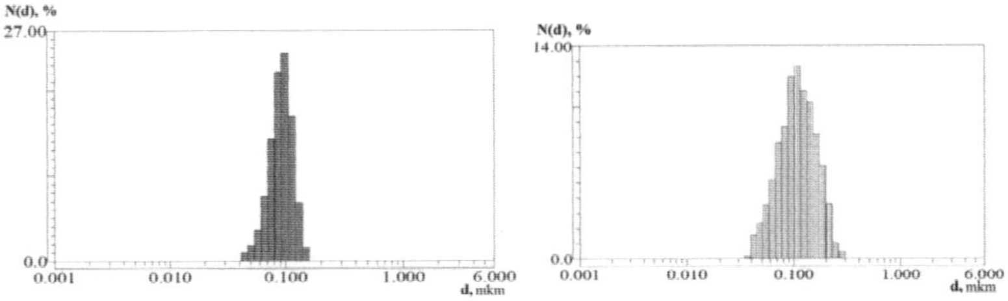

Fig. 1. The histogram of the particle diameters distribution in colloidal system, which was obtained using laser irradiation (intensity 10^6 W/cm^2, scanning speed 100 μm/s) of the monolithic target (a) and porous (b).

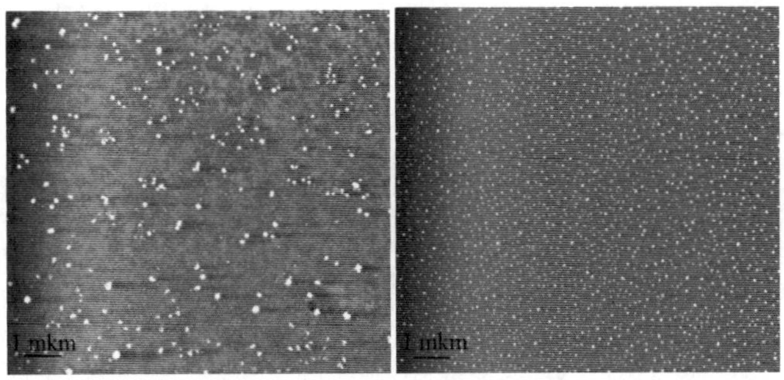

Fig. 2. REM-images of spray-jet deposited colloidal particles, obtained during the ablation of the porous (a) and monolithic (b) targets.

3. The colloidal particles deposition

Images of the particles shown in fig. 2 correspond to the evaporation process of the ethanol colloidal drop with nanoparticles inside. The experiment was carried out in a temperature controlled chamber with the temperature of 20 C. As it can be seen from fig. 2 there are uniformly deposited particles on the base area of the drop after its evaporation.

To control the deposition of the nanoparticles from a small colloidal drop we have used the methods described in [8, 10, 11]. We have deposited silicon nanopaticles using the sputtering small colloidal drops technology on the surface of the transparent substrate.

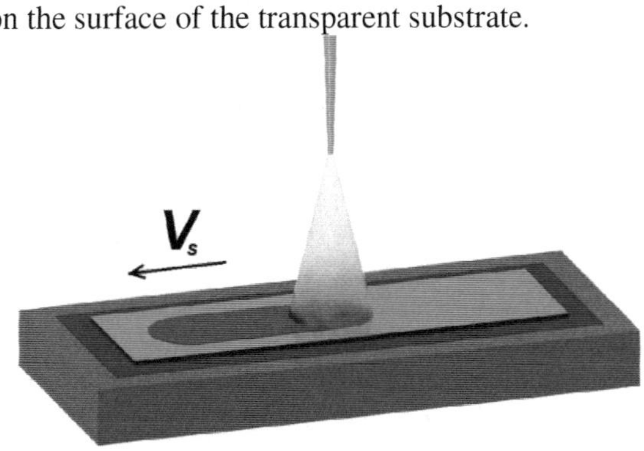

Fig.3. The colloidal particles deposition scheme.

Colloidal systems have been sprayed through a capillary with a diameter of 100 μm (see fig. 3).

The system was pressurized by the pneumatic pump, and the pressure was about 5 bar. The capillary was positioned perpendicular to the surface of the substrate at a distance of 100 mm. The glass substrate was fixed on the surface of thermostabilized one-coordinate table with a temperature of 20 ° C. The table movement speed (V_s) was ranged from 0.1 mm/s to 1 mm/s. The deposited droplets took a spherical shape, because of the Bond condition

$$B_o = \frac{g(\rho-\rho_m)d^2}{\sigma} \ll 1, \quad (1)$$

where **g** – the gravity acceleration index, **ρ** – the density of the drop material, $\boldsymbol{\rho_m}$ – the density of the volume, in which is a drop, **σ** – the surface tension, d – the droplet diameter.

In this case the drop evaporation may be calculated using Maxwell diffuse model. Then the changing of the drop profile and height in time may be described by following equation:

$$h(r,t) = \sqrt{\left(\frac{h(0,t)^2+R^2}{2h(0,t)}\right)^2 - r^2} - \frac{R^2-h(0,t)^2}{2h(0,t)}, \quad (2)$$

where $h(0,t)$ – the height of drop in center, R – the drop base radius.

4. The measurement of the optical properties of the particles using a dark-field microscopy

The measurements of nanoparticle optical properties were carried out using dark field microscope [5], where only scattered light is collected. As we can see from fig. 4a nanoparticles fabricated from the crystalline Si targets can resonantly scatter light of different frequencies.

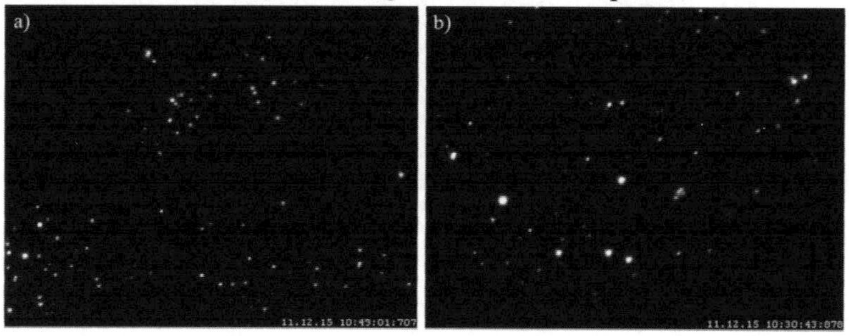

Fig. 4 A dark-field (scattered light) image of the deposited colloidal systems of monolithic (a) and porous (b) silicon.

It is also important that in the case of the irradiation of the porous silicon target a large amount of nanoparticles does not support the resonant optical response in the visible range fig. 4b, because of their low refractive index. The different color presentation of the dark-filed images of the nanoparticles in fig. 4a is connected with excitation of electric and magnetic multipole resonances. Depending on the nanoparticle size these resonances are realized at different spectral points. The larger nanoparticle the large resonant wavelength [4,11]. Q-factors of the multipole resonances are decreased with decreasing of the nanoparticle dielectric permittivity as a result dark-filed images of nanoparticles with low refractive index are presented by white color like in fig. 4b.

5. Conclusion

It has been experimentally demonstrated that the cw-laser ablation can be applied for synthesis of silicon nanoparticles supporting resonant optical responses in the visible spectral range. Crystalline and porous silicon targets have been considered. The proposed method allows the silicon nanoparticles with an average size of 100 nm, the dispersion of the sizes used depends on the target structure. In the latter case resulting porous nanoparticles had relatively low refractive index and did not show strong resonances. In the case of crystalline nanoparticles, the strong electric and magnetic multipole resonances have been demonstrated. These obtained results have important implications to the realization of optical metasurfaces, with many functional properties, in the visible spectral range.

The article was supported by the Ministry of Education and Science of the Russian Federation (state project no. 16.5592.2017/VU), RFBR grants 16-32-60067 mol_a_dk, 17-42-330928 r_a.

References

[1] U. Zywietz, A.B. Evlyukhin, C. Reinhardt, B.N. Chichkov, *Nat. Commun.* 5, 3402 (2014)
[2] A.B. Evlyukhin, S.M. Novikov, U. Zywietz, R.L. Eriksen, C. Reinhardt, S.I. Bozhevolnyi, B.N. Chichkov, *Nano Lett.* 12 (7), 3749 (2012)
[3] D.V. Abramov, A.A. Antipov, S.M. Arakelian, K.S. Khorkov, A.O. Kucherik, S.V. Kutrovskaya, V.G. Prokoshev *Laser Phys*, 24, 6886356 (2014).
[4] U. Zywietz, C. Reinhardt, A.B. Evlyukhin, T. Birr, B.N. Chichkov, *Appl. Phys.* A 114, 45 (2014)
[5] P. Spinelli, M.A. Verschuuren, A. Polman, Nat. Commun. 3, 692 (2012)
[6] U. Zywietz, M.K. Schmidt, A.B. Evlyukhin, C. Reinhardt, J. Aizpurua, B.N. Chichkov, *ACS Photonics* 2(7), 913 (2015)
[7] A.A. Antipov, S.M. Arakelyan, T.A. Vartanyan, T.E. Itina, S.V. Kutrovskaya, A.O. Kucherik, I.V. Sapegina, *Opt Spectrosc.* 119 (1), 119 (2015)
[8] M. Boutinguiza, R. Comesana, F. Lusquinos, A. Riveiro, J. del Val, J. Pou, *Appl. Surf. Sci.* 336, 108 (2014)
[9] S. Arakelian, S. Zimin, S. Kutrovskaya, A. Kucherik, A. Makarov, A. Osipov, *Laser Phys.* 24, 074002 (2014)
[10] S.M. Arakelyan, V.P. Veiko, S.V. Kutrovskaya, A.O. Kucherik, A.V. Osipov, T.A. Vartanyan, T.E. Itina, J. *Nanopart. Res.* 18, 1 (2016)
[11] A.B. Evlyukhin, C. Reinhardt, A. Seidel, B.S. Luk'yunchuk, B.N. Chichkov, *Phys. Rev.* B 82, 045404 (2010)

Fabrication of porous nanostructures for Raman signal amplification

E V Mitsai[1], S A Syubaev[1], A A Kuchmizhak[1,2]

[1] Institute of Automation and Control Processes FEB RAS, 5 Radio Str., 690041 Vladivostok, Russia
[2] Far Eastern Federal University, 8 Sukhanova Str., 690090 Vladivostok, Russia

mitsai@dvo.ru

Abstract. Pulsed-laser dry printing of noble-metal microrings with a tunable internal porous structure, which can be revealed via an ion-beam etching post-procedure, was demonstrated. Average size of the pores inside the microrings were shown to be tuned in a wide range by varying the incident pulse energy and a nitrogen doping level controlled in the process of magnetron deposition of the gold film in the appropriate gaseous environment. The fabricated porous microrings were shown to provide many-fold near-field enhancement of incident electromagnetic fields, which was confirmed by mapping of the characteristic Raman band of a nanometer-thick covering layer of Rhodamine 6G molecules and supporting calculations. The proposed laser-printing/ion-beam etching approach is demonstrated to be a unique tool aimed at designing and fabricating multifunctional plasmonic structures and metasurfaces for spectroscopic bioidentification based on surface-enhanced Raman scattering and photoluminescence detection schemes.

Surface-enhanced Raman scattering (SERS) is an ultrasensitive non-invasive spectroscopic technique based on a label-free identification of different molecules placed in the vicinity of plasmonic-active nanostructured metallic substrates. Intensity of the characteristic Raman signal defining the specific vibrational signatures of individual molecules is usually very weak, but it can be significantly increased near nanotextured surfaces or nanostructures generating localized, strongly enhanced plasmon-mediated electromagnetic fields. Multiple attempts were undertaken to increase the efficiency of SERS-active nanotextured substrates in terms of the achieved maximal enhancement factor inside a single "hot spot" as well as the number of "hot spots" per individual nanostructure.

A variety of nanotextured structures, predominantly having a large surface-to-volume ratio and generating dense hot spots were fabricated and tested as versatile SERS substrates. Specifically, porous materials, nanostructures, and nanoparticles, routinely reaching uniform SERS enhancement sufficient to overcome the single-molecule detection limit independently of excitation/detection conditions, are of growing interest. Meanwhile, to design sensitive elements for advanced biosensors, along with desirable porosity, it is also important to control the overall size and shape of such porous templates as well as to arrange them into ordered arrays at a specific point on a substrate with micrometer-scale lateral accuracy.

In this work, an ion-beam assisted, liquid-free nanosecond (ns) laser printing of isolated plasmonic ring-shaped nanostructures with pronounced porosity is demonstrated. The proposed procedure for fabricating the isolated porous rings includes two consecutive steps schematically illustrated in figure 1. First, a 50-nm thick glass- or Si-supported Au film is irradiated with single second-harmonic (532 nm), 7-ns pulses delivered by a Nd:YAG laser system (Brio, Quantel) focused with a dry objective lens of

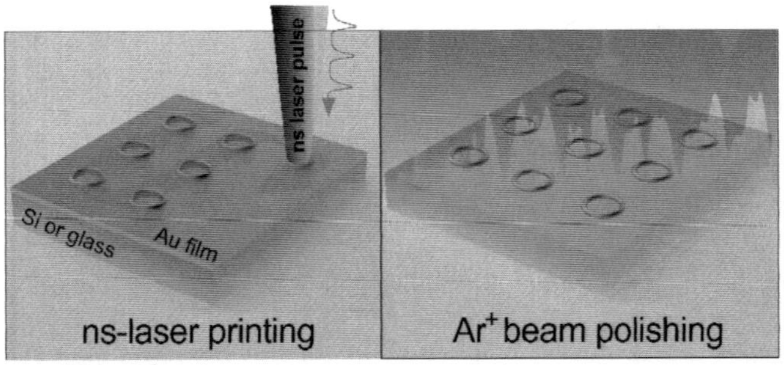

Figure 1. Schematic of the two-step fabrication procedure for printing of the isolated porous plasmonic rings.

numerical aperture NA = 0.6. In all our experiments, we used the same focusing conditions, while the only incident pulse energy was varied by means of an adjustable attenuator. In a certain range of incident pulse energies [1], such single-pulse ns-laser ablation of the metal film leaves a through hole surrounded by a smooth resolidified rim typically 2.5–3 times thicker, compared to the initial thickness of the deposited Au film. The second step involves removal of the unprocessed parts of the film using a liquid-free etching with an accelerated argon-ion beam (IM4000, Hitachi) at fixed values of an acceleration voltage of 3 kV, a discharge current of 105 μA, and a gas flow of 0.1 cm^3/min. This parameter set provides a relatively slow rate of 1 nm/s to avoid film melting. The removal of the residual, initially deposited 50-nm thick Au film leaves the isolated microrings on the substrate surface.

Main geometric parameters of the microrings produced with the developed two-step procedure were shown to be tuned by varying the thickness of the deposited film and the incident pulse energy [1, 2]. In the present study, we demonstrate that an additional degree of tunability of the internal structure of the produced microrings can be provided by a minor variation in the chemical composition of the sputtered metal film via its controlled magnetron deposition in appropriate gaseous environments. To deposit the Au film on a glass substrate, a commercial magnetron sputter was used (Quorum Technologies). Typically, argon is used as a discharge gas for such magnetron sputtering. Then, ns-laser printing followed by the subsequent ion-beam polishing usually results in formation of smooth microrings with a relatively small amount of defects.

Surprisingly, detailed imaging of the structures produced with the single-shot ns-pulse ablative patterning of an Au film sputtered in a mixture of the argon and nitrogen (50/50 wt. %) reveals dense pores inside the fabricated microrings. The nanosized (appox. 60-nm diameter) spherical-shape pores initially located inside the resolidified metal rim can be further visualized through the ion-polishing post-procedure. Moreover, even more pronounced porosity with a significantly larger averaged pore size was found for such structures produced on the surface of the Au film deposited in the nitrogen atmosphere [Figure 2(a)]. The statistical study of the distribution of the pore size inside the microrings produced at the same pulse energy indicates the clear correlation between the nitrogen concentration in the deposition chamber and the average nanopore diameter in the produced microrings [figure 2(a)].

Subsurface boiling around nitrogen-doped cracks and N_2-reach areas of the film appears to be the key mechanism responsible for the formation of the densely packed nanopores inside the molten rim [3]. As known, gold can interact with nitrogen, producing gold-nitride phase Au_xN, upon Au target bombardment with accelerated nitrogen ions [4, 5]. Also, Au_xN films were fabricated, using ablative pulsed laser deposition in a N_2 containing atmosphere. Mass and X-ray inspections show that increased amount of nitrogen corresponds to case of film deposition in nitrogen ambient.

The printed microrings, with their well-controlled micrometer-scale diameter and pronounced nanotexture governed by the unveiled multiple circumferentially spaced pores, are expected to provide strong near-field enhancement of incident electromagnetic waves in visible and IR spectra ranges, making them promising for chemo- and biosensing applications. In this respect, the porous rings arranged into the ordered array can be considered as multifunctional substrates, providing enhancement of both photoluminescence and SERS signals from adsorbed analyte molecules commonly pumped in

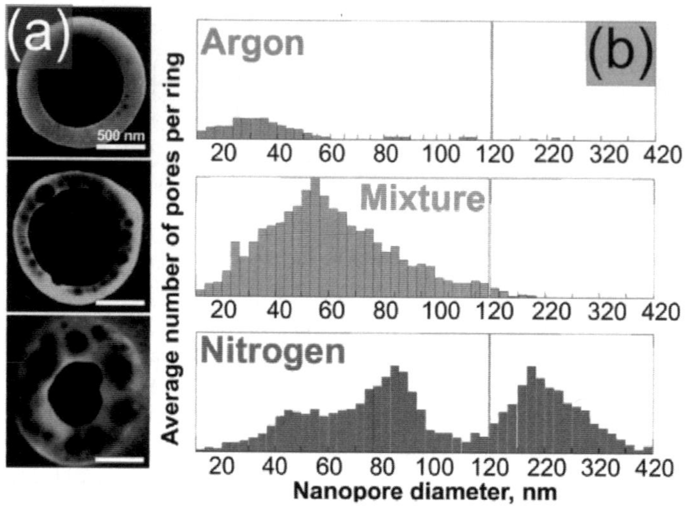

Figure 2. (a) Series of normal-view false-color SEM images of microrings produced on the 100-nm thick Au films deposited in the argon (top), argon-nitrogen (50/50 wt. %) gas mixture (middle), and nitrogen gas atmospheres (bottom) via magnetron sputtering. All three microrings were produced at the same pulse energy E = 17 nJ. (b) Averaged distribution of nanopore diameters inside the fabricated microrings. The statistics for each film type was accumulated over 50 microrings of the same size produced under similar experimental conditions.

the optical range, while giving also possibility to utilize near-to-mid infrared wavelengths to excite main dipolar localized plasmon resonances of the microrings.

We restrict our studies of biosensing performance for the laser-printed porous microrings by measuring their Raman response from a self-organized monolayer of Rhodamine 6G (R6G) molecules and mapping surface intensity distribution of their specific main bands. A commercial Raman microscope (Alpha, WiTec) was used to determine the SERS performance of the fabricated microrings. An ethanol solution of R6G with a concentration of 10^{-5} M was drop-cast on the sample surface, and after complete drying, it was rinsed with distilled water producing a monolayer. All Raman spectra were measured at the 1-mW excitation laser power and the accumulation time of 0.5 s per point. Similar parameters were used for mapping of the surface intensity distribution of the main R6G Raman bands near the several smooth and porous microrings within the 2x2 μm^2 square sample area with 0.2-μm sampling step. The build-in software (WiTec Project) was used to obtain the 2D maps related to each spectrally narrow Raman band.

Typical Raman spectra acquired from the R6G layer on both smooth and porous microrings, having almost the same outer diameter and the wall thickness, reveal all main Raman bands characteristic for R6G molecules, while these "fingerprints" cannot be identified on the glass substrate even under 10 times increased accumulation time [figure 3(a)]. This measurement indicates strong SERS performance of the fabricated structures. It points out a clear plasmonic contribution to the SERS enhancement, rather than the concentration effect. Mapping the surface distribution of the Raman intensity at the 1361-cm^{-1} band of the R6G molecules, covering the smooth microring, reveals the position of the local electromagnetic "hot spots", which were found to depend on the polarization direction of the exciting laser source [orange arrows in figures 3(b)-3(e)].

According to our supporting FDTD calculations [1] such 532-nm, normal-incident linearly polarized laser irradiation provides low-efficiency excitation of the smooth microring walls with quite moderate 5-fold enhancement of the squared electric field amplitude [figure 3(d)], while the corresponding hot spots, following the linear polarization direction, are distributed non-uniformly along the ring circumference, yielding similar distribution of the SERS signal [figure 3(b)]. On the contrary, according to our FDTD calculation performed for the porous microring, nanosized surface features produce

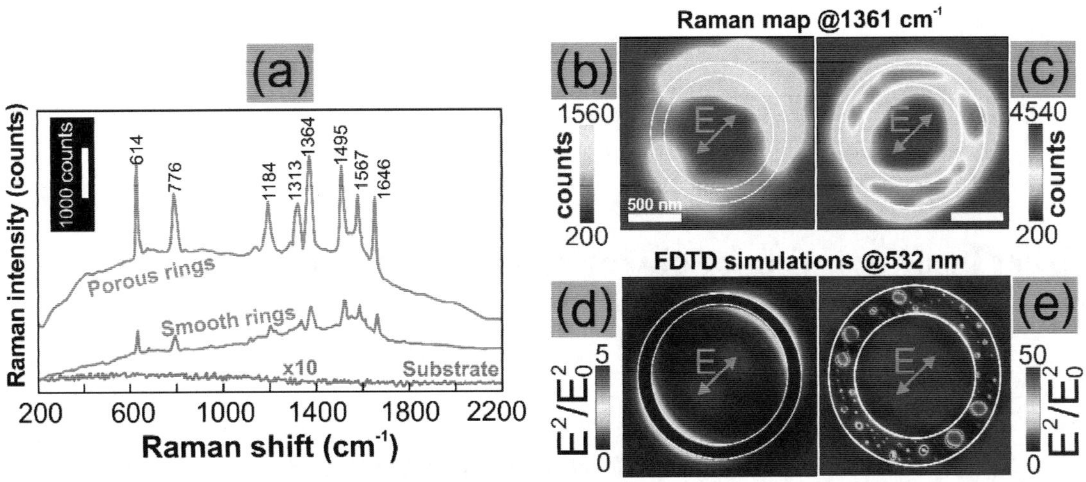

Figure 3. Biosensing performance of smooth and porous microrings presented as R6G Raman spectra measured on the smooth (brown curve) and porous microrings (purple curve) as well as on the glass substrate after complete removal of the Au film via the ion polishing post-procedure. (b) and (c) SERS maps measured for the 1361-cm^{-1} Raman band of the R6G molecules nanolayer covering the smooth (b) and porous microrings (c). These maps were found to provide SERS enhancement according to the characteristic E-field amplitude distribution near the smooth (d) and porous microrings (e) calculated for the normally incident, linearly polarized 532-nm laser radiation. The orange arrows indicate the polarization direction. The white circles show the actual dimensions of the microrings.

multiple strongly enhanced E-fields, homogeneously distributed along the microring surface independently of the polarization direction [figure 3(e)] and generally correlating with the acquired SERS map [figure 3(c)].

In conclusion, in this work, we have fabricated porous microrings from thin gold films deposited in nitrogen and nitrogen-argon mixed atmospheres, by their ns-laser ablation and the following ion-beam etching post-procedure, and studied their SERS performance. The proposed approach allows us to fabricate both smooth and porous microrings with variable geometric parameters as well as tunable inner structure. The developed laser-printing/ion-beam etching procedure is demonstrated to be a versatile tool for fabricating multifunctional plasmonic structures and metasurfaces for spectroscopic bioidentification based on the combination of SERS and photoluminescence detection schemes on a single sensing substrate.

Acknowledgements

Authors acknowledge the partial support from RF Ministry of Science and Education (Contract No. MK-3287.2017.2) through the Grant of RF President and Russian Foundation for Basic Research (Grant No. 17-02-00571) as well as Far East Program.

References

[1] Kuchmizhak A, Gurbatov S, Vitrik O, Kulchin Y, Milichko V, Makarov S and Kudryashov S 2016 *Sci. Rep.* **6** 19410
[2] Kuchmizhak A, Gurbatov S, Kulchin Y and Vitrik O 2015 *Opt. Commun.* **356** 1
[3] Kuchmizhak A, Gurbatov S, Kulchin Y and Vitrik O 2014 *Opt. Express* **22** 19149
[4] Šiller L, Peltekis N, Krishnamurthy S, Chao Y, Bull S and Hunt M 2005 *Appl. Phys. Lett.* **86**, 221912
[5] Brieva A, Alves L, Krishnamurthy S and Šiller L 2009 *J. Appl. Phys.* **105**, 054302

Direct laser printing of chiral plasmonic nanojets by vortex beams

S A Syubaev

Institute for Automation and Control Processes FEB RAS, 5 Radio Str., 690041 Vladivostok, Russia

trilar@bk.ru

Abstract. Donut-shaped laser radiation, carrying orbital angular momentum, namely optical vortex, was recently shown to provide vectorial mass transfer, twisting transiently molten material and producing chiral micro-scale structures on surfaces of different bulk materials upon their resolidification. In this paper, we show that nanosecond laser vortices can produce chiral nanoneedles (nanojets) of variable size on thin films of such plasmonic materials, as silver and gold films, covering thermally insulating substrates. These results suggest optical interference of the incident and reflected laser beams as a source of complex surface intensity distributions in metal films, possessing spiral components and driving both center-symmetric and spiral thermocapillary melt flows to yield in frozen nanoneedles with their pre-determined spiral nanocarving.

1. Introduction

Chirality is a specific inherent feature, typically occurring in natural living organisms at almost all length scales from snails and sea shells to chiral molecules and DNA. Artificially designed nanoscale chiral-shaped structures, mimicking their natural analogues owing to unique ways of interaction with optical radiation, possess remarkable properties as circular dichroism, enhancement of nonlinear signals, highly directional emission and photoactivity. Meanwhile, modern methods of fabrication such chiral structures are rather expensive and small-scale (low-throughput) or limit the number of the available materials as well as the achievable size of the produced structures, that triggers, in its turn, search for alternative technologies. Recently, another alternative pioneering approach, employing optical radiation with specially designed intensity-, polarization- or phase states for fast fabrication of chiral structures, was demonstrated. More importantly, vortex laser pulses, carrying simultaneously orbital (OAM) and spin (SAM) angular momenta, were demonstrated to twist transiently molten metal, producing chiral-shaped micron-size needles [1]. Based on their experiments, the authors revealed that mass transfer and handedness of the produced surface structures are associated with corresponding phase helicity (or OAM), while the polarization helicity (or SAM) accelerates/decelerates the movement of the molten material.

In this work, we explore formation of twisted nanojets under direct translative ablation of Ag films of variable thickness, using nanosecond (ns) laser vortex pulses generated from circularly polarized pulses by an S-waveplate. We suggest an alternative explanation of the chiral nanoneedle formation on the Ag film surface based on a combination of straight and twisted thermocapillary flows of the molten material, which are triggered by the characteristic spiral-like intensity distribution, originating from the interference of the incident donut-shape vortex beam and its replica reflected from transiently evolving molten metal surface.

2. Experimental details

A vortex beam, carrying OAM $l = \pm 1$ (per photon, in units of ħ), was generated by transferring second-harmonic radiation from a Nd:YAG laser system (Brio, Quantel: central wavelength – 532 nm, pulse duration – 4 ns, repetition rate – 20 Hz, maximal pulse energy – 50 mJ) through a Glan-Taylor polarizer and a quarter-waveplate (Fig. 1), to produce a circularly-polarized beam, and finally through a commercially available radial polarization converter (S-waveplate, Altechna). Then, generated ns-laser vortex pulses were focused onto the sample surface (Fig. 1(A)) by means of different microscope objectives (Xirox: NA = 0.45, 50x; Nikon Plan Fluor: NA = 0.3, 15x and NA = 0.65, 50x), yielding in donut-shape intensity distributions (Fig. 1(B)) with the outer optical diameter Dvortex of ~2.05 μm, ~2.7 μm and ~3.9 μm for NA = 0.65, 0.45 and 0.3, respectively.

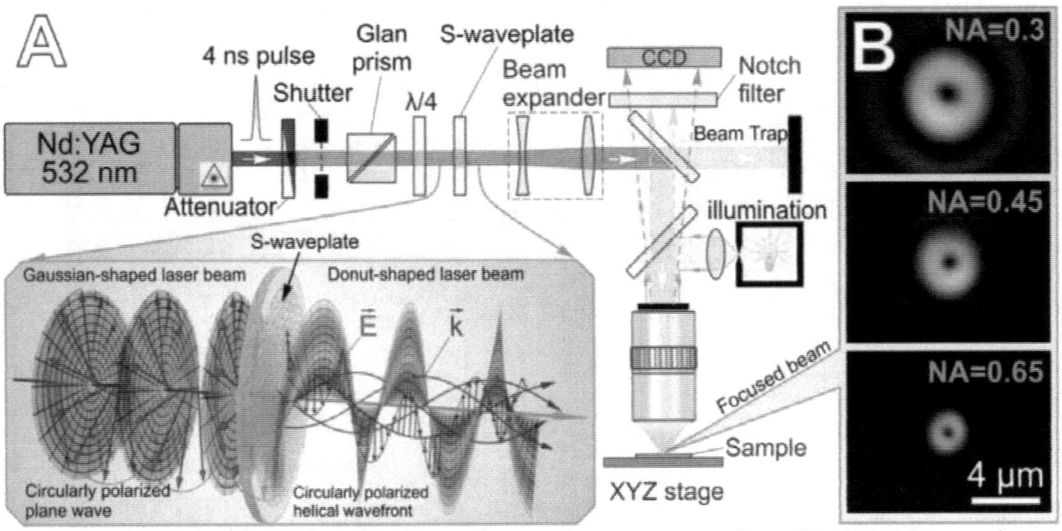

Fig. 1. (A) Schematic of the experimental setup for nanostructuring with the nanosecond vortex beams. (B) Vortex-beam intensity distributions measured in the focal plane of the microscope objectives with NA = 0.3 (top), NA = 0.45 (middle) and NA = 0.65 (bottom).

3. Results and discussion

Series of energy-resolved SEM images (Figs. 2(A)-2(D)) illustrates the formation and evolution of the twisted nanojets produced under single-pulse irradiation of the 500-nm thick Ag film, covering the silica glass substrate, with ns-laser vortex pulse focused at NA = 0.3. Besides the evident twisted shape of the produced nanojets with their clockwise rotational symmetry (the sign of vortex helicity), expectedly coinciding with the vortex pulse handedness (Figs. 1(C)-1(E)), their formation process appears to be very similar to that for common nanojets fabricated under single-pulse ablation of noble (semi-noble) metal films with Gaussian-shape (zero-OAM) ns-laser pulses.

The main energy-resolved steps include (i) accumulation of the molten material at the spatial center of the beam through thermocapillary forces and melt flows, which rapidly thin the peripheral part, (ii) formation of the liquid nanojet, which undergoes a Rayleigh-Plateau hydrodynamic instability, resulting in appearance and ejection of the molten droplets and, finally, (iii) formation of the through hole by breaking the significantly thinned area, surrounding the nanojet, at increased pulse energy. The tipped twisted nanojets are expectedly fabricated for typical pulse energies sufficient to trigger the ejection of all molten droplets from the resolidifying tip. Under such condition, the produced twisted nanojet can be characterized by very fine tips with the typical averaged curvature radius R_{tip}~12 ± 6 nm, as it was shown by the high-resolution SEM imaging (see inset in Fig. 2(C)).

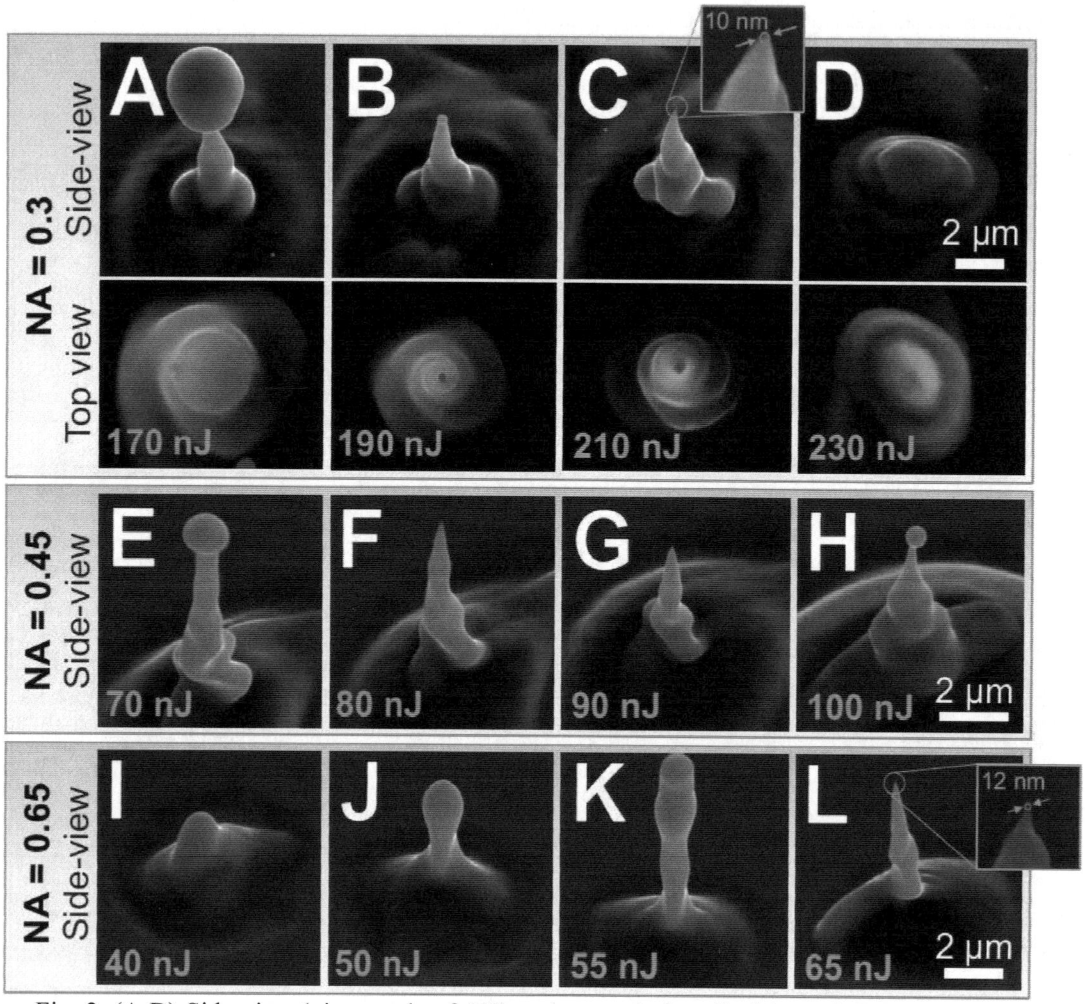

Fig. 2. (A-D) Side-view (view angle of 40°) and normal-view (in false color) SEM image pairs of twisted Ag nanojets produced under single-pulse ablation of 500-nm thick Ag film with the vortex pulses focused at NA = 0.3. Similar side-view (view angle of 40°) SEM images showing twisted nanojets fabricated on the surface of the same Ag film with single vortices focused with microscope objective at NA = 0.45 (E-H) and NA = 0.65 (I-L). The pulse energy in each image row increases from left to right. Insets in figures (C) and (L) show the magnified view of the nanojet tip indicating the typical curvature radii smaller than 12 nm.

Similarity of our present observations and the previous extensive experience in fabrication of thin-film nanojets, using short- and ultrashort laser pulses, provokes us to compare formation of the non- and chiral nanoneedles. For Ag films studied in this paper, both Gaussian- and donut-shaped beams produce nanojets, having non- and chiral shapes, respectively. In particular, this indicates that for almost similar general mechanism, underlying the formation of the nanojets on the noble-metal film surface and attributed to the temperature-gradient driven thermocapillary flow of the molten material, strong rotational movement appears under vortex-pulse irradiation, twisting molten material in the direction, coinciding with the vortex helicity sign. The helical thermocapillary melt flows within the evolving nanojet and surrounding microbump area possibly can originate from the corresponding characteristic spiral-like (or more complex) intensity distribution. For the noble-metal films considered in this study, the origin of such spiral-shaped intensity distribution and corresponding temperature profile on the metal film surface can be explained in terms of "pure" optical processes as interference of the incident donut-shaped vortex beam with the zero-OAM beam, having a spherical (or more complex) wavefront. It is clear that appearance of such perturbed reflected wave, providing the desired "spiral" component of the corresponding surface intensity distribution, can be associated with the transient nanosecond evolution of the metal film surface, which can be triggered via its heating, melting and thermal expansion by the incident donut-shaped laser pulse. Scalar diffraction theory calculations show that structure of a vortex beam reflected from a metal film can be perturbed during its reflection from a film, containing multiple submicroscale inhomogeneities with average size more than 20 nm. In this case, interference of the incident vortex beam and the reflected speckle-modulated zero-OAM one produces the spiral-like intensity distribution on the metal film surface, which handedness coincides with that one of the incident vortex beam (Fig. 3). To the contrary, similar calculation performed for a smooth metal mirror shows a common annular-shape interference pattern, originating from two counter-propagating vortex beams, retaining with opposite helicity and the similar OAM (this intensity distribution is equal to that one presented in left-most image of the Fig. 3). As any substantial surface evolution of the initially smooth metal film starts after electron-phonon relaxation (typical timescale - few picoseconds for both noble metal film and Si), the "heating spiral" as well as the helical shape of the nanojets are expected to be missing for ultra-fast energy deposition with the pulse durations shorter, than this characteristic time.

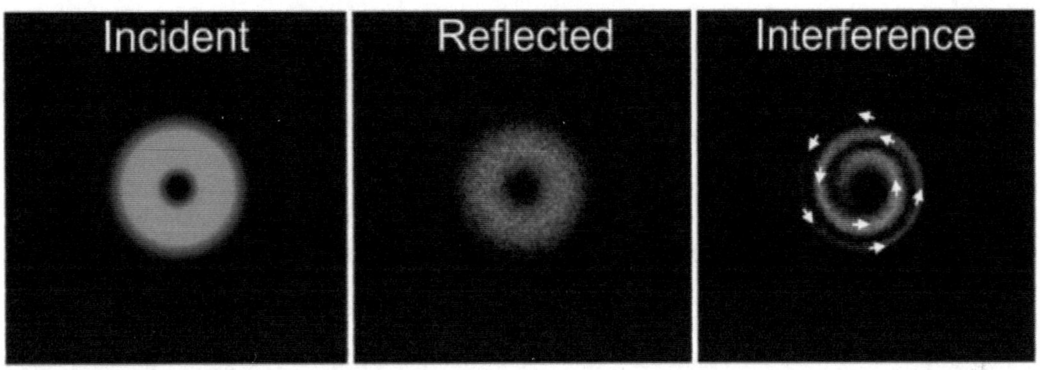

Fig. 3. Intensity distributions of the incident focused first-order vortex beam (left), distorted beam reflected from the rough metal surface (middle) and their interference pattern (right). The white arrows in the right-most image indicate the direction of the intensity gradient, which leads to the appearance of a "thermal spiral".

4. Conclusions

To conclude, donut-shaped nanosecond vortex pulses generated from circularly polarized ones by an S-waveplate, were found to produce twisted nanojets under single-pulse translative ablation of Ag films. We suggest an alternative explanation of the chiral nanoneedle formation on the silver film based on a combination of straight and twisted thermocapillary flows of the molten material, which are triggered by the characteristic spiral-like intensity distribution originating from interference of the incident donut-shape vortex beam and its replica reflected from the transiently evolving molten metal surface.

5. Funding

The work was supported by Russian Foundation for Basic Research (Projects No.17-02-00571-a) and FASO Russia (the Far East program).

References

[1] K. Toyoda, K. Miyamoto, N. Aoki, R. Morita, and T. Omatsu, "Using optical vortex to control the chirality of twisted metal nanostructures," Nano Lett. 12(7), 3645–3649 (2012).

IOP Conf. Series: Journal of Physics: Conf. Series **951** (2017) 012026 doi:10.1088/1742-6596/951/1/012026

Effects of mass defect in atomic clocks

A V Taichenachev[1,2], V I Yudin[1,2,3]

[1] Institute of Laser Physics SB RAS, Novosibirsk 630090, Russia
[2] Physics Department, Novosibirsk State University, Novosibirsk 630090, Russia
[3] Department of Physics and Technology, Novosibirsk State Technical University, Novosibirsk 630073, Russia

taichenachev@laser.nsc.ru, viyudin@mail.ru

Abstract. We consider some implications of the mass defect on the frequency of atomic transitions. We have found that some well-known frequency shifts (such as gravitational and quadratic Doppler shifts) can be interpreted as consequences of the mass defect, i.e., without the need for the concept of time dilation used in special and general relativity theories. Moreover, we show that the inclusion of the mass defect leads to previously unknown shifts for clocks based on trapped ions..

1. Introduction

At the present time, atomic clocks are most precise scientific devices. The principle of operation of these quantum instruments is based on modern methods of laser physics and high-precision spectroscopy. In this way, the unprecedented value of fractional instability and uncertainty at the level of 10^{-18} has already been achieved with the goal of 10^{-19} on the horizon [1]. Frequency measurements at such a level could have a huge influence on further developments in fundamental and applied physics. In particular, we can foresee tests of quantum electrodynamics and cosmological models, searches for drifts of the fundamental constants, new types of chronometric geodesy, and so on (see, for example, review [2]). However, this level of experimental accuracy requires a comparable level of theoretical support, which would account for systematic frequency shifts of atomic transitions due to different physical effects. Thus, modern atomic clocks are also at the point of interweaving different areas of theoretical physics.

In this paper we develop the mass defect concept with respect to atomic clocks. Historically, considerations of the mass defect have been connected with nuclear physics, where the mass defect explains the huge energy emitted due to different nuclear reactions. However, a quite unexpected result is that this effect has a direct relation to frequency standards, where it leads to shifts in the frequencies of atomic transitions.

The main idea of our approach is as follows. Let us consider an arbitrary atomic transition between the ground (g) and excited (e) states with unperturbed frequency $\omega_0 = (E_e^{(0)} - E_g^{(0)})/\hbar$, where $E_e^{(0)}$ and $E_g^{(0)}$ are the unperturbed energies of the corresponding states (see in Fig.1). Using Einstein's famous formula, $E=Mc^2$, which links the mass M and energy E of a particle (c is the speed of light in vacuum), we can find the rest masses of our particle, M_g and M_e, for the states (g) and (e), respectively:

$E_g^{(0)} = M_g c^2$ and $E_e^{(0)} = M_e c^2$. The fact that $M_g \neq M_e$ is the essence of the so-called mass defect. In our case, the connection between M_g and M_e reads:

$$M_e c^2 = M_g c^2 + \hbar\omega_0 \;\Rightarrow\; M_e = M_g + \hbar\omega_0 / c^2 \tag{1}$$

We show that the relationship (1) allows us to reinterpret some well-known systematic frequency shifts (e.g. those connected with the time dilation effects) [3,4]. Moreover, our approach predicts some new shifts previously unconsidered, to our best knowledge, in the scientific literature.

2. Gravitational shift

As the first example, let us show how the mass defect allows us to formulate a very simple explanation of the gravitational redshift even under a classical description of the gravitational field (as classical potential U_G). Indeed, because the potential energy of a particle in a classical gravitational field is equal to the product MU_G (where $U_G<0$), we can write the energy of j-th state $E_j(U_G)$ as:

$$E_j(U_G) = M_j c^2 + M_j U_G = M_j c^2 (1 + U_G / c^2),\ (j = \text{g}, \text{e}) \tag{2}$$

Using eqs.(1) and (2), we find the frequency of the transition (g)→(e) in the gravitational field:

$$\omega = \omega_0 (1 + U_G / c^2). \tag{3}$$

This expression coincides to the leading orderwith the well-known formula from the general relativity theory:

$$\omega = \omega_0 \sqrt{1 + 2U_G / c^2} \approx \omega_0 (1 + U_G / c^2) \tag{4}$$

in the limit of weak gravitational field $|U_G|/c^2 \ll 1$.

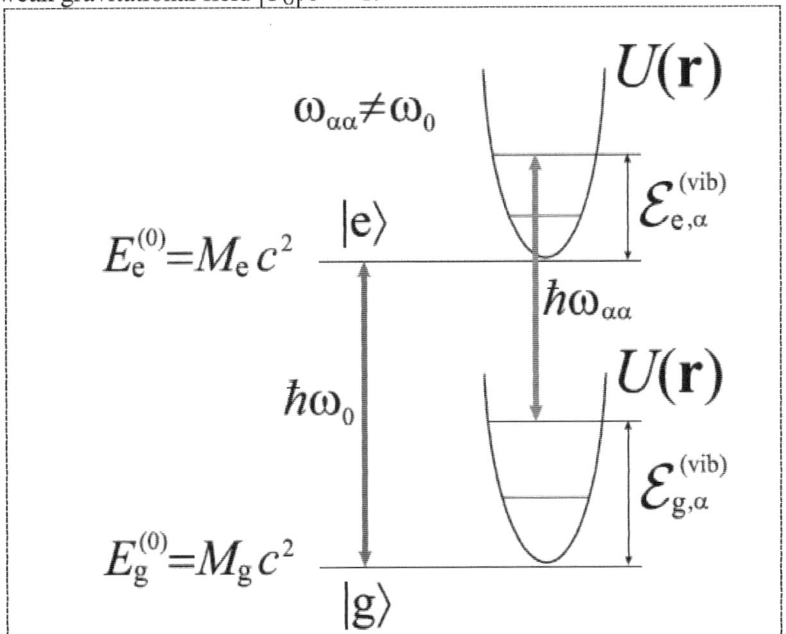

Figure 1. Scheme of atomic transition $(g) \leftrightarrow (e)$. Also it is shown the quantization of energy levels on translational degrees of freedom in the confining potential $U(\mathbf{r})$, where $\omega \neq \omega_0$ due to the mass defect.

3. Motion-induced shifts for atoms (ions) trapped in a confining potential

The second example concerns frequency shifts in the presence of a confining potential $U(\mathbf{r})$, which we take to be the same for both states (g) and (e). Such a situation occurs both for clocks based on neutral atoms in optical lattice at magic wavelength and those based on trapped ions. In this case, we have the standard task of quantizing the energy levels with translational degrees of freedom:

$$\hat{H}_j\left|\Psi_{j,\alpha}(\mathbf{r})\right\rangle = E^{(\mathrm{vib})}_{j,\alpha}\left|\Psi_{j,\alpha}(\mathbf{r})\right\rangle,\ \hat{H}_j = \hat{\mathbf{p}}^2/2M_j + U(\mathbf{r}), \tag{5}$$

where the Hamiltonian $\hat{H}_j$ describes the translational motion of the particle in the j-th internal state (j=g,e), the wavefunction $\left|\Psi_{j,\alpha}(\mathbf{r})\right\rangle$ corresponds to the α-th vibrational level (α=0,1,2,...) and $\mathbf{r}$ is coordinate of atomic center-of-mass. Thus, taking into account the translational motion, the atomic wave function is described by the pair products $|j\rangle\otimes\left|\Psi_{j,\alpha}(\mathbf{r})\right\rangle$. Because of the mass defect ($M_g \neq M_e$), the energy levels for the lower and upper states differ: $E^{(vib)}_{g,\alpha} \neq E^{(vib)}_{e,\alpha}$. Consequently, the frequency $\omega_{\alpha\alpha}$ between corresponding levels of trapped particle is different from the unperturbed frequency, ω_0, with a value $\Delta\omega$ (see Fig.1): $\Delta\omega = \omega_{\alpha\alpha} - \omega_0 = \left(E^{(vib)}_{e,\alpha} - E^{(vib)}_{g,\alpha}\right)/\hbar$. Let us now estimate this value. For this purpose, we write the Hamiltonian for upper state in the following form:

$$\hat{H}_e = \hat{H}_g + \Delta\hat{H};\ \Delta\hat{H} = \frac{\hat{\mathbf{p}}^2}{2M_e} - \frac{\hat{\mathbf{p}}^2}{2M_g} = -\frac{\hbar\omega_0}{2M_e M_g}\hat{\mathbf{p}}^2. \tag{6}$$

where the operator ΔH can be considered as a small perturbation. Then using standard perturbation theory, we obtain the following estimate for the relative value of the shift

$$\frac{\Delta\omega}{\omega_0} \approx -\frac{1}{2c_2}\frac{\left\langle\Psi_{j,\alpha}\left|\hat{\mathbf{p}}^2\right|\Psi_{j,\alpha}\right\rangle}{M_g^2} \tag{7}$$

We note that this expression coincides with a well-known relativistic correction, which is the quadratic Doppler shift due to the time dilation effect for moving particle [4]. Indeed, at the present time the following explanation is conventionally used. In accordance with special relativity, the tick rate Δt in the moving (with velocity $\mathbf{v}$) coordinate system changes with respect to the tick rate $\Delta t'$ in motionless (laboratory) coordinate system by the law: $\Delta t = \Delta t'(1-\mathbf{v}^2/c^2)^{1/2}$. As a result, an atomic oscillation with eigenfrequency ω_0 is perceived by an external observer to be shifted to $\omega = \omega_0(1-\mathbf{v}^2/c^2)^{1/2}$. In the nonrelativistic limit, ($\mathbf{v}^2/c^2 << 1$), we have: $\omega \approx \omega_0(1-\mathbf{v}^2/2c^2) = \omega_0[1-(\mathbf{p}/M)^2/2c^2]$ (where $\mathbf{p}$ is momentum of particle). Then, if we take into account quantum considerations through the replacement $\mathbf{p}\to\hat{\mathbf{p}} = -i\hbar\nabla$, we obtain the expression for frequency shift (7).

4. Previously unconsidered field-induced shifts for trapped ions

Besides the reinterpretation of some well-known shifts, the mass defect concept predicts additional contributions for field-induced shifts that have not been previously discussed in the scientific literature. We emphasize that these additional shifts are associated with translational degrees of freedom and they vanish if we will not take into account the mass defect.

We will consider a trapped ion (with charge eZ_i) in the presence of additional weak electric field with potential $\phi_{\mathrm{add}}(\mathbf{r})$, which describes all controlled and uncontrolled fields except the trapping potential $U(\mathbf{r})$. Let us show how the additional potential $\phi_{\mathrm{add}}(\mathbf{r})$ will perturb the vibrational structure formed by $U(\mathbf{r})$. For this purpose we will use vibrational eigenfunctions $\left|\Psi_{j,\alpha}(\mathbf{r})\right\rangle$, which describe a spatial localization of the ion in the internal states $|j\rangle$ (j=g,e) due to the trap potential $U(\mathbf{r})$, as a basis for perturbation theory on the small additional interaction $U_{\mathrm{add}}(\mathbf{r}) = eZ_i\phi_{\mathrm{add}}(\mathbf{r})$.

The center of the ion localization $\mathbf{r}_0$ in the trap is determined by averaging: $\mathbf{r}_0 = \left\langle\Psi_{j,\alpha}\left|\mathbf{r}\right|\Psi_{j,\alpha}\right\rangle$. In this case, we will use the following Taylor series of the operator $U_{\mathrm{add}}(\mathbf{r})$ at the point of ion localization $\mathbf{r}_0$:

$$\mathrm{U}_{\mathrm{add}}(\mathbf{r}) = eZ_i\phi_{add}(\mathbf{r}_0) - (\mathbf{d}\cdot\mathbf{E}_{add}(\mathbf{r}_0)) + (\mathbf{Q}\cdot\mathbf{W}_{add}(\mathbf{r}_0)) + ..., \tag{8}$$

where $\mathbf{E}_{add}(\mathbf{r}_0) = -\nabla\phi_{add}(\mathbf{r})\big|_{\mathbf{r}_0}$ and $\mathbf{W}_{add}(\mathbf{r}_0) = \frac{1}{6}\nabla\otimes\nabla\phi_{add}(\mathbf{r})\big|_{\mathbf{r}_0}$ are the electric vector and electric tensor, respectively, at the point $\mathbf{r}_0$. The operators **d** and **Q** correspond to the mesoscopic dipole and quadruplole moments of ion cloud in a trap:

$$\mathbf{d} = eZ_i(\mathbf{r}-\mathbf{r}_0)$$
$$Q_{kl} = eZ_i\left\{3(\mathbf{r}-\mathbf{r}_0)_k(\mathbf{r}-\mathbf{r}_0)_l - \delta_{kl}|\mathbf{r}-\mathbf{r}_0|^2\right\} \qquad (9)$$

As we see, that apart from the well-known electronic dipole and quadrupole moments the charge spatial distribution connected with motion of ion in a trap leads to mesoscopic multipole moments coupled to the weak additional electric field and its gradients.

Let us consider now the first order shift:

$$\Delta_{j,\alpha} = \frac{\langle\Psi_{j,\alpha}|U_{add}|\Psi_{j,\alpha}\rangle}{\hbar} = \frac{(\mathbf{Q}_{j,\alpha}\mathbf{W}_{add}(\mathbf{r}_0))}{\hbar}, \qquad (10)$$

where $\mathbf{Q}_{j,\alpha} = \langle\Psi_{j,\alpha}|\mathbf{Q}|\Psi_{j,\alpha}\rangle$. The shift of the clock transition frequency reads

$$\delta_\alpha = \Delta_{e,\alpha} - \Delta_{e,\alpha} = \frac{(\Delta\mathbf{Q}_\alpha \cdot \mathbf{W}_{add}(r_0))}{\hbar}. \qquad (11)$$

Here the residual quadrupole moment caused by the mass defect

$$\Delta\mathbf{Q}_\alpha = \langle\Psi_{e,\alpha}|\mathbf{Q}|\Psi_{e,\alpha}\rangle - \langle\Psi_{g,\alpha}|\mathbf{Q}|\Psi_{g,\alpha}\rangle \qquad (12)$$

has nonzero value because the wavefunctions describing the translational motion in the excited and ground states are different in general case. Estimate of order of magnitude gives $|\Delta Q_\alpha| \approx eZ_i R^2\, \hbar\omega_0 / Mc^2$, where R is the size of ion cloud and $M \approx M_{e,g}$. Though the expression (12) contains a very small multiplier, $\hbar\omega_0 / Mc^2 << 1$, the size of ion localization R significantly exceeds the Bohr radius a_0. Indeed, $R \sim 10^2$–10^3 a_0 even for the deeply-cooled ion to the lowest vibrational level in the confined potential $U(\mathbf{r})$ (i.e., for the quantum limit of cooling), and $R \sim 10^4 a_0$ for the upper vibrational states, which are populated if the ion is laser-cooled to the usual Doppler temperature (mK range). As a result, the quadrupole shift of the clock transition, modified by the mass defect, can be metrologically significant for modern and future optical frequency standards. For example, let us consider an atomic clock based on the transition $^1S_0 \to {}^3P_0$ in the ion $^{27}Al^+$ [5]. Because of the zero electronic angular momentum for the clock transition, $J_g = J_e = 0$, the quadrupole moment, associated with internal degrees of freedom, is very small ($|Q| \sim 10^{-6} e\, a_0^2$, see Ref. [6]). However, for the residual quadrupole moment caused by mass defect we have the estimate $|\Delta Q_\alpha| \sim 2\times(10^{-2} - 10^{-6})a_0^2$ for $R \sim 10^2$–10^4 a_0. Such values of the quadrupole moment may be important for atomic clocks with fractional uncertainty at the level of 10^{-18}–10^{-19}.

5. Conclusion

We have considered some manifestations of the mass defect in atomic clocks. As a result, some well-known systematic shifts, previously interpreted as the time dilation effects in the frame of special and general relativity theories, can be considered as a consequence of the mass defect in quantum mechanics. Furthermore, our approach has predicted a series of previously unknown shifts for ion clocks. These results are important for high precision optical atomic clocks and can be interesting for theoretical quantum physics.

Acknowledgments
This work was supported by the Russian Scientific Foundation (No. 16-12-00052). V.I.Yudin was also supported by the Ministry of Education and Science of the Russian Federation (No. 3.1326.2017/4.6), and Russian Foundation for Basic Research (No. 17-02-00570)..

IOP Conf. Series: Journal of Physics: Conf. Series **951** (2017) 012026 doi:10.1088/1742-6596/951/1/012026

References

[1] Schioppo M et al.2017 *Nature Photonics* **11** 48.
[2] Ludlow A D et al. *Rev. Mod. Phys.* 2015 **87** 637
[3] Einstein A *Annal. Physik* 1905 **17** 891
[4] Chou C W et al. *Science* 2010 **329** 1630
[5] Chou C W et al. *Phys. Rev. Lett.* 2010 **104** 070802
[6] Beloy K et al. 2017 *Phys. Rev. A* **95** 043405

Mathematical modelling of thin films growth and calculation of coefficients reflection, transmission and absorption waves

A V Istratov and M N Gerke

Stoletovs Vladimir State University, 87 Gorky str., Vladimir 600000, Russia

E-mail: aleksgreene@gmail.com

Abstract. Progress in nano- and microsystem technology is directly related to the development of thin-film technologies. At the present time, thin metal films can serve as the basis for the creation of new instruments for nanoelectronics. One of the important parameters of thin films affecting the characteristics of devices is their optical properties. That is why the island structures, whose optical properties, can change in a wide range depending on their morphology, are of increasing interest. However, despite the large amount of research conducted by scientists from different countries, many questions about the optimal production and use of thin films remain unresolved.

1. Introduction

The quality of many famous materials can be enhanced by using nanoparticles and atomic processing. Nanotechnology make it possible to create lighter, thinner and stronger composite (mixed, compound) materials.

To study nanostructures, it's important to know not only their mass or arrangement of atoms, but what they consist of. Determine the chemical composition of the samples (i.e. the content atoms of various elements in them) are allowed spectroscopic methods that use various instruments for studying the spectra of radiation, absorption, reflection, scattering.

As is known, ordinary light (visible radiation) is a collection of electromagnetic waves of different lengths (~ 400-760 nm) perceived by the human eye. Our eye perceives different colors not because objects have some abstract property "color", but because they are able to absorb and reflect electromagnetic waves of some length.

The analysis of the spectra of optical transmission and reflection plays a primary role among various methods of investigating quantum-size structures.

The characteristics of a number of optical elements, which include interference coatings, including a metal layer, are significantly influenced by the parameters of thin metal layers (refractive index, main absorption index and thickness). To this group of optical elements are mirrors, both metal and metal-dielectric, attenuating optical filters for a wide spectral range, gradient attenuators (shaders) and metal-dielectric narrow-band filters. The characteristics of each of the listed elements (transmission, reflection) are to some extent influenced by the optical parameters of the metal layers, which in turn depend on the deposited material and the resulting morphology of the sample surface.

The morphology of the surface of thin films is one of the most important parameters. That is why there is an increasing interest in the structure of islands, their roughness and thickness. Today, obtaining thin films is unthinkable without the use of mathematical modeling, numerical methods and complex programs. In this study, the degree of influence of morphology (particle diameter, number of layers, etc.) on the optical properties of a deposited thin film of bimetallic clusters was studied.

2. Computer modelling

For computer modeling of the growth of metal thin film, a mathematical model of the "cellular automaton" type was used, which has recently been widely used in modeling processes in nanotechnologies.

The growth model of a granular film in the form of a deterministic cellular automaton consists in depositing particles on a substrate randomly. Each particle hits random points on the substrate and is fixed there, increasing at a given point the thickness of the resulting film by one relative unit.

For research the optical properties of modeled structures, the Finite Difference Time Domain method (FDTD) was used, which is one of the most popular methods of numerical solution of Maxwell's equations. Since introduction in 70th years of the previous century this method became popular due to it certain advantages:

- Simplicity of explicit numerical scheme;
- High parallel efficiency;
- Easiness of complex geometry generation;
- Ability to handle dispersive and nonlinear media;
- Natural description of impulsive regimes.

FDTD includes various numerical techniques and options, such as algorithm for dispersive and nonlinear media modeling, different mesh types, simulation results postprocessing etc.

We list the "main participants" of the FDTD numerical experiment:

- Material bodies, whose optical properties we are investigating.
- The source of the electromagnetic wave. The simplest way to specify a source is to specify the time dependence of in Maxwell's equations. This type of source is usually used in simulating dipoles. To generate a plane wave, another type of source that is implemented using the total field / scattered field method is more convenient.
- Detectors that remove fields on the grid during the entire numerical experiment. Detectors are not responded to by any real bodies in space, by which is meant only that we write the values of the fields at some points inside the computational volume into a file. Upon completion of the calculation based on this file, you can restore the course of the numerical experiment.

The medium (material) of objects is described by dielectric permittivity, conductivity, magnetic permeability and magnetic losses. Their default values are (1, 0, 1, 0). Unlike frequency domain methods, the dielectric permittivity of dispersive materials in tabular form can't be directly replaced by the FDTD scheme. Instead, it can be approximated with the help of several members of Drude or Lorentz in the form:

$$\varepsilon(\omega) = \varepsilon_\infty - \sum_{p=1}^{N_D} \frac{\Delta\varepsilon_p \omega_p^2}{i\omega\gamma_p + \omega_p} + \sum_{p=1}^{N_L} \frac{\Delta\varepsilon_p \omega_p^2}{\omega_p^2 - 2i\omega\gamma_p - \omega^2} \quad (1)$$

The parameters for the Drude and Lorenz constructors are the plasma frequency ω_0 and attenuation γ and the value $\Delta\varepsilon$. The frequency and damping of the plasma are measured in radians per unit time FDTD.

The Total Field/Scattered Field method is used to simulate an infinitely distant plane wave source. It's based on the linearity of Maxwell's equations and the following superposition principle. FDTD difference equations can be independently applied both for the total field and for the incident or scattered fields, which allows us to break the computational volume into the field of the total field and the region of the scattered field. They are separated by a virtual boundary, which serves to generate a plane wave in the region of the total field. The difference equations used to calculate the field components in the grid nodes adjacent to this boundary differ from the original ones by the presence of additional components that take into account the value of the field of the incident wave.

To simulate the departure of a wave from the computational volume to infinity, absorbing boundary conditions are required. One of the best implementations is the use of a thin layer of special material called the Perfectly Matched Layer (PML) along the boundary. Thus, this material absorbs all waves incident on the boundary of the computational volume, regardless of the angle of incidence. To model infinite periodic structures, periodic boundary conditions are used in one or several directions.

In order to measure the fields inside the computational volume, detectors are needed. They read the values of the fields at the points where they are placed, without affecting them at all. The values on the detectors are obtained by interpolation at nearby grid nodes. In the course of a numerical experiment, detectors record their readings in binary files. At the end of calculations, based on these binary files, readable text files are created.

In EMTL there are different types of detectors which can be divided into two groups: universal and special detectors. At the same time, it is customary in EMTL to operate not with separate detectors, but with sets of detectors.

A universal detector records the history of the signal in the form of a table with columns ***t***, ***x***, ***y***, ***z***, ***Ex***, ***Ey***, ***Ez***, ***Hx***, ***Hy***, ***Hz*** for time, coordinates and values of fields ***E*** and ***H***.

A special detector creates input files with time, coordinates and E and H field values for plotting and 3d views. Special detectors include: NearToFar - used to convert fields from near field to distant; RTA - is used to measure the passage, reflection and absorption of a signal from a structure; Flux - used to measure energy flows through surfaces.

Typical scenario of FDTD experiment includes following steps:

- User specifies calculated volume and mesh resolution, optical properties and geometry of the structure, boundary conditions (typically, periodic or absorbing), wave source and set of points where field values should be recorded (detectors).
- Source generates finite time width impulse impinging on structure. Its propagation and scattering is recorded by detectors and possibly transformed to the frequency domain. Total exit of the radiation through absorbing boundaries determines the simulation time.
- Recorded field values are processed (for example, energy flux integrating through the chosen surface) to get optical characteristics of the structure.

3. Results of modelling

Using the FDTD method: a plane wave propagates along the Z-axis and automatically calculates the energy flux for the transmitted (***T***) waves using the data from the detectors.

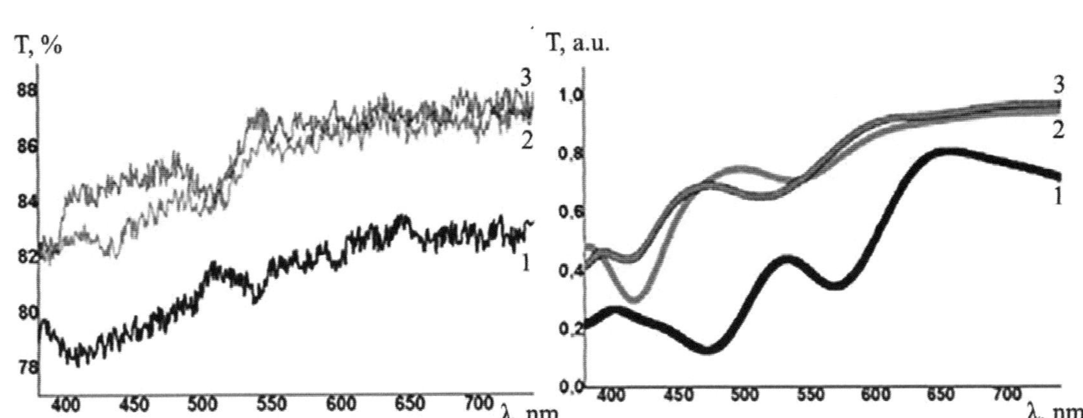

Figure 1. Transmission spectra of the deposited films: **1** – Au:Ag(1:1), a particle diameter (D) is 50nm, one layer, the distance between particles (gap) – is 5nm; **2** – Au:Ag(1:1) , D is 10nm and five layers, gap is 4nm; **3** – Au:Ag(1:1), D is 10nm and five layers, gap is 2nm.

IOP Conf. Series: Journal of Physics: Conf. Series **951** (2017) 012027 doi:10.1088/1742-6596/951/1/012027

Using the data from the detectors, it's also possible to calculate the coefficients of the reflected (***R***) waves. Absorption (***A***) is obtained as $\boldsymbol{A = 1 - R - T}$.

Figure 1 shows the transmission spectra of deposited bimetallic films and transmission spectra obtained with the help of FDTD of simulated thin films.

The calculated transmission spectra are in qualitative agreement with the experimental data.

References

[1] Aderhold J, Davydov V Yu, Fedler F, Klausing H, Mistele D, Rotter T, Semchinova O, Stemmer J and Graul J 2001 *J. Cryst. Growth* **222** 701
[2] Arakelyan S M, Veiko V P, Kutrovskaya S V *et al.* 2016 *J. Nanopart. Res.* **18** 155
[3] Antipov A A, Arakelyan S M, Vartanyan T A, Itina T E, Kutrovskaya S V, Kucherik A O, Sapegina I V 2015 *Optics and Spectroscopy* **119** 119
[4] Antipov A A, Arakelyan S M, Kutrovskaya S V *et al.* 2012 *Bulletin of the Russian Academy of Sciences: Physics* **76** 611
[5] Abramov D V, Antipov A A, Arakelian S M *et al.* 2014 *Laser Phys.* **24** 074010
[6] Meilikhov E Z 1999 *JETP* **115** 1484
[7] Arakelian S, Kucherik A, Kutrovskaya S 2016 *Journal of Russian Laser Research* **37** 494
[8] Abramov D, Arakelian S, Kochuev D et al. 2016 *Nanosystems: Physics, Chemistry, Mathematics* **7** 220
[9] Abramov D V, Arakelian S M, Makov S A *et al.* 2013 *Technical Physics Letters* **39** 14
[10] Abrikosov A A 2009 *Principles of metal theory.* (Moscow: FIZMATLIT) p 600
[11] Arakelian S, Kutrovskaya S, Kucherik A *et al.* 2016 *Proc. of SPIE* **9884** 988425
[12] Emel'yanov V I, Koroteev N I 1981 *Sov. Phys. Usp.* **24** 864
[13] Konov V I 2015 *Qantum Electron.* **45** 1043
[14] Kucherik A O, Arakelian S M, Garnov S V *et al.* 2016 *Quantum Electron.* **46** 627
[15] Kucherik A, Arakelian S, Vartanyan T *et al.* 2016 *Optics and Spectroscopy* **121** 263
[16] Shi L, Rohringer P, Suenaga K *et al.* 2016 *Nature Materials* **15** 634
[17] Ravagnan L, Siviero F, Lenardi C *et al.* 2002 *Phys. Rev. Let.* **89** 285506
[18] Rozhkova N N, Gribanov A V, Khodorkovskii M A 2007 *Dimond and related materials* **16** 2104
[19] Kao T-Y, Reno J L, Hu Q 2016 *Nature Photonics* **10** 541
[20] Hu Y, Wang Z, Weng Z *et al.* 2016 *Applied Optics* **55** 3226

IOP Conf. Series: Journal of Physics: Conf. Series **951** (2017) 012028 doi:10.1088/1742-6596/951/1/012028

Modelling of the tunnelling effect in granulated metallic nanostructures

A V Istratov and A O Kucherik

A.G and N.G. Stoletov Vladimir State University (VSU), 87 Gorki st., Vladimir, Russia, 600000

E-mail: aleksgreene@gmail.com

Abstract. Obtaining thin films of today is unthinkable without use of mathematical modeling, numerical methods and complex programs. In this regard, the practical importance of this calculations is that it can be used to investigate the conductivity of nano-sized granular structures that expands the diagnostic capabilities of thin films, opens up new perspectives in the creation of new devices based on thin-film technology, allow to predict their properties.

1. Introduction

The achievements of nanotechnologies are based on fundamentally new scientific knowledge about the nature of the structure materials, and, consequently, fundamentally new technologies and design principles. As the particle size of a substance decreases, its physical and chemical properties can change substantially.

The characteristic dimensions of semiconductor structures at the modern micro- and nanoelectronics are 10^{-7}-10^{-8} m. Such a range of linear dimensions of elements is a fundamental physical barrier behind which all the properties of a solid body, including electrical conductivity, change drastically.

Dimensional effects in solids are a phenomenon observed when the geometric dimensions of an object are comparable with one or other of the lengths that determine the course of physical processes (for example, the mean free path of the charge carrier, the de Broglie wavelength, etc.). Depending on the size of the sample being examined, classical and quantum dimensional effects are distinguished, which can affect practically any properties of the substance. As a rule, for nanometer objects, where the particle sizes are comparable with the de Broglie wavelength of an electron, the quantum size effects determine such properties of matter as heat capacity, electrical conductivity, optical properties, and the like.

The most striking representative of quantum size effects is the tunnel effect, a phenomenon that plays an important role in nanotechnology. The tunnel effect is the microparticle overcoming a potential barrier between two conductors with a small cross-section. Using this effect, you can control the movement of individual electrons. According to the basic principles of quantum mechanics, microparticles (in particular electrons) can pass through an insulator (dielectric) from one conductor to another that is a tunnel transition.

One-electron devices are promising nanoelectronic devices based on the effect of discrete tunneling of individual electrons and providing ultra-low levels of energy consumption at ultra-low operating voltages. The first devices based on the tunnel effect: tunnel diodes, transistors, sensors, thermometers

IOP Conf. Series: Journal of Physics: Conf. Series **951** (2017) 012028 doi:10.1088/1742-6596/951/1/012028

for measuring ultra-low temperatures, and finally, scanning tunneling microscopes, which laid the foundation for modern nanotechnology.

Since physical models of colored plastic balls don't accurately reflect the real properties of nanoparticles, computer models are usually used in which one can set the real laws of quantum physics. Based on a powerful mathematical apparatus, computer modeling plays a key role in the development of nanosystems.

2. Computer modelling

Algorithm, which simulates the electrical properties of thin granular films, is realized in several stages:

2.1. Simulation of a granular film.

The sedimentation of particles on the surface of the substrate is carried out by random sedimentation method. Random sedimentation is the simplest model of surface growth - a cellular automaton that includes simple deterministic deposition rules. The precipitated particles fall into randomly chosen points on the surface, are fixed there and do not have the ability to diffuse into neighboring points. A particle that hits a certain point on the surface increases its thickness by one relative unit. Mathematically, this model is similar to the model of Brownian motion of particles and is intensively studied by analytical and numerical methods. The disadvantage of deterministic cellular automata is that the state at the next instant of time is uniquely determined by the state at the previous moment of time.

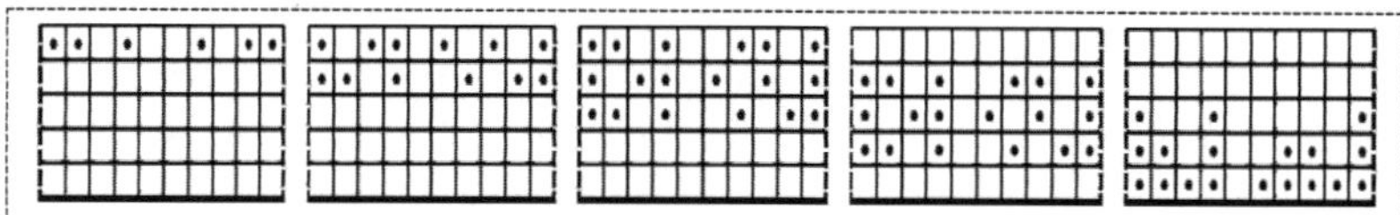

Figure 1. The random sedimentation model

The proposed model for the growth of thin-film surfaces is as follows:

a) The surface of a thin film in the model is a rectangular grid;

b) The new particles are assumed to be identical and their sedimentation locations are chosen randomly, and for one relative unit of time within the given model, more than one particle does not precipitate into one cell of the surface;

c) The rate of particle sedimentation is determined by the number of particles that hit the surface in one step of modeling the sedimentation of particles;

d) After the step of modeling the sedimentation of film particles, a transition to the next.

2.2. Modelling tunnel effect

To simulate a metallic granular nanostructure with a potential difference applied to the ends, a one-electron mathematical model was used in the form of a three-dimensional stochastic cellular automaton. The metal particles are represented as spheres located on the substrate at a distance S from each other. The electric potential is applied to the ends of the film, and the charge accumulates in the particles until its magnitude is sufficient for making the tunnel by the electrons to the subsequent particles in the direction of the positive potential: this flow of electrons moving in the same direction is electric amperage.

In the Euclidean metric, the distance between the pairs of neighboring particles in question is not the same, therefore, the type of neighborhood (common face, edge or vertex) must be taken into account. A cellular automaton, implemented in the Moore neighborhood in the three-dimensional case.

The three-dimensional Moore neighborhood in the simulated space consists of 26 neighboring cubic cells in which the particles are located and which have a current particle with the cell: common faces (6 units) are neighbors of the first order; common ribs (8 units) - neighbors of the second order; common vertices (8 units). Obviously, the distance between the cell centers of the particle in question and the neighboring one, having a common edge, is √2 times larger than the distance between the centers of the cells that have a common face. If they have only a common vertex, then the distance between the centers of the given cells will increase by √3 times.

Calculation of the probability of the tunnel effect. The probability is calculated for each pair of particles: the current particle and its nearest neighbor are not more than the third order:

$$P_i = \begin{cases} \dfrac{\operatorname{arcctg}\left(8\cdot\left(\sqrt{n_i}\cdot(S+2\cdot R)-2\cdot R-\dfrac{U\cdot(x_i-x_0)}{\sqrt{n_i}}-\left(\dfrac{q_i-q_0}{e_0}\right)+0.5\right)\right)}{\pi}, \\ 0, \text{ if } \dfrac{q_i}{e_0}=1 \text{ or } \dfrac{q_0}{e_0}=-1 \end{cases} \quad (1)$$

Where $\boldsymbol{n_i}$ - the order of the neighborhood of the i-th neighboring particles; $\boldsymbol{S}$ - the distance between the nearest particles along the axes; $\boldsymbol{R}$ - the radius of the deposited particles; $\boldsymbol{U}$ - the voltage at the ends of the simulated thin film, collinear with the X-axis; $\boldsymbol{x_0}$ - the x-coordinate of the current particle center; $\boldsymbol{x_i}$ - the x-coordinate of the i-th neighboring particle center; $\boldsymbol{q_0}$, $\boldsymbol{q_i}$ - the charges of the particles under consideration; $\boldsymbol{e_0}$ - the charge of the electron.

3. Computer modelling

The main quantitative morphological parameters characterizing the inhomogeneity of thin films: thickness and surface roughness film (standard deviation of the film thickness), particle size (diameter of spheres) and gap (the distance between the particles).

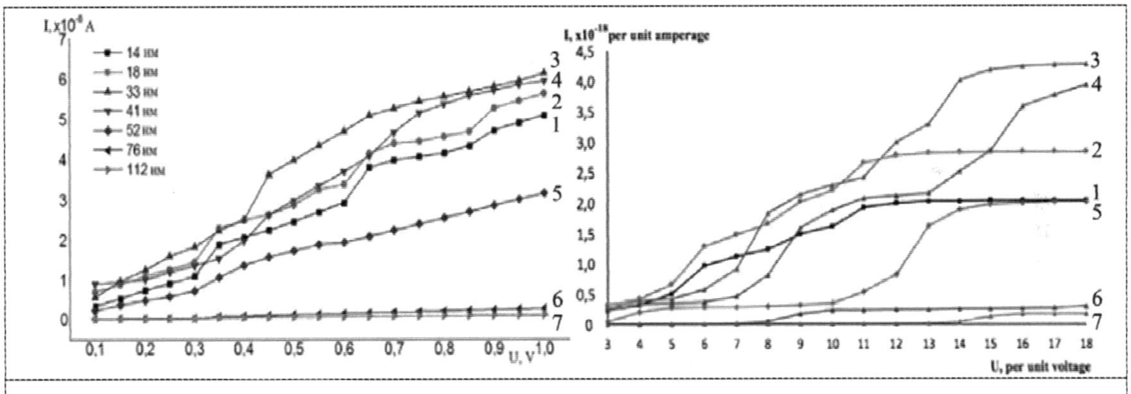

Figure 2. Current-voltage characteristics of the deposited and simulated gold and silver films: **1** – a particle diameter (D) is 4nm, the distance between the particles (gap) is 2nm, the surface roughness (W) about 7nm; **2** – D is 4nm, gap is 2nm, W about 8.5nm; **3** – D is 4nm, gap is 3nm, W about 13.5nm; **4** – D is 6nm, gap is 3nm, W about 16.5nm; **5** – D is 9nm, gap is 4nm, W about 21nm; **6** – D is 10nm, gap is 10nm, W about 34nm; **7** – D is 14nm, gap is 15nm, W about 53nm.

The volt-ampere characteristics of the deposited films and the results of modeling the tunnel effect using the model of a stochastic cellular automaton are shown in Figure 2.

The calculated volt-ampere characteristics are in qualitative agreement with the experimental one. The discrepancy in numerical values can be explained by the geometry of the calculation of features: in the model, the particles lie strictly on each other (because of the need to use a computational grid) and a clearly defined interval $\boldsymbol{S}$.

IOP Conf. Series: Journal of Physics: Conf. Series **951** (2017) 012028 doi:10.1088/1742-6596/951/1/012028

References

[1] Genov D A, Sarychev A K, Shalaev V M and Wei A 2004 *Nano lett.* **4** 153-158
[2] Destouches N, Crespo-Monteiro N, Vitrant G, Lefkir Y, Reynaud S, Epicier T, Liu Y, Vocanson F and Pigeon F 2014 *J. Mater. Chem.* C **2** 6256
[3] Persson B N J. and Liebsch A 1983 *Phys. Rev.* B **28** 4247-54
[4] Baba K, Okuno T and Miyagi M 1993 *App. Phys. Lett.* **62** 437
[5] Pereira R M S, Borges J, Peres F C R, Pereira P A S, Smirnov G V, Vaz F, Cavaleiro A and Vasilevskiy M I 2015 *J. Nanophoton.* **9** 0937961-12
[6] Antipov A A, Arakelyan S M, Vartanyan T A, Itina T E, Kutrovskaya S V, Kucherik A O and Sapegina I V 2015 *Optics and Spectroscopy* **119** 119-123
[7] Bakhti S, Biswas S, Hubert C, Reynaud S, Vocanson F and Destouches N 2014 *J. Phys. Chem.* C **118** 7494-7500
[8] Antipov A A, Arakelyan S M, Kutrovskaya S V, Kucherik A O, Makarov A A, Nogtev D S and Prokoshev V.G. 2012 *Bulletin of the Russian Academy of Sciences: Physics* **76** 611-617
[9] Arakelyan S M, Veiko V P, Kutrovskaya S V, Kucherik A O, Osipov A V, Vartanyan T A and Itina T E 2016 *J Nanopart Res* **18** 15501-12
[10] Abramov D V, Antipov A A, Arakelian S M, Khor'kov K S, Kucherik A O, Kutrovskaya S V and Prokoshev V G 2014 *Laser Phys* **24** 0740101-9
[11] Mie G 1908 *Ann. Phys. Leipzig* **25**, 377-452
[12] Xu Y-l and Gustafson B A S 2001 *JQSRT* **70** 419
[13] SCHOTT refractive Index Database, http://refractiveindex.info/
[14] Antipov A A, Arakelian S M, Kutrovskaya S V, Kucherik A O and Vartanian T 2014 *Optics and Spectroscopy* **116** 324-327
[15] Maximova K, Aristov A, Sentis M and Kabashin A 2015 *Nanotechnology* **26** 065601
[16] Antipov A A., Arakelyan S M., Kutrovskaya S V. and Kucherik A O 2014 *Technical Physics Letters* **6** 529–532

IOP Conf. Series: Journal of Physics: Conf. Series **951** (2017) 012029 doi:10.1088/1742-6596/951/1/012029

Studies of photon echo from exciton ensemble in (In,Ga)As quantum dots

Ia.A. Babenko[1*], I.A. Yugova[1], S.V. Poltavtsev[1,2], M. Salewski[2], I.A. Akimov[2,3], M. Kamp[4], S. Höfling[4], D.R. Yakovlev[2,3] and M. Bayer[2,3]

[1] Spin Optics Laboratory, 198504 St. Petersburg, Russia

[2] Experimentelle Physik 2, Technische Universität Dortmund, 44221 Dortmund, Germany

[3] Ioffe Physical-Technical Institute, Russian Academy of Sciences, 194021 St. Petersburg, Russia

[4] Technische Physik, Universität Würzburg, D-97074 Würzburg, Germany

* e-mail: babenkoyana@mail.ru

Abstract. Photon echo from trions and excitons in (In,Ga)As/GaAs quantum dots has been studied theoretically and experimentally. Theoretical analysis allowed us to distinguish photon echo signals from excitons and trions measured in the same range of wavelength using different polarization configurations of laser excitation. The theoretical predictions are in good agreement with the experimental data.

1. Introduction

Photon echo from trions and excitons in (In,Ga)As/GaAs quantum dots has been studied theoretically and experimentally. Theoretical analysis allowed us to distinguish photon echo signals from excitons and trions measured in the same range of wavelength using different polarization configurations of laser excitation. The theoretical predictions are in good agreement with the experimental data.

The interaction between material excitations (excitons and trions) in semiconductor nanostructures and light can be very effective due to high oscillator strength of the excitations. Recently it has been found that using the photon echo (PE) technique, one can transfer an information contained in the optical field into a spin system where it is decoupled from the optical field and may persist for a long time [1].

Semiconductor quantum dots (QDs) are considered to be promising for storing an information about the optical excitation since spin relaxation of electrons and holes in these nanostructures is characterized by significantly long times [2]. The inhomogeneity of the QD ensemble and spectral overlap of the exciton and trion transitions hamper investigations and analysis of these structures.

In the present work we describe the theoretical model developed for the photon echo signal from exciton ensemble in (In,Ga)As/GaAs QDs. Theoretical results complemented by the results for trions from Ref. [1] are compared with the experimental data.

2. Results

We study the dynamics of PE from an ensemble of the (In,Ga)As/GaAs semiconductor QDs excited by the picosecond laser pulses. Several polarization protocols have been used in the experiment: the linear polarization of the optical pulses parallel to the QD axes *x* or *y* (*z*-axis is the structure grows axis) denoted as H and V, respectively, and the linear polarization parallel to the tilted 45° with respect to the *x*- and *y*-axes (denoted D and X, correspondingly). Transient PE signal was measured using heterodyne detection [4]. Experimental signals measured in different polarization configurations are shown in Fig. 1.

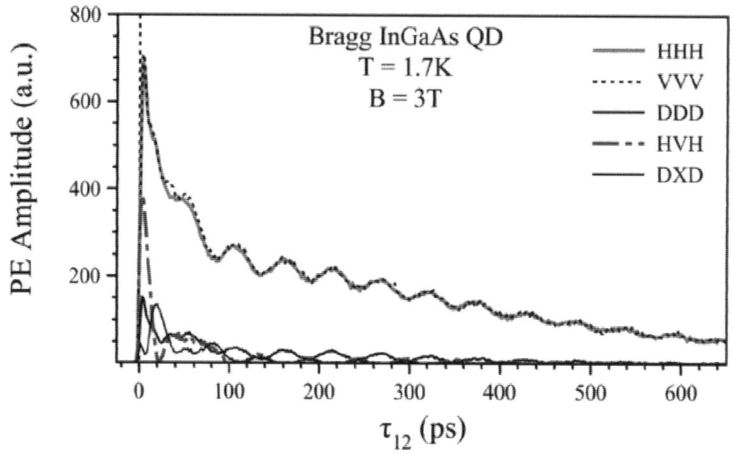

Fig. 1. The delay dependence of the experimental PE signals. Symbols HHH etc. in the legend denote polarization of the first and second pulses and of the detected signal, respectively

The theoretically calculated exciton PE signals for different polarization configurations are described by the following expressions:

$$P_{HHH}^{exc} \propto e^{-\frac{2\tau_{12}}{T_2}}\left[cos^2\left(\Omega_p\tau_{12}\right)+\frac{\delta_0^2}{4}\frac{sin^2\left(\Omega_p\tau_{12}\right)}{\Omega_p^2}\right], \tag{1}$$

$$P_{VVV}^{exc} \propto e^{-\frac{2\tau_{12}}{T_2}}\left[cos^2\left(\Omega_m\tau_{12}\right)+\frac{\delta_0^2}{4}\frac{sin^2\left(\Omega_m\tau_{12}\right)}{\Omega_m^2}\right], \tag{2}$$

$$P_{HVH}^{exc} = P_{VHV}^{exc} = 0, \tag{3}$$

$$P_{DDD}^{exc} \propto \frac{1}{4}e^{-\frac{2\tau_{12}}{T_2}}\left[\left(cos\left(\Omega_p\tau_{12}\right)+cos\left(\Omega_m\tau_{12}\right)\right)^2+\frac{\delta_0^2}{4}\left(\frac{sin\left(\Omega_p\tau_{12}\right)}{\Omega_p^2}+\frac{sin\left(\Omega_m\tau_{12}\right)}{\Omega_m^2}\right)^2\right], \tag{4}$$

$$P_{DXD}^{exc} \propto \frac{1}{4}e^{-\frac{2\tau_{12}}{T_2}}\left[\left(cos\left(\Omega_p\tau_{12}\right)-cos\left(\Omega_m\tau_{12}\right)\right)^2+\frac{\delta_0^2}{4}\left(\frac{sin\left(\Omega_p\tau_{12}\right)}{\Omega_p^2}-\frac{sin\left(\Omega_m\tau_{12}\right)}{\Omega_m^2}\right)^2\right]. \tag{5}$$

Here $\Omega_p \equiv \sqrt{\left(\omega_L^e+\omega_L^h\right)^2+\left(\delta_0/\hbar\right)^2}$, $\Omega_m \equiv \sqrt{\left(\omega_L^e-\omega_L^h\right)^2+\left(\delta_0/\hbar\right)^2}$.

In the expressions above ω_L and ω_L^h are electron and hole Larmor precession frequencies respectively, τ_{12} is the delay time between pulses, δ_0 is an isotropic exchange interaction constant, the exponential factor describes relaxation processes after the pulses action. Indices HHH etc. again denote polarization of the first and second pulses and of the detected PE signal, respectively.

The theoretical trion PE signals for different polarization configurations can be obtained according to the theoretical model provided in Ref. [1]:

$$P^{tr}_{HHH} = P^{tr}_{VVV} \propto \frac{1}{4} e^{-\frac{2\tau_{12}}{T_2}}, \tag{6}$$

$$P^{tr}_{HVH} \propto e^{-\frac{2\tau_{12}}{T_2}} cos\left(\left(\omega_L^e - \omega_L^h\right)\tau_{12}\right), \tag{7}$$

$$P^{tr}_{VHV} \propto e^{-\frac{2\tau_{12}}{T_2}} cos\left(\left(\omega_L^e + \omega_L^h\right)\tau_{12}\right), \tag{8}$$

$$P^{tr}_{DDD} \sim \frac{1}{4} e^{-\frac{2\tau_{12}}{T_2}} \left[cos\left(\left(\omega_L^e - \omega_L^h\right)\tau_{12}\right) + cos\left(\left(\omega_L^e + \omega_L^h\right)\tau_{12}\right) - 2cos\left(\omega_L^e \tau_{12}\right) - 2cos\left(\omega_L^h \tau_{12}\right) - 2\right], \tag{9}$$

$$P^{tr}_{DXD} \sim \frac{1}{4} e^{-\frac{2\tau_{12}}{T_2}} \left[cos\left(\left(\omega_L^e - \omega_L^h\right)\tau_{12}\right) + cos\left(\left(\omega_L^e + \omega_L^h\right)\tau_{12}\right) + 2cos\left(\omega_L^e \tau_{12}\right) + 2cos\left(\omega_L^h \tau_{12}\right) - 2\right], \tag{10}$$

The theoretically obtained dependencies of exciton and trion PE signals calculated for direct and tilted polarizations on time delay are demonstrated in Fig. 2.

Theoretical analysis of the obtained dependencies for excitons (1-5) and of the similar theoretical dependencies for trions (6-10) described in Ref. [1] allows us to propose the polarization protocols of optical excitation to separate experimentally the contributions of the exciton and the trion PE signals. For example, in HHH configuration, the exciton signal has an oscillating character, while the trion signal has a decaying character without any oscillations. For the HVH configuration, the exciton signal is zero, while the trion signal shows cosine-like oscillations.

Similar behavior is observed for the DDD and DXD polarizations when g_h=0 (see expressions 4, 5, 9, 10): the exciton DDD signal is oscillating while the trion DDD signal is decaying; the exciton DXD signal is zero, while the trion DXD signal oscillates.

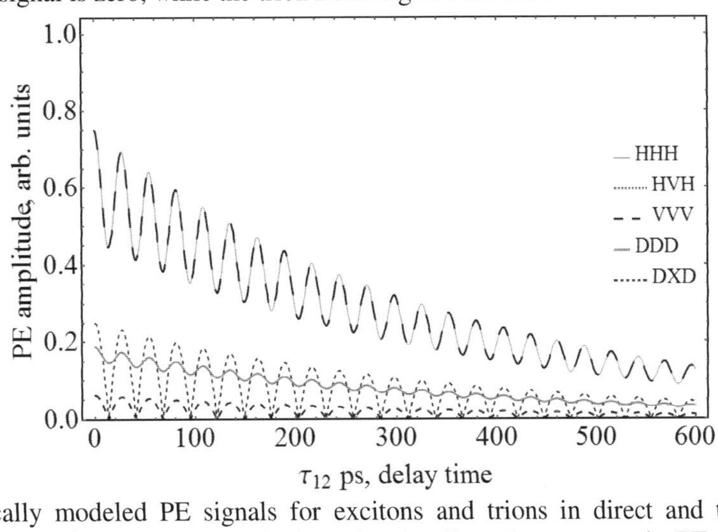

Fig. 2. Theoretically modeled PE signals for excitons and trions in direct and tilted polarization protocols. The thin gray and the black dashed overlapping lines demonstrate the PE signal in the HHH and VVV polarization geometries, the black dotted line demonstrates the PE signal in the HVH geometry, the gray solid line demonstrates PE signal in the HVH geometry, and the black short dashed line demonstrates the PE signal in DXD geometry respectively. The parameters are: electron g-factor

$|g_e| = 0.44$, magnetic field B = 3 T, the dephasing time T_2 = 700 ps, the isotropic exchange interaction constant for excitons δ_0 and hole g-factor are assumed to be zero. We assume that the fraction of excitons are two times larger than trions

Since the PE signals from the excitons and trions demonstrate different character for the parallel and perpendicular linear polarizations of laser pulses, as already discussed above, it is interesting to understand the behavior of the PE signals in dependence on the different angles between linear polarizations. The dependencies of the PE signals from excitons and trions on the angle between the linear polarizations of the laser pulses are shown in Fig. 3.

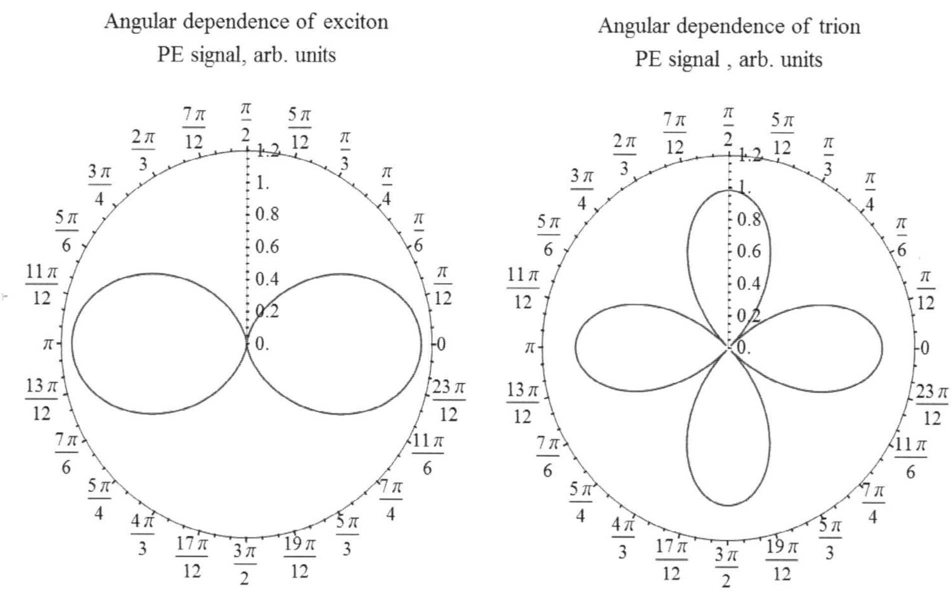

Fig. 3. Dependencies of the PE signals for excitons and trions on the angle between the linear polarizations of exciting pulses

As seen, the angular dependencies demonstrate very different character that allows us to distinguish the excitons and trions. This difference shows the way to experimentally distinguishing of the excitons and trions in PE signals.

3. Conclusion

Finally, the theoretical possibility of the experimental separation of the exciton and trion contributions in PE signal by appropriate choice of the exciting pulses polarization configurations have been shown in the present work. The most demonstrative separation of the exciton and trion contributions in the PE signals on the sample with a close to a zero hole *g*-factor has been analyzed.

Acknowledgement

This work was carried out in the framework of the joint Russian-Greek project supported by Ministry of Science and Education of Russian Federation (project RFMEFI61617X0085).

References

[1] L. Langer, S. V. Poltavtsev, I. A. Yugova et al, Nature Photonics 219, 851-857 (2014)
[2] A. Greilich, D. R. Yakovlev, A. Shabaev et al, Science, 313, 341-345 (2006)
[3] S.V. Poltavtsev, M. Salewski M., Yu. V. Kapitonov, et al, Phys. Rev. B 93, 121304(R) (2016)
[4] L. Langer, S. V. Poltavtsev, I. A. Yugova, et al, Phys. Rev. Lett. 109, 157403 (2012)

Nonlinear bifurcation diagrams of the current states of microcavity exciton-polaritons in the lattice

I Yu Chestnov[1], A V Yulin[2] and O A Egorov[3]

[1]Vladimir State University named after A. G. and N. G. Stoletovs, Gorkii St. 87, 600000, Vladimir, Russia
[2]ITMO University, Kronverksky pr. 49, 197101, St. Petersburg, Russia
[3]Technische Physik, Physikalisches Instität and Wilhelm Conrad Röntgen-Center for Complex Material Systems, Universität Würzburg, Am Hubland, D-97074 Wüzburg, Germany

E-mail: igor_chestnov@mail.ru

Abstract. We study nonequilibrium properties of exciton-polaritons in an incoherently driven semiconductor microcavity with an embedded weak-contrast periodic lattice. Within the framework of mean-field theory, we describe nonlinear eigenstates of coherent exciton polaritons having the structure of Bloch waves and identify nonlinear current states at the boundaries of Brillouin zone. We demonstrate a profound modification of the eigenfrequency-momenta bifurcation diagrams of the condensate evoked by exciton-exciton interactions. The discussed phenomenon occurs for the states possessing high degree of symmetry, i.e. in particular, for the states lying at the centre or at the boundary of Brillouin zone.

1. Introduction

Periodic media have been attracting attention of physicist for many decades. One of the reasons of such interest is the rich dispersion properties typical for these media. The modification and control of the dispersive properties are among the fundamental problems of current material science, photonics, condensed matter and solid state physics. In particular, manipulation of a band structure of a given physical system is a vital practical task of the design of new micro- and nanostructures possessing unusual properties [1]. Using of nonlinear properties of the medium for managing of its dispersion is one of the promising approaches for this task. For instance, it has been revealed for both atomic Bose-Einstein condensates (BECs) [2] and for solid-state polaritonic systems [3] that the periodical modulation can support self-localized states of the quantum particles known as gap solitons even under repulsive nonlinear interactions between them. So, the nontrivial nonlinear dynamics and coherent properties of the periodic media was always the fascination problem.

A semiconductor microcavity supporting exciton polaritons is a very promising platform for the investigation of the spontaneous coherence build-up in periodic lattices [3-5]. Unlike the conservative bosonic systems, in particular, atomic BEC [6], the exciton-polaritons are subject to rapid radiative decay. Thus the population of polariton state has to be maintained by an external pumping. The direct way to it is the use of nonresonant pumping implying creation of a reservoir of hot excitons, which can relax in energy to form a polariton condensate. This nonequilibrium open-dissipative behaviour represents an important intrinsic feature of the exciton-polaritons in semiconductor microcavities and imprint the nonlinear dynamics in both homogeneous [7] and periodically modulated configurations

[8]. In the paper we consider spatially modulated exciton-polariton systems focusing our attention on the nonlinear states possessing nonzero current. For the sake of clarity and simplicity we restrict our consideration to a 1D case.

Note that the nonlinear energy renormalization of conservative atomic BEC placed in 1D optical lattice have been discussed before [9-11]. In particular, it was shown that two-body (atom-atom) interactions lead to the formation of the loops in the dispersion characteristics of the nonlinear states (so-called "swallow tails") both at the centre and at the edges of BZ. The most intriguing feature of the obtained loop structure is connected with a non-zero flow appearing at the edges of BZ. The existence of such nonlinear current-currying states was interpreted as a consequence of superfluid behaviour of a condensate in periodic potential [10]. The linear stability properties of these nonlinear Bloch waves have been intensively studied [12-14]. The regions of both the energetic and dynamical instabilities have been identified for such essentially conservative systems. The novelty of our paper is related to the open-dissipative nature of the polariton condensate which shows a number of interesting effects.

In such class of problems analytical methods are suitable only for special kinds of a potential such as elliptic [12] or Kronig-Penney [13, 14] potentials. That is why in this paper we use the numerical methods of finding nonlinear Bloch waves which are appropriate for the strongly nonlinear case when perturbative approaches fail.

2. The model

To describe polariton condensate properties we use an open driven-dissipative model that accounts for the particle exchange between the condensate and the excitonic reservoir formed by a non-resonant pump [12]. This approach is based on a mean field theory and assuming the spontaneous formation of the exciton-polariton condensate, we consider an open-dissipative Gross-Pitaevskii (GP) model, that describes incoherently pumped condensate coupled to an excitonic reservoir. The polariton order parameter $\Psi(x,t)$ is described by a GP-type equation

$$i\hbar\frac{\partial\Psi}{\partial t}=\left[-\frac{\hbar^2}{2m}\frac{\partial^2}{\partial x^2}+V_0\cos(\beta x)+g_c|\Psi|^2+\frac{i\hbar}{2}(Rn-\gamma_c)+g_R n\right]\Psi, \qquad (1)$$

and the exciton reservoir density $n(x,t)$ is governed by the rate equation:

$$\frac{\partial n}{\partial t}=P-\gamma_R n-Rn|\Psi|^2, \qquad (2)$$

Here γ_c and γ_R are the condensate and the reservoir dumping rates, g_c and g_R are coefficients of nonlinearity due to polariton-polariton and polariton-reservoir interaction. The reservoir of incoherent polaritons is formed by an external pump P. The parameter of coupling between reservoir and condensate is denoted as R. The value of V_0 characterizes depth of the lattice, $\beta=2\pi/l$ is the lattice wavevector, and l – the lattice period. We expect that nonlinearity affects polariton band structure if the term $g_c|\Psi|^2$ is comparable or higher than the potential depth V_0.

In this paper we use following parameters, which agree with experimental data [15]: $\gamma_c=0.33\ \mathrm{ps^{-1}}$, $\gamma_R=0.495\ \mathrm{ps^{-1}}$, $g_c=6\times10^{-3}\ \mathrm{meV\mu m^{-2}}$, $g_R=2g_c$, $R=0.01\ \mathrm{ps^{-1}\mu m^{-2}}$, $m=0.568\ \mathrm{meVps^{-2}\mu m^{-2}}$, $l=8\ \mathrm{\mu m}$.

3. Steady-states and frequency-momenta bifurcation diagrams of the condensate in the lattice

3.1. Searching for the condensate stationary states

The problem of the nonlinear states in the periodic lattice is solved as follows. Imposing periodic boundary conditions and restricting ourselves with the class of solutions to the functions having the same periodicity as lattice we search the stationary solutions which are parameterized by the quasi-momentum, the index of the mode and the average particle density. The obtained solutions resemble the familiar Bloch states and formally obey the Bloch theorem [16]. However the investigated solutions satisfy the nonlinear problem and possess several distinctive features which will be discussed latter. That is why we address these states as the nonlinear Bloch states [13].

In the case of weak nonlinearity the problem can be approached perturbatively when the solution is sought in the form of the expansion over the basis of the Bloch modes calculated for the corresponding linear problem and treating the nonlinearity as a perturbation. Naturally, within this approach the modes with quasi-momentum lying at the edges of Brillouin zone (BZ) are necessarily currentless. However this result cannot be generalized for highly nonlinear systems like exciton-polariton condensate. In this case strong Kerr-like nonlinear properties of condensate caused by many-body interactions could essentially affect the properties of the nonlinear Bloch modes.

We are looking for the solution of equations (1) and (2) in the general form

$$\Psi(x,t) = \psi(x)e^{ikx-i\mu t}, \qquad n(x,t) = f(x), \tag{3}$$

where the condensate wavefunction is represented in the standard form of the structural function ψ and the exponent with the quasi-momentum k and eigenfrequency (chemical potential) μ.

The key property of the open-dissipative model (1), (2) is that a nontrivial solution $\Psi \neq 0$, $n \neq 0$ can be formed when the pump exceeds a threshold value P^{th}. For the spatially homogeneous system ($V_0 = 0$) equations (1) and (2) possesses solution in the form of nonlinear plain waves (3) $\Psi = \Psi_0 e^{-i\mu_0 t}$ and $n_0 = \gamma_c / R$ above the threshold $P > P^{\text{th}} = \gamma_R \gamma_c / R$. In this case the chemical potential is $\hbar\mu_0 = g_c |\Psi_0|^2 + g_R n_0$ and the condensate density $|\Psi_0|^2 = \left(P - P^{\text{th}}\right)/\gamma_c$ increases linearly with the pump P.

The situation is more complicated when the condensate is placed in a periodic lattice. Then we expand $\Psi(x)$ and $n(x)$ in Fourier series on spatial harmonics of the external periodic potential:

$$\psi(x) = \sum_{j=-N}^{N} \psi_j e^{i\cdot j\beta x}, \quad n(x) = \sum_{j=-N}^{N} n_j e^{i\cdot j\beta x}, \tag{4}$$

with normalization condition $I = \sum_{j=-N}^{N} |\psi_j|^2$, where the value of I corresponds to the condensate density averaged over lattice cell, i.e. $I = \frac{1}{l}\int_{-l/2}^{l/2} |\Psi(x)|^2 dx$. We restrict ourselves with $2N+1$ harmonics where N is a maximum number of spatial harmonic. This approach is similar to the method used in [11] to determine the energy band structure of atomic BEC localized in an optical lattice. Unlike the latter case the exciton-polariton condensate is strongly affected by the spatial modulation of excitonic reservoir [12], and that is accounted by the second expansion in equation (4).

Substituting (4) into (1) and (2) and collecting terms with equal exponential factors one obtains $4N+2$ coupled equations with $4N+3$ unknowns (including the eigenfrequency μ). The missing equation can be found from the condition of the stationarity of condensate density:

$$\frac{\partial |\Psi|^2}{\partial t} = 0. \tag{5}$$

The used approach allows to determine condensate wavefunction and its energy in every particular point of BZ simultaneously, i.e. to determine the full bifurcation digram of the solution.

The number of energy bands described by (4) and (3) is $2N+1$. However we are primarily interested in the lowest energy levels on which polariton condensation occurs in current experiments [4, 5]. In [8] the three-mode approximation (*N=1*) was used to describe momentum space oscillations of condensate in a lattice. This approximation adequately describes polariton dynamics for the case of weak-contrast lattices and for relatively low pumping rates. However it becomes inaccurate in a strongly nonlinear regime, i.e. far above the condensation threshold. Indeed, with the increase of condensate density $|\Psi|^2$ the nonlinearity leads to effective coupling between different harmonics ψ_j. Thus, more Fourier components are required to provide sufficient precision of the simulations. We have solved equations (1) and (2) for different values of N. Our numerical analysis has shown that for the moderate value of potential depth ($V_0 / \hbar = 0.5\ \mathrm{ps}^{-1}$) the shape of the solutions belonging to the low energy bands almost does not depend on N for $N > 5$. It means that the major contribution to wavefunctions is provided by spatial harmonics with low wavevectors. All numerical calculations in the present paper have been made for $N = 9$ which guaranties a sufficient precision of our numerical analysis.

3.2. Stability of the nonlinear Bloch states

Stability is a crucial characteristic of the nonlinear Bloch states discussed above. To prove the dynamical stability of the solutions we use the standard Bogoliubov-de Gennes approach [9, 17], which implies perturbation of the solution (3) in the form

$$\Psi(x) = \mathrm{e}^{ikx - i\mu t} \sum_j \left[\psi_j(x) + \delta\psi_j(x) \right] \mathrm{e}^{\mathrm{i} \cdot j\beta x}, \tag{6}$$

$$n(x) = \sum_j \left[n_j(x) + \delta n_j(x) \right] \mathrm{e}^{\mathrm{i} \cdot j\beta x}, \tag{7}$$

where small perturbations of the condensate and the reservoir steady states are taken in the form $\delta\psi_j(x) = u_j \mathrm{e}^{\mathrm{i}\omega t + \mathrm{i}qx} + v_j^* \mathrm{e}^{-\mathrm{i}\omega t - \mathrm{i}qx}$ and $\delta n_j(x) = \eta_j \mathrm{e}^{\mathrm{i}\omega t + \mathrm{i}qx} + \eta_j^* \mathrm{e}^{-\mathrm{i}\omega t - \mathrm{i}qx}$. Here ω is eigenfrequency of small excitation with quasi momentum q. Substituting this solution into equations (1) and (2) and keeping only those terms which are linear in respect to small perturbations we obtain the linearized system of equations for all u_j, v_j and η_j. Stability analysis implies the solution of the eigenvalue problem for this system. The Bloch state is unstable if it possesses frequency ω with a negative imaginary part for any values of q. Since analytical analysis is impossible in our case, we perform linear stability analysis numerically.

3.3. Frequency-momenta bifurcation diagrams

The results of numerical solution of equations (3)-(5) are presented below. As expected, in the vicinity of the pumping threshold, where condensate density is small and nonlinear effects are negligible, the frequency-momenta diagrams of exciton-polariton condensate, i.e. $\mu(k)$, resembles a dispersion of a single particle in periodic potential [16] – see figure 1. Note that it is characterized by zero derivatives at the BZ edge, $k = \beta / 2$, meaning that in the linear case all the Bloch waves at the zone boundary possess zero group velocity $v_g = \partial\mu / \partial k = 0$, i.e. carry no current. This phenomenon can be interpreted as the first-order Bragg scattering of the condensate on the periodic lattice. In this case the states at the boundary of BZ are separated by the energy gap which is approximately equal to V_0 – see [8].

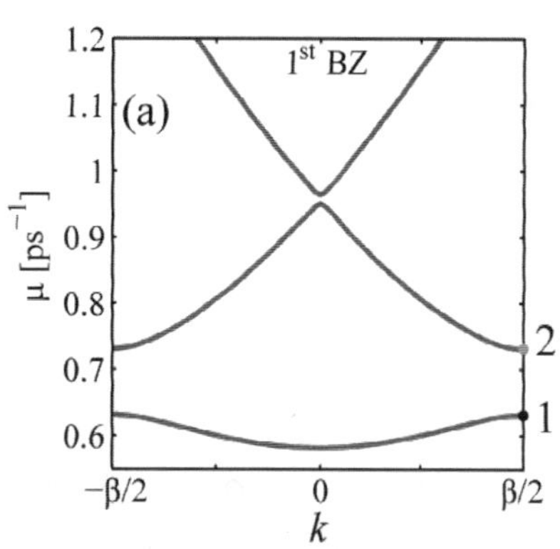

Figure 1. Frequency-momentum diagram of the exciton-polariton condensate in a one-dimensional periodic lattice ($V_0 / \hbar = 0.1\ \mathrm{ps}^{-1}$) in the vicinity of the condensation threshold $P_{\mathrm{th}} = 16.335\ \mu\mathrm{m}^{-2}\mathrm{ps}^{-1}$.

When pumping P increases, the condensate density $|\Psi|^2$ increases as well, making "extra" nonlinear solutions of equations (1) and (2) possible to appear. The properties of the solutions of a Bloch-type are governed by the periodic lattice. That is why the nonlinear Bloch states appear due to the interplay between periodic potential and nonlinearity, i.e., in particular, when the nonlinear mixing between the condensate harmonics ψ_j (mediated by the term $g_c|\Psi|^2$ in equation (1)) becomes comparable with the linear scattering of them by the potential. Thus the nonlinear regime for nonlinear Bloch waves can be reached either under high pumping rates or for weak-contrast lattices (small values of V_0) under moderate level of pumping.

Besides the two-particle repulsive interaction between coherent polaritons the incoherent reservoir also contributes to the nonlinear renormalization of the band structure. Indeed, owing to the reservoir-polariton scattering (the term proportional to g_R in equation (1)) the energy of the Bloch waves experiences an additional nonlinear blue shift. Moreover, the density-dependent saturation of the gain results in the spatial modulation of the reservoir which induces additional spatially modulated potential. All aforementioned nonlinear mechanisms affect substantially both shape and stability of the Bloch waves resulting in the modification of the energy dispersion typical for the linear system.

Note that the conception of the energy band structure is not relevant for a nonlinear dissipative system such as exciton-polaritons in a lattice. Actually, the particle number is a free parameter in the conservative system parametrizing the family of the solutions with different k. However in the driven dissipative case the condensate population is fixed by the balance of gain and losses which is essentially affected by the condensate distribution. Thus the relevant parameter which parameterizes solutions is the pumping strength P which is a key control parameter and could be smoothly varied in experiment. In this case all the states belonging to the same band have different spatial shapes and thus possess different densities. So, here we search the Bloch solutions possessing different wavectors k for the fixed value of pumping P. The obtained dependences of the condensate eigenenergy versus momentum, $\mu(k)$, differ from the convenient dispersion and can be interpreted as frequency-momenta bifurcation diagram. For instance, the value $\partial\mu/\partial k$ is not a group velocity in the open-dissipative system.

Modification of the energy-momenta diagram near the BZ edge is shown in figure 2 for the case of gradual reduction of the potential contrast from $V_0/\hbar = 0.5\ \text{ps}^{-1}$ for panel (a) down to $0.075\ \text{ps}^{-1}$ for panel (d). All the diagrams are plotted for the fixed pumping $P = 50\ \mu\text{m}^{-2}\text{ps}^{-1}$ that is sufficiently above condensation threshold.

The case shown on the panel (a) corresponds to the deep potential $V_0/\hbar = 0.5\ \text{ps}^{-1}$ when the influence of nonlinear effects should be suppressed. However even in this case the energy-momenta diagrams substantially differs from the linear case shown in figure 1. In particular, the chemical potential of the condensate at the center of the first band ($k = 0$) is larger than at BZ edge. Such a contraintuitive result should be attributed to the driven-dissipative nature of exciton-polariton system. Actually, the average population of the particular Bloch state is determined by the balance of losses and effective gain which is determined by the spatial overlap of the condensate distribution $|\Psi(x)|^2$ with the distribution of the hot excitons $n(x)$. Since all the states of the same band have different spatial structures they possess different densities and thus gain different blueshifts. In particular, the steady-state distribution of reservoir $n(x) = P/\left(\gamma_R + R|\Psi|^2\right)$ is noted for the local depletion which implies that the value of n is minimal at the maxima of the condensate density. Therefore, the overlap integral $\int_{-l/2}^{l/2} |\Psi(x)|^2 n(x)dx$ responsible for the effective gain (see equation (1)) is lower for the Bloch states with high density contrast. Thus, these states are less populated than the states with a smooth density profile.

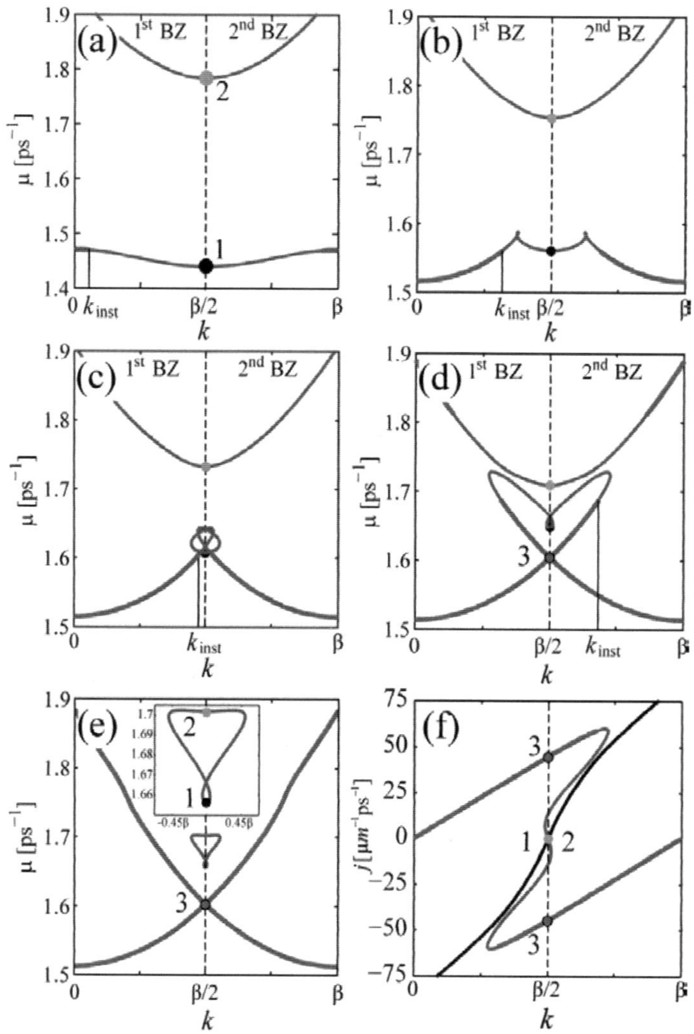

Figure 2. (a)-(d) eigenfrequency-momenta bifurcation diagrams, $\mu(k)$, near the boundary of BZ (vertical dashed line) in the extended zone representation. For all panels the pumping rate is $P = 50\ \mu m^{-2} ps^{-1}$ $\mu m^{-2} ps^{-1}$ whereas the depth of periodic potential gradually decreases for (a)-(d). In particular, (a) – $V_0 / \hbar = 0.5\ ps^{-1}$; (b) – $V_0 / \hbar = 0.3\ ps^{-1}$; (c) – $V_0 / \hbar = 0.2\ ps^{-1}$; (d) – $V_0 / \hbar = 0.1\ ps^{-1}$; (e) – $V_0 / \hbar = 0.075\ ps^{-1}$. Red (thick) sections correspond to the regions of dynamical stability while k_{inst} denotes momentum above which the states of the lower band becomes dynamically unstable. The inset to the panel (e) shows the magnified region of the closed loop below it. Panel (f) the averaged current $\langle j \rangle$ distribution for the parameters corresponding to panel (d). The black curve corresponds to the upper band. The blue curve (with the red sections) corresponds to the first band.

IOP Conf. Series: Journal of Physics: Conf. Series **951** (2017) 012030 doi:10.1088/1742-6596/951/1/012030

In particular, the lower state at the BZ boundary (labeled as '1' in figure 1(a)) possesses cosine-like shape and its density vanishes at the potential maxima. At the same time the state at the zone centre has a more smooth density distribution. Its density minima also locate under the potential maxima but are always above zero since the condensates located at the neighboring lattice cites are in-phase (cf. with [18]). So, the state at the BZ centre is usually more populated than the state at the zone boundary. Thus it possesses the stronger blueshift. Besides the state '1' acquires a strong redshift for the case high-contrast lattice since the energy gap is proportional to V_0. So, the interplay of the aforementioned effects leads to the unusual shape of the energy band shown in figure 2(a).

The shape of the first band modifies for the smaller potential depth. Unlike the case of conservative BEC system [9-11] the energy loops emerge from the first band not at the BZ edge but at some k located inside Brillouin zone – see panel (b) of figure 2. Reduction of the potential depth V_0 allows nonlinearity to increase its impact. The loops move towards each other and tie into a knot of a rather complicated shape at the zone boundary – panel (c). With the further decrease of the potential depth the edges of this knot stretch towards the second band forming a typical loop structure shown in panel (d) of figure 2. The shape of this loop resembles the "swallow tail" structure [13-18] but with the twisted upper edge.

In the limit of small but not vanishing potential depths ($V_0 / \hbar = 0.075$ ps^{-1}, panel (e)) the first and the second bands merge. In this case the nonlinear effects completely dominate over the contribution of the external periodic potential in the sense that the energy gap, which is a key feature of a linear periodic potential, disappears. Finally, the $\mu(k)$ dependence approaches a parabolic shape implying that the lattice becomes effectively screened by the nonlinear interactions.

The stability analysis of the nonlinear Bloch states states was also performed. The domains of dynamical stability are shown by thick red curves in figure 2. The region of existence of the stable states grows with the reduction of V_0. Besides as the influence of the nonlinearity increases the second energy band also stabilizes – see figure 2(d). Finally, all the nonlinear Bloch states stabilize as the bands merge and approach parabolic shape.

A remarkable feature of the observed loops of frequency-momenta bifurcation diagrams (see figures 2(c) and 2(d)) is the existence of states possessing nonzero current $j = \frac{i\hbar}{2m}\left(\frac{\partial \Psi^*}{\partial x}\Psi - \frac{\partial \Psi}{\partial x}\Psi^* \right)$ even at the edges of BZ. The state is currentless if the value of current averaged over the lattice period $\langle j \rangle = \int_{-l/2}^{l/2} j(x)dx$ is zero. Figure 2(f) illustrates the current-momenta diagram corresponding to the typical loop structure shown in the panel (d). The red curve corresponds to the nonlinear Bloch states belonging to the first band while the black one corresponds to the second band. Note that the red curve crosses BZ boundary five times. There are four current states but only two of them are dynamically stable – points '3' in figure 3(f). These states correspond to equivalent oppositely directed polariton flows and thus they are frequency degenerate – point '3' in figure 3(d). Our numerical simulations reveal that nonlinear current states possessing smaller absolute value of the average current are always unstable (see [19, 20] for the details).

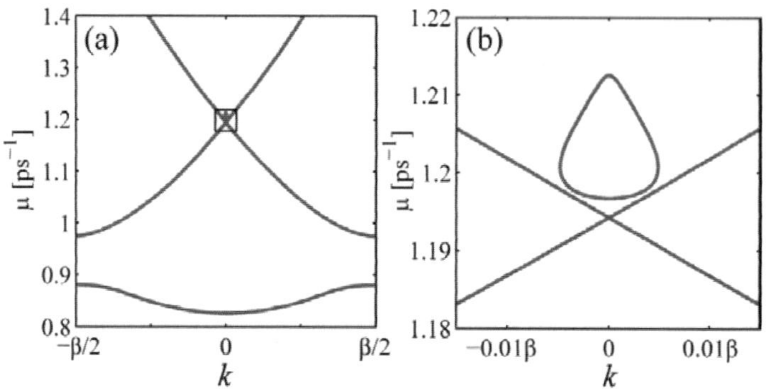

Figure 3. (a) the typical shape of a frequency-momenta bifurcation diagram with a closed gap between the second and the third band. (b) the magnified region framed with the black square on the panel (a). The parameters are: $P = 25\ \mu m^{-2} ps^{-1}$ and $V_0 / \hbar = 0.1\ ps^{-1}$.

Besides these "fluid states" the conventional currentless Bloch solutions at the BZ boundaries also survive even in the strongly nonlinear regime. For instance, these nonlinear Bloch modes belong to the closed curve on the dispersion relation shown in the inset to figure 2(d). The lower and the upper states of this curve [marked by '1' and '2'] are direct counterparts of the highly symmetric Bloch waves known for the single-particle band structure, which nest on the top of the first and on the bottom of the second Bloch bands respectively – see figures 1 and 2(a). We have found that all the states on this closed curve are unstable and thus they do not exhibit in condensate dynamics.

The shape of the second energy band in the vicinity of $k = 0$ also needs to be discussed. In the atomic BEC systems the emergence of swallow tails was predicted from the top of the second band. Our simulations reveal that in our case the nonlinear Bloch states appear only in the very narrow region close to the zone centre, see figure 3. Like at the BZ edges the gap between the second and the third bands closes in the strongly nonlinear limit resulting in formation of the closed loop shown in figure 3(b). Note that all the states belonging to this loop are unstable.

4. Conclusion

We have performed the theoretical analysis of the properties of the nonlinear Bloch states of exciton-polariton condensate confined in 1D periodic potential. The interplay between the periodic potential and the nonlinear interactions results in essential modification of the energy landscape of the condensate states possessing the same periodicity as the lattice. The relevant eigenfrequency-momenta bifurcation diagrams of the nonlinear Bloch states were plotted for the fixed values of pumping rate P which is a key control parameter of a driven-dissipative polariton system. The main peculiarity of these diagrams is the loop-like structures emerging at the first and the second bands. We have shown that the main manifestations of the periodic potential, which are typical for the linear problem, i.e. energy gap and currentless states at the Brillouin zone boundary, becomes suppressed in the limit of strong nonlinearity. Formation of the nonlinear current states at the BZ boundary was also described. Such nonlinear current of polariton condensate in a lattice can be experimentally observed either with the specifically designed localized pumping [19] or via spontaneous formation in the presence of complex periodic potential [20].

Acknowledgment

This work was supported by the RFBR Grants No. 16-32-60102, 15-59-30406, 15-52-52001 and partially by the President of Russian Federation for state support of young Russian scientists, Grant MK-2988.2017.2. The work was also supported by Ministry of Education and Science of the Russian

Federation, project no. 16.1123.2017/4.6. A.V.Y. acknowledges the support by the Russian Federation Grant 074-U01 through ITMO Early Career Fellowship scheme.

References

[1] Mousavi S H, Khanikaev A B and Wang Z 2015 Topologically protected elastic waves in phononic metamaterials *Nature Commun.* **6** 8682

[2] Eiermann B, Anker T, Albiez M, Taglieber M, Treutlein P, Marzlin K-P and Oberthaler M K 2004 Bright Bose-Einstein Gap Solitons of Atoms with Repulsive Interaction *Phys. Rev. Lett.* **92** 230401

[3] Tanese D, Flayac H, Solnyshkov D, Amo A, Lemaitre A, Galopin E, Braive R, Senellart P, Sagnes I, Malpuech G and Bloch J 2013 Polariton condensation in solitonic gap states in a one-dimensional periodic potential *Nature Commun.* **4** 1749

[4] Cerda-Mendez E A, Krizhanovskii D N, Wouters M, Bradley R, Biermann K, Guda K, Hey R, Santos P V, Sarkar D and Skolnick M S 2010 *Phys. Rev. Lett.* **105** 116402

[5] Lai C, Kim N, Utsumomiya S, Roumpos G, Deng H, Fraser M, Byrnes T, Recher P, Kumada N, Fujisawa T and Yamamoto Y 2007 *Nature* **450** 529

[6] Pethick C J and Smith H 2008 *Bose-Einstein Condenstaion in dilute gases* (Cambridge University Press)

[7] Liew T C H, Egorov O A, Matuszewski M, Kyriienko O, Ma X and Ostrovskaya E A, 2015 Instabilityinduced formation and nonequilibrium dynamics of phase defects in polariton condensates *Phys. Rev.* B **91** 085413

[8] Ma X, Chestnov I Yu, Charukhchayan M V, Alodjants A P and Egorov O A 2015 Oscillatory dynamics of nonequilibrium dissipative exciton-polariton condensates in weak-contrast lattices *Phys. Rev.* B **91** 214301

[9] Wu B and Niu Q 2003 Superfluidity of Bose–Einstein condensate in an optical lattice: Landau–Zener tunnelling and dynamical instability *New Journal of Physics* **5** 104

[10] Mueller E J 2002 Superfluidity and mean-field energy loops: Hysteretic behavior in Bose-Einstein condensates *Phys. Rev.* A **66** 063603

[11] Machholm M, Pethick C J and Smith H 2003 Band structure, elementary excitations, and stability of a Bose-Einstein condensate in a periodic potential *Phys. Rev.* A **67** 053613

[12] Bronski J C, Carr L D, Deconinck B and Kutz J N 2001 Bose-Einstein Condensates in Standing Waves: The Cubic Nonlinear Schrödinger Equation with a Periodic Potential *Phys. Rev. Lett.* **86** 1402

[13] Seaman B T, Carr L D and Holland M J 2005 Effect of a potential step or impurity on the Bose-Einstein condensate mean field *Phys. Rev.* A **71** 033622

[14] Danshita I and Tsuchiya S 2007 Stability of Bose-Einstein condensates in a Kronig-Penney potential *Phys. Rev.* A **75** 033612

[15] Roumpos G, Fraser M D, Loffler A, Hofling S, Forchel A and Yamamoto Y 2011 Single vortex–antivortex pair in an exciton-polariton condensate *Nature Physics* **7** 129

[16] Ashcroft N W and Mermin N D, *Solid State Physics* (Holt-Saunders, Philadelphia)

[17] Smirnov L A, Smirnova D A, Ostrovskaya E A and Kivshar Y S 2014 Dynamics and stability of dark solitons in exciton-polariton condensates *Phys. Rev.* B **89** 235310

[18] Ohadi H, Gregory R L, Freegarde T, Rubo Y G, Kavokin A V, Berloff N G and Lagoudakis P G 2016 Nontrivial phase coupling in polariton multiplets *Phys. Rev.* X **6**(3) 031032

[19] Chestnov I Yu, Yulin A V, Alodjants A P and Egorov O A 2016 Nonlinear Bloch waves and current states of exciton-polariton condensates *Phys. Rev.* B **94**(9) 094306.

[20] Nalitov A V, Liew T C H, Kavokin A V, Altshuler B L and Rubo Y G 2017 Spontaneous polariton currents in periodic lateral chains *Phys. Rev. Lett.* **119**(6) 067406.

Quantum fluctuation and nonlinear properties of exciton polaritons in semiconductor microcavities

S S Demirchyan[1], T A Khudaiberganov[1], I Yu Chestnov[1], A P Alodjants[2] and S M Arakelian[1]

[1]Vladimir State University named after A. G. and N. G. Stoletovs, Gorkii St. 87, 600000, Vladimir, Russia
[2]ITMO University, Kronverksky pr. 49, 197101, St. Petersburg, Russia

E-mail: sevakdemirchyan@gmail.com

Abstract. The impact of quantum fluctuations on the properties of quantum well excitons excited inside 0D semiconductor microcavity possessing strong light confinement is analysed. The analysis takes into account the exciton-exciton repulsion and the presence of a reservoir. Analytical solutions are obtained for the generalized Glauber P-function of the system in the adiabatic limit. Destabilization of the homogeneous nonlinear solution belonging to the lower bistability branch in the presence of quantum noise is predicted.

1. Introduction

One of the most important and promising trends in nonlinear photonics is the investigation of the quantum properties of optical radiation interacting with a nonlinear medium. These researches have many important applications in quantum information and precision measurements, as well as in material science [1], biophotonics [2], etc. The usual approach for such problems is based on the study of quantum fluctuations of Hermitian quadratures or of the number of photons of the field interacting with classically (or quasiclassically) described nonlinear medium [3]. However, in many problems of practical importance, in particular for engineering of nonclassical light sources, taking into account quantum fluctuations of the medium is required [4]. The choice of such media has recently been substantially broadened to include Bose-Einstein concentrates, which represent a new quantum state of matter observed both in atomic gases [5] and in solid-state micro- and nanostructures [6].

Semiconductor microcavities having quantum wells inside them appear to be convenient systems for observing Bose-Einstein condensation of exciton polaritons, which are Bose quasi-particles and are consequently capable of macroscopically filling ground state via the Bose-stimulation effect [7]. An important feature of the microcavity exciton polaritons that will be treated in this paper is the strong exciton-exciton repulsion resulting in the appearance of the Kerr-type exciton nonlinearity [8,9]. The magnitude of the repulsion strength is notably high. Experiments reveal that it exceeds the purely photonic nonlinearity on several orders of magnitude. Since the nonlinearity magnifies the effects of noise semiconductor microcavity systems are ideal candidates for studying of a phenomena of quantum fluctuations in the hybrid light-matter system. Another important feature of semiconductor microcavity systems is a strong dissipation inherent in both the photon and exciton components of the polariton state which also leads to the appearance of fluctuations according to the fluctuation-dissipative theorem [10]. Besides, strong dissipation results in requirement of pumping to maintain the steady-state. In this paper we consider the case of external laser pumping, which tuned closeto both exciton and photon resonances and coherently feeds the ground state. The combination of the discussed peculiarities causes bistability of the steady-state solution for the exciton field. This phenomenon manifests itself in the formation of the *S*-shape loop in the pump/signal dependence which has been studied in fairly great detail in the classical limit [11]. In the present paper we investigate whether the purely quantum treatment of the system, which also takes quantum noise into account, affects properties of the steady-state solutions.

2. The model

We consider exciton polaritons in a pillar microcavity under coherent near-resonant continuous-wave pumping (figure 1(a)). The basic model is described below. The Hamiltonian of a system of excitonic (b) and photonic (a) boson fields with frequencies ω_{ex} and ω_{ph}, respectively, is

$$H = \hbar\omega_{ex}a^+a + \hbar\omega_{ph}b^+b + \hbar g\left(b^+a + a^+b\right) + \hbar\chi b^{+2}b^2 + i\hbar\left(Ea^+ - E^*a\right) + \\ + a^+\Gamma_{c1} + a\Gamma_{c1}^+ + b^+\Gamma_{c2} + b\Gamma_{c2}^+. \quad (1)$$

The quantity $2g$ in Eq. (1) is the Rabi splitting frequency, χ describes the angular exciton–exciton scattering, and the external homogeneous pumping is given in the form $E = E_p e^{-i\delta t}$, where E_p is the pumping amplitude of the photon mode, and δ is its frequency (the quantity $I_p = \left|E_p\right|^2$ is proportional to the pumping intensity). Last four terms describe the interaction of the photons and excitons with the corresponding thermal reservoirs, described by operators Γ_{c1} and Γ_{c2}, respectively. Note, that since the light is tightly confined inside the micropillar cavity which is under consideration the kinetic energy terms are not included in the Hamiltonian (1).

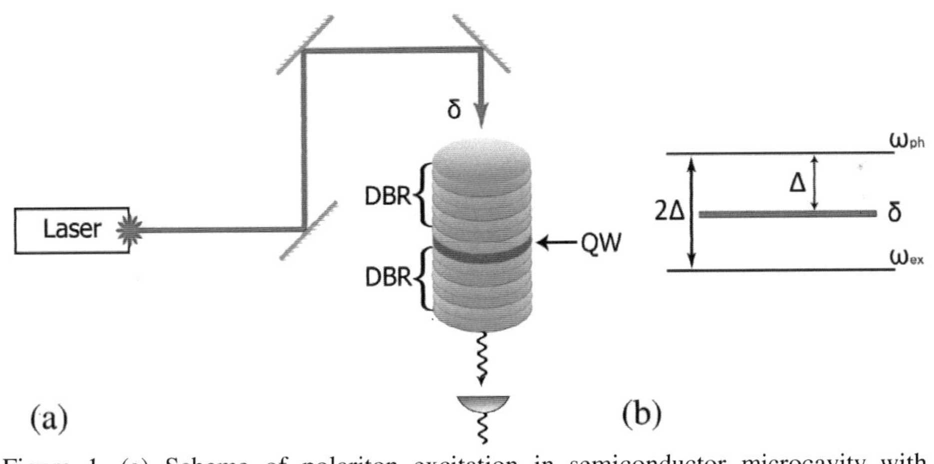

Figure 1. (a) Scheme of polariton excitation in semiconductor microcavity with embedded quantum well (QW) sample. (b) Energy levels of photon and exciton modes and energy level of pumping laser.

To take the dissipation into account, we write the standard master equation for the density matrix ρ:

$$\dot{\rho} = -i\omega_{ph}\left[a^+a,\rho\right] - i\omega_{ex}\left[b^+b,\rho\right] - ig\left[b^+a,\rho\right] - ig\left[a^+b,\rho\right] - i\chi\left[b^{+2}b^2,\rho\right] + \\ + E\left[a^+,\rho\right] - E^*\left[a,\rho\right] + \gamma_{ph}(2a\rho a^+ - \rho a^+a - a^+a\rho + 2n_{th1}[[a,\rho],a^+]) + \\ + \gamma_{ex}(2b\rho b^+ - \rho b^+b - b^+b\rho + 2n_{th2}[[b,\rho],b^+]), \quad (2)$$

IOP Conf. Series: Journal of Physics: Conf. Series **951** (2017) 012031 doi:10.1088/1742-6596/951/1/012031

where γ_{ph} and γ_{ex} describe the damping of the photon and exciton modes, respectively. In Eq. (2) and what follows, a dot over a symbol corresponds to the time derivative. Here we neglect the thermal noise, since $\hbar\omega_{ph,ex} \gg kT$ and the population of the exciton and photon reservoirs is small. We first consider the exciton–photon system of Eqs. (1) in the mean-field approximation, replacing the operators with their mean values i.e., setting $\alpha = \langle a \rangle$, $\beta = \langle b \rangle$ and $\langle \Gamma_{c1,c2} \rangle = 0$. Going to slowly varying amplitudes, we restrict ourselves to a consideration of the solution in the form $\alpha \to \alpha e^{-i\delta t}$ and $\beta \to \beta e^{-i\delta t}$, where the incident laser field has the frequency $\delta = (\omega_{ex} + \omega_{ph})/2$ (figure 1(b)). Then we get:

$$\begin{aligned} \dot{\alpha} &= -(i\Delta + \gamma_{ph})\alpha + E_p - ig\beta, \\ \dot{\beta} &= -(-i\Delta + \gamma_{ex})\beta - ig\alpha - 2i\chi|\beta|^2\beta, \end{aligned} \tag{3}$$

where $\Delta = (\omega_{ph} - \omega_{ex})/2$. We shall search for the steady-state solution $\dot{\alpha} = \dot{\beta} = 0$ of the system of Eqs. (3). In this regime, the exciton-mode population $n_{ex} = |\beta|^2$ is determined from the parametric equation:

$$I_p = 4\frac{(\gamma_{ph}^2 + \Delta^2)\chi^2}{g^2} n_{ex}^{\ 3} - 4\frac{(\gamma_{ph}^2 + \Delta^2)c}{g^2} \chi n_{ex}^{\ 2} + \frac{(\gamma_{ph}^2 + \Delta^2)}{g^2}(c^2 + d^2) n_{ex}, \tag{4}$$

where we have introduced the following notation $c = \Delta + \Delta g^2 (\Delta^2 + \gamma_{ph}^2)^{-1}$, $d = \gamma_{ex} + g^2\gamma_{ph}(\Delta^2 + \gamma_{ph}^2)^{-1}$. Equation (4) determines the condition for the existence of a bistability loop for the exciton-mode population i.e., the presence of two stable stationary states for the same pump intensity. It is easy to show that this requires the condition $c > \sqrt{3}d$ to be satisfied.

We now turn to a study of the quantum noise for the system described by the Hamiltonian (1). It is fairly difficult to solve the main kinetic equation with the nonlinearity expressed in operator form. Therefore, it is worthwhile to go to commuting quantities, or *c*-numbers [12]. One method of solving this problem is to diagonalize the density matrix with respect to coherent states by means of the P-function – the so-called P-representation, or Glauber representation [3]. Using the Glauber representation, we can go from a consideration of the exciton and photon modes in the form of quantum oscillators to a model of stochastic oscillators with definite displacements and diffusion matrices. The Fokker-Planck equation, corresponding to the density-matrix equation, is obtained using standard methods based on the Glauber representation. We use the approach based on the so-called P-function – an inversely ordered representation of the density matrix. Unlike other distribution functions, it is unbounded and can be nonanalytical in the case of nonclassical states. In the same way, unlike its classical analog, the probability density, it can be negative – for example, when describing nonclassical photon statistics [13]. Therefore, we use a nondiagonal, generalized P-representation. For that we represent the density operator in the form $\rho = \int P(\{x\})\hat{\Lambda}(\{x\})d\sigma(\{x\})$, where $\{x\} = (\alpha, \alpha^+, \beta, \beta^+)$ and $\hat{\Lambda}(\alpha, \alpha^+, \beta, \beta^+) = \frac{|\alpha\rangle|\beta\rangle\langle\beta^{+*}|\langle\alpha^{+*}|}{\langle\alpha^{+*}|\langle\beta^{+*}||\beta\rangle|\alpha\rangle}$ is the projection operator, while $d\sigma(\alpha, \alpha^+, \beta, \beta^+)$ is the integration volume. Note that now α and β have the sense of *c*-numbers i.e., the eigenvalues of the photon and exciton creation and annihilation operators, respectively.

Using the standard operator algebra [12,14], we arrive at the Fokker–Planck equation:

$$\frac{\partial P}{\partial t}=[-\frac{\partial}{\partial\alpha}\left(-\left(i\omega_{ph}+\gamma_{ph}\right)\alpha+E-ig\beta\right)-\frac{\partial}{\partial\beta}\left(-\left(i\omega_{ex}+\gamma_{ex}\right)\beta-ig\alpha-2i\chi\beta^{+}\beta^{2}\right)+ \\ +\frac{\partial^{2}}{\partial\beta^{2}}\left(-i\chi\beta^{2}\right)+\gamma_{ph}n_{th1}\frac{\partial^{2}}{\partial\alpha\partial\alpha^{+}}+\gamma_{ex}n_{th2}\frac{\partial^{2}}{\partial\beta\partial\beta^{+}}+H.c.]P, \tag{5}$$

where H.c. corresponds to the Hermite-conjugate terms. One of the methods for solving this problem is to obtain the exact stochastic Langevin equations in the Ito calculus by converting the Fokker–Planck equation given by Eq. (5) and consider small fluctuations around the steady-state. It allows finding the noise spectrum and the correlation functions of the field. But we consider the more rigorous solution of the steady-state Fokker–Planck equation (5) beyond the limits of linearization with respect to fluctuations. In this case, we assume that the potentiality condition is satisfied [15]: $\partial_i V_j=\partial_j V_i$, where $V_x=\sum_{\nu}\left(D^{-1}\right)_{x\nu}\left(2A_\nu+\sum_l\frac{\partial D_{\nu l}}{\partial q_l}\right)$. Here A is drift and D is diffusion matrix. We next use the adiabatic approximation [16], which is valid since the dissipation of the ground photon mode of the micropillar cavity is usually more fast than the quenching of the exciton mode, $\gamma_{ph}\gg\gamma_{ex}$. After some easy computations, we get that the potentiality condition is satisfied if $\partial V_1/\partial\beta^{+}=\partial V_2/\partial\beta=-2$. Integrating the generalized force V in order to find the potential function, we arrive at an expression for the quasi-probability distribution in the steady-state:

$$P_{ss}\left(\beta,\beta^{+}\right)=\beta^{-2-i\frac{\gamma}{\chi}}\beta^{+^{-2+i\frac{\gamma^{*}}{\chi}}}\exp\left(-\mu E_{p}\frac{1}{\beta}-\mu^{*}E_{p}^{*}\frac{1}{\beta^{+}}+2\beta\beta^{+}\right), \tag{6}$$

where $\gamma=\gamma_{ex}+g^{2}\gamma_{ph}\left(\gamma_{ph}^{2}+\Delta^{2}\right)^{-1}-i\left(\Delta+\Delta g^{2}\left(\gamma_{ph}^{2}+\Delta^{2}\right)^{-1}\right)$ and $\mu=\chi^{-1}\left(g\gamma_{ph}-i\Delta g\right)\left(\gamma_{ph}^{2}+\Delta^{2}\right)^{-1}$. It can be seen from this that ordinary integration cannot be carried out in the complex plane since the potential in this case diverges, $\beta\beta^{+}\to\infty$. This means that, instead of the usual diagonal Glauber-Sudarshan representation, it is necessary to use the generalized P-representation, implying the integration region $\beta^{+}\neq\beta^{*}$, so that the distribution function falls off at the edges. After some easy computation, we get the normalization integral and the moments of the distribution function. The ratio of the moment to the normalization integral corresponds to the expression for the correlation function (with zero delay):

$$G^{(IJ)}=\left\langle b^{+I}b^{J}\right\rangle=\frac{\int\beta^{+I}\beta^{J}P_{ss}\left(\beta^{+},\beta\right)d\mu}{\int P_{ss}\left(\beta^{+},\beta\right)d\mu}=-\frac{\left(-\mu E_{p}\right)^{J}\left(-\mu E_{p}^{*}\right)^{I}\Gamma\left(\frac{i\gamma^{*}}{\chi}\right)\Gamma\left(-\frac{i\gamma}{\chi}\right)}{\Gamma\left(I+\frac{i\gamma^{*}}{\chi}\right)\Gamma\left(J-\frac{i\gamma}{\chi}\right)}\times \\ \times\frac{{}_0F_2\left(I+\frac{i\gamma^{*}}{\chi},\mathrm{J}-\frac{i\gamma}{\chi},2\mu^{2}\left|E_{p}\right|^{2}\right)}{{}_0F_2\left(\frac{i\gamma^{*}}{\chi},-\frac{i\gamma}{\chi},2\mu^{2}\left|E_{p}\right|^{2}\right)}. \tag{7}$$

This is the basic equation, since it can be used to obtain expressions for the mean of the exciton-annihilation operator and the first-order corrector, respectively:

$$\left\langle\hat{b}\right\rangle=\frac{\chi\mu E_{p}\,{}_0F_2\left(\frac{i\gamma^{*}}{\chi},-\frac{i\gamma}{\chi}+1,2|\mu|^{2}\left|E_{p}\right|^{2}\right)}{i\gamma\,{}_0F_2\left(\frac{i\gamma^{*}}{\chi},-\frac{i\gamma}{\chi},2|\mu|^{2}\left|E_{p}\right|^{2}\right)}, \tag{8}$$

$$\left\langle \hat{b}^{+}\hat{b} \right\rangle = \frac{\chi^2 |\mu|^2 |E_p|^2 \, {}_0F_2\left(\frac{i\gamma^*}{\chi}+1, -\frac{i\gamma}{\chi}+1, 2|\mu|^2 |E_p|^2 \right)}{|\gamma|^2 \, {}_0F_2\left(\frac{i\gamma^*}{\chi}, -\frac{i\gamma}{\chi}, 2|\mu|^2 |E_p|^2 \right)}. \tag{9}$$

These quantities are plotted in figure 2, which also shows for comparison the exciton-mode population, determined in terms of semiclassical analysis. Let us compare the predictions of the quantum-mechanical and semiclassical approaches. In particular, as pointed out above, the semiclassical theory predicts that the exciton population is bistable. Bistability can be experimentally observed in the form of an abrupt change of state of the system as the pumping increases above the certain threshold (point 2 in figure 2). According to quasiclassical treatment of this event the jump occurs when the frequency of the ground state of the system is nonlinearly shifted into resonance with the pump. In this case the system switches to the state $2'$ which is more populated. With the further increase of pumping amplitude the exciton population then increases along the upper steady-state bistability branch.

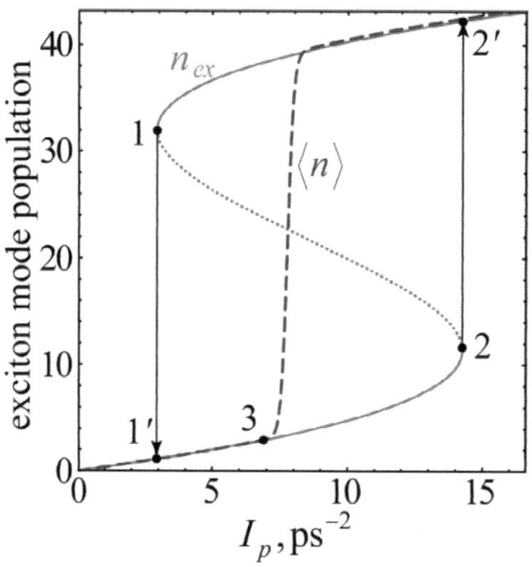

Figure 2. (a) The semiclassical value n_{ex} (solid curve with dotted section) and quantum-mechanical mean value $\langle b^{+}b \rangle = \langle n \rangle$ (dashed curve) of the exciton population as a function of the optical pump I_p. The parameters are $\Delta = 0.5$ ps^{-1}, $\hbar\chi = 0.05$ meV , $\hbar g = 1$ meV , $\gamma_{ex} = 0.01$ ps^{-1} and $\gamma_{ph} = 0.1$ ps^{-1}.

An analysis of the solution of the Fokker–Planck equation shows that the quantum fluctuations that occur around the steady state can force the system out of the basin of attraction of stable steady-state solution of the lower bistability branch into the region of attraction of the upper stable branch. The state of the system in this case will follow the dashed (purple) curve in figure 2 as the pumping increases, and, after it reaches point 3 in figure 2, the mean exciton population is observed to sharply increase all the way to the upper bistability branch. However, when the parameters of the system are such that bistability is not observed at any pumping value, the quantum-mechanical and semiclassical descriptions coincide. Thus, in the region where the semiclassical description (dashed curve) predicts bistability, no bistability is observed when quantum fluctuations are taken into account. It should be pointed out that the section of the lower branch between points 3 and 2 of figure 2 is metastable in this case.

However it is important to note that whether the bistability phenomenon is observed in experiment depends on the characteristic time of random transition from one branch to the other, which in turn is determined by the fluctuation amplitude. However, if the pump field increases in a time interval shorter than the given random transition time, a hysteresis loop will be observed.

3. Conclusion

The effect of quantum fluctuations on the properties of exciton polaritons localized in a semiconductor pillar microcavity is analyzed, taking into account the nonlinear interaction between excitons and in the presence of quantum noise. Analytical solutions are obtained for the generalized Glauber P-function in the adiabatic limit. It was found that the presence of quantum noise leads to the metastability of the homogeneous solution related to the lower branch of the bistability.

Acknowledgment

This work was supported by the RFBR Grants No. 16-32-60102, 15-59-30406, 15-52-52001 and partially by the President of Russian Federation for state support of young Russian scientists, Grant MK-2988.2017.2. The work was also supported by Ministry of Education and Science of the Russian Federation, project no. 16.1123.2017/4.6.

References

[1] Bachor H A and Ralph T C 2004 *A guide to experiments in quantum optics* (Weinheim: Wiley-VCH)

[2] Lambert N, Chen Y N, Cheng Y C, Li C M, Chen G Y and Nori F 2013 Quantum biology *Nat. Phys.* **9** 10–18

[3] Walls D F and Milburn G J 2008 *Quantum optics* (Berlin: Springer-Verlag)

[4] Heshami Kh, England D G, Humphreys P C, Bustard P J, Acosta V M, Nunn J and Sussman B J 2016 Quantum memories: emerging applications and recent advances *J. Mod. Opt.* **63**(20) 2005–28

[5] Dalfovo F, Giorgini S, Pitaevskii L P and Stringari S 1999 Theory of Bose-Einstein condensation in trapped gases *Rev. Mod. Phys.* **71** 463–512

[6] Deng H, Haug H and Yamamoto Y 2010 Exciton-polariton Bose-Einstein condensation *Rev. Mod. Phys.* **82** 1489–1537

[7] Kasprzak J, Richard M, Kundermann S, Baas A, Jeambrun P, Keeling J M J, Marchetti F M, Szymanska M H, André R, Staehli J L, Savona V, Littlewood P B, Deveaud B and Dang L S 2006 Bose-Einstein condensation of exciton polaritons *Nature* **443** 409–14

[8] Ciuti C, Savona V, Piermarocchi C, Quattropani A and Schwendimann P 1998 Role of the exchange of carriers in elastic exciton-exciton scattering in quantum wells *Phys. Rev.* B **58**(12) 7926–33

[9] Wouters M and Carusotto I 2007 Parametric oscillation threshold of semiconductor microcavities in the strong coupling regime *Phys. Rev.* B **75**(7) 075332

[10] Gardiner C W and Zoller P 2004 *Quantum noise: a handbook of Markovian and non-Markovian quantum stochasctic methods with applications to quantum optics* (Berlin: Springer-Verlag)

[11] Yulin A V, Egorov O A, Lederer F and Skryabin D V 2008 Dark polariton solitons in semiconductor microcavities *Phys. Rev.* A **78** 061801

[12] Lax M 1968 *Fluctuation and Coherence Phenomena in Classical and Quantum Physics* (New York: Science Publishers Inc.)

[13] Carmichael H J and Walls D F 1976 A quantum-mechanical master equation treatment of the dynamical Stark effect *J. Phys. B: At.Mol. Phys.* **9** 1199–1219

[14] Drummond P D and Gardiner C W 1980 Generalised P-representations in quantum optics *J. Phys.* **13** 2353–68

[15] Drummond P D, McNeil K J and Walls D F 1981 Non-equilibrium transitions in sub/second harmonic generation *Opt. Acta* **28**(2) 211–25

[16] Gardiner C W 1984 Adiabatic elimination in stochastic systems. II. Application to reaction diffusion and hydrodynamic-like systems *Phys. Rev.* A **29** 2814–22

Photon statistics of semiconductor microcavity polaritons

T A Khudaiberganov, I Yu Chestnov, S S Demirchyan

Vladimir State University named after A. G. and N. G. Stoletovs, Gorkii St. 87, 600000, Vladimir, Russia

E-mail: igor_chestnov@mail.com

Abstract. Quantum statistical properties of the coherently driven semiconductor microcavity are studied. The unique quantum solution for the two-mode system of coupled excitons and photons was obtained in the adiabatic approximation. This solution predicts both bunching and antibunching regimes for the photonic field in the domain of parameters where the single mode polariton approximation fails.

1. Introduction

Investigation of the quantum properties of light interacting with nonlinear medium represents a fundamental task of modern quantum optics and many-body physics. In the problem of such kind statistical properties of light are of special interest. Nowadays engineering of photon sources possessing non-classical statistics is essential in many practically important applications. The main principle utilized in such devises is a government the light properties by the interaction with the medium which is usually considered as classical in the sense that its fluctuations are negligible. However the influence of fluctuations of the purely quantum medium represents an intriguing problem. The appropriate candidates for such quantum media are Bose-Einstein condensates. These exotic states of matter were observed both in atomic and solid-state semiconductor structures. While atomic condensates are obtained at rather low temperatures the exciton polariton condensates formed in semiconductor microcavities with a wide band gap (like GaN and ZnO) exist even at room temperatures. Quantum well excitons are distinguished by the strong nonlinearity evoked by the Coulumb interaction. Thus a rich dynamical and statistical behaviour triggered by quantum fluctuations are expected in microcavity system. So, semiconductor systems supporting interaction of quantum well excitons with a microcavity mode represent a robust platform for engineering of light sources with unusual statistics.

Here we study the system consisted of a 0D semiconductor microcavity with embedded quantum well irradiated by a coherent laser field which frequency is close both to photon and to exciton resonances of the structure. We perform an analysis of the influence of quantum fluctuations on the properties of the photonic mode paying special attention to nonlinear interaction between the excitons. Analytical solutions for this problem were found using generalized *P*-function [1] in the adiabatic limit [2]. It has been also demonstrated that a presence of quantum noise leads to metastability of the steady-state solution of the lower branch of bistability loop [3]. Non-classical statistics (both antibunching and bunching regimes) of photons were predicted.

2. The model

The system we are working with may arise in a variety of contexts. It consists of a single photonic mode interacting with a nonlinear medium. For concreteness we consider a semiconductor micropillar cavity of a small radius which fundamental mode frequency ω_{ph} is tuned closely to the exciton resonance ω_{ex} of the embedded quantum well. Since the light is tightly confined inside the pillar both excitonic and photonic modes are devoid of spatial degrees of freedom. It implies that, unlike in planar semiconductor microcavities, the exciton wavevector $\vec{k}$ normal to the structure growth axis has the fixed value, say $k=0$. To sustain the nontrivial steady-state population of both photonic and excitonic modes an external quasiresonant laser driving is applied. Thus the Hamiltonian of the two coupled modes under coherent driving reads

$$\hat{H}_{\mathrm{C}}=\hbar\omega_{\mathrm{ph}}\hat{\phi}^{+}\hat{\phi}+\hbar\omega_{\mathrm{ex}}\hat{\chi}^{+}\hat{\chi}+\hbar\omega_{\mathrm{R}}\left(\hat{\chi}^{+}\hat{\phi}+\hat{\phi}^{+}\hat{\chi}\right)+\hbar\alpha\hat{\chi}^{+2}\hat{\chi}^{2}+\mathrm{i}\hbar\left(E\hat{\phi}^{+}-E^{*}\hat{\phi}\right). \quad (1)$$

Here $\hat{\phi}$ ($\hat{\phi}^{+}$) and $\hat{\chi}$ ($\hat{\chi}^{+}$) annihilates (creates) photon and exciton respectively. The quantity ω_{R} corresponds to the strength of the coherent coupling between the excitonic and the photonic modes and corresponds to the half of Rabi splitting. The quantity α sets the strength of two-body exciton-exciton interaction. The coherent CW driving E is spatially homogeneous and intense. So we take it as a classical force directly exciting the photon field, $E=E_{\mathrm{d}}\mathrm{e}^{-\mathrm{i}\omega_{\mathrm{d}}t}$ with the frequency ω_{d} which is close but not necessary equals to the photonic resonance of the micropillar. Thus, the value of modulus square $\left|E_{\mathrm{d}}\right|^{2}$ is proportional to the intensity of the driving laser.

The considered system is open in the sense that it interacts with the environment which acts as a sink for both photon and exciton fields balancing the driving gain. Such an interaction also leads to the rise of noise according to the driving-dissipative theorem. Accounting of this noise implies treating the environment as a thermal baths that is why we address the noise of such kind as "thermal". Since the current experiments with exciton-polariton supporting microstructures are carried out under rather low temperatures (about units of Kelvins for GaAs-based systems, for instance), the level of thermal noise is rather low. Actually, the noise strength is proportional to the thermal occupation of the bath which exponentially falls as the ratio $\hbar\omega_{\mathrm{ph,ex}}/k_{\mathrm{B}}T$ increases. So we neglect by the effect of thermal noise since $\hbar\omega_{\mathrm{ex}}\sim\hbar\omega_{\mathrm{ph}}\gg k_{\mathrm{B}}T$. We also exclude the factor of noise originating from the apparatus such as fluctuations of the driving intensity [4]. Thus, the only source of noise that we account for is the so-called quantum noise which influence is expected to be amplified by the strong nonlinear interactions in the exciton subsystem.

Nevertheless, the impact of the environment can not be completely neglected since it evokes fast damping of both modes. The presence of the damping affects essentially dynamical and steady-state properties of the system. We treat losses of the exciton-photon system in the Born-Markov approximation which leads to the following master equation for the density operator:

$$\frac{\partial\rho}{\partial t}=\frac{1}{\mathrm{i}\hbar}\left[\hat{H}_{C},\rho\right]+\gamma_{\mathrm{ph}}\left(2\hat{\phi}\rho\hat{\phi}^{+}-\rho\hat{\phi}^{+}\hat{\phi}-\hat{\phi}^{+}\hat{\phi}\rho\right)+\gamma_{\mathrm{ex}}\left(2\hat{\chi}\rho\hat{\chi}^{+}-\rho\hat{\chi}^{+}\hat{\chi}-\hat{\chi}^{+}\hat{\chi}\rho\right), \quad (2)$$

where γ_{ph} and γ_{ex} are the damping rates for photonic and excitonic modes respectively.

The semiclassical treatment of the equations (2) and (1) implies the mean field description. It operates with the average values instead of operators, $\chi=\langle\hat{\chi}\rangle$ and $\phi=\langle\hat{\phi}\rangle$, and postulates factorization of the averaged products as the products of averaged, in particular $\langle\chi^{+}\chi\chi\rangle\to|\chi|^{2}\chi$. The corresponding dynamical equations in the frame rotating with the driving frequency are

$$\frac{\partial}{\partial t}\phi=-\left(\mathrm{i}(\Delta-\Omega)+\gamma_{\mathrm{ph}}\right)\phi+E_{\mathrm{d}}-\mathrm{i}\omega_{R}\chi, \quad (3a)$$

$$\frac{\partial}{\partial t}\chi=-\left(-\mathrm{i}(\Delta+\Omega)+\gamma_{\mathrm{ex}}\right)\chi-\mathrm{i}\omega_{\mathrm{R}}\phi-2\mathrm{i}\alpha|\chi|^{2}\chi, \quad (3b)$$

where $\Delta = (\omega_{ph} - \omega_{ex})/2$ is a half of the exciton-photon detuning and $\Omega = \omega_d - (\omega_{ph} - \omega_{ex})/2$ is detuning of the driving field from the central frequency between the exciton and the photon resonances.

The system (3) is known to sustain a couple of stable steady-state solutions, $\partial\chi/\partial t = 0$, $\partial\phi/\partial t = 0$ [5], reflecting the bistable optical response of the microcavity which manifests itself in a hysteretic behaviour of the output light intensity [3] – see figure 1(a). Originating from competition between the losses and the positive nonlinear feedback of the microcavity excitons [4] the optical bistability occurs within the definite range of the driving field intensities.

The existence of the bistable regime is determined by both the exciton-photon detuning and the frequency of the driving field [5, 6]. The bistability domain on the parameter plane of Δ and Ω is shown on the figure 1(b) – see [7]. Color of the filling corresponds to the value of the threshold driving intensity I^{th} above which the upper stable solution exists (see the panel (b)).

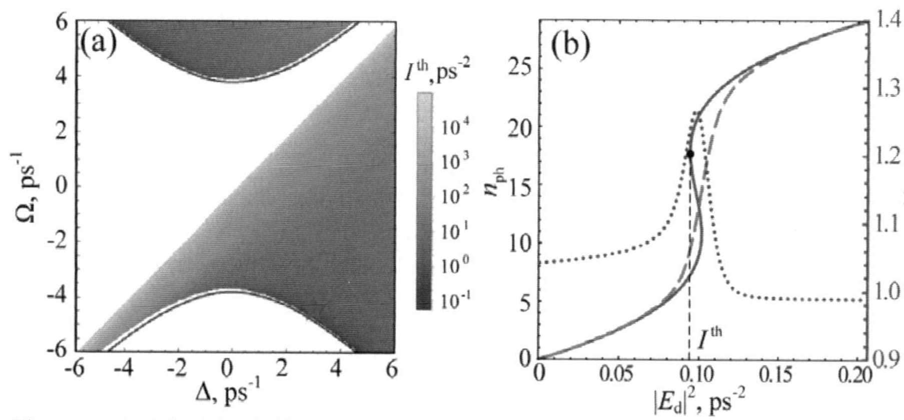

Figure 1. (a) The bistability map on the parameter plane of Δ and Ω. Shaded region corresponds to the existence of bistability. Color of the filling corresponds to the lower threshold of the driving field intensity within the bistability loop (see panel (b)). The parameters are $\hbar\alpha = 0.001$ meV, $\hbar\omega_R = 2.5$ meV, $\gamma_{ex} = 0.01$ ps^{-1} and $\gamma_{ph} = 0.1$ ps^{-1}. (b) Population of the photonic mode predicted within the mean field approach (blue curve) and by the unique quantum solution (4) (red dashed curve) for $\Delta = 3.1$ ps^{-1} and $\Omega = -4.85$ ps^{-1} (blue star on the panel (a)). The green dotted line corresponds to the second order coherence function of the photon field for zero delay.

There are two domains where the bistability exists. The shape of lower boundaries of these domains can be explained with the use of polariton states which are linear eigenstates of the Hamiltonian (1) arising from the strong coupling between excitons and photons. Dispersions of the two polariton branches are $\omega_{LP,UP} = \pm\sqrt{g^2 + \Delta^2}$, where LP (UP) denotes the lower (upper) polariton state – blue solid lines in figure 1(a). The bistability occurs if the driving frequency is blue detuned in respect to the polariton resonance. Actually, the increase of the driving intensity leads to the blue shifting of the energy of the state until it locks in resonance with the driving. This results in switching from the lower to the upper state of the bistability loop. So, the bistability domains lie above both polaritonic branches and are shifted upwards on the amount of frequency which is proportional to the level of losses, see [3].

In the context of mean field theory the upper and the lower states of the bistability loop are dynamically stable [7]. It implies that these states live infinitely long in the presence of small perturbations of the amplitudes of the exciton and the photon fields. However, generally, this statement can't be extended to the presence of noise. If the noise, either purely quantum or originating from the apparatus, is strong, it forces the system to switch from one of the stable branches to the other. These stochastic switches are clearly observed experimentally [4,8,9].

Besides stochastic dynamics, the statistical properties of light reemitted by the microcavity can be also affected by quantum noise. Within the mean field approach both excitonic and photonics modes are supposed to be in the coherent state since they inherit its property from the coherent driving field. This assumption seems reasonable under high level of pumping when quantum fluctuations are negligible. However it should be verified for an arbitrary intensity of the coherent driving and both within and out of bistability existence domain shown in figure 1(b).

Both of previously mentioned phenomena have been already treated in the context of dissipative phase transition in a microcavity polariton system [10]. Experimental studies of the statistics of the microcavity radiation under quasiresonant driving reveal [8,9] both random switching between bistability branches and non-classical second-order correlations of the emitted photonic field. The obtained results demonstrate good coincidence with theoretical predictions. Note that the developed approaches treated a single mode problem when the driving is tuned closely to the low polariton branch and the upper polariton state can be carefully omitted. In this paper we, on the contrary, does not restrict ourselves with the lower polariton branch and use a complete two-component model of the coupled excitonic and photonic fields.

3. Quantum statistics of light in semiconductor microcavity

To get access into quantum properties of the exciton-photon system beyond mean field we use an approach of positive P-distribution developed by P. Drummond [1]. Converting the master equation (2) into partial differential equation for the distribution we get the Fokker-Planck equation for the P-function. The latter appears when expressing the density operator ρ in the basis of coherent states of the photonic and the excitonic fields, see [2,7] for more details. In order to make the diffusion matrix positively definite the phase space should be doubled [1]. The obtained Fokker-Planck equation is written for the quasiprobability Glauber-Sudarshan P-function which phase space dimension is eight. This equation can be solved with the method of potentials. However to satisfy the condition of potentiality a so-called adiabatic approximation is necessary. It implies that excitonic and photonic modes possess greatly unequal dumping rates. Although the values of γ_{ph} and γ_{ex} vary in a wide range, the condition $\gamma_{\mathrm{ph}} \gg \gamma_{\mathrm{ex}}$ is satisfied for a typical micropillar structure. In this case the strongly dissipative photonic mode can be adiabatically eliminated that leads to a reduced Fokker-Planck equation yielding the following steady-state solution for the quasi-probability distribution in the steady-state:

$$P_{\mathrm{ss}} = \chi^{-(\mathrm{i}\Upsilon/\alpha+2)}(\chi^{+})^{\mathrm{i}\Upsilon^{*}/\alpha-2}\cdot \mathrm{e}^{2\chi\chi^{+}-(\sigma E_{\mathrm{d}}/\chi+\sigma^{*}E_{\mathrm{d}}^{*}/\chi^{+})}, \tag{4}$$

where $\Upsilon = \gamma_{\mathrm{ex}} + \omega_{\mathrm{R}}{}^{2}\gamma_{\mathrm{ph}}\Big/\left(\gamma_{\mathrm{ph}}^{2} + (\Delta-\Omega)^{2}\right) - \mathrm{i}\left(\Delta+\Omega+(\Delta-\Omega)\omega_{\mathrm{R}}{}^{2}\Big/\left(\gamma_{\mathrm{ph}}^{2}+(\Delta-\Omega)^{2}\right)\right)$ and $\sigma = \omega_{\mathrm{R}}\alpha^{-1}\left(\gamma_{\mathrm{ph}} - \mathrm{i}(\Delta-\Omega)\right)\Big/\left(\gamma_{\mathrm{ph}}^{2}+(\Delta-\Omega)^{2}\right)$. Note that χ and χ^{+} have a meaning of c-numbers, while the asterisk demotes complex conjugate algebraic numbers. Solution (4) gives the values of normally ordered quantum averages $G^{(\mathrm{mn})} = \left\langle (\chi^{+})^{\mathrm{m}}\chi^{\mathrm{n}}\right\rangle$ which are found as moments of the P-distribution:

$$G^{(\mathrm{mn})} = -\frac{(-\sigma E_{\mathrm{d}})^{\mathrm{n}}(-\sigma^{*}E_{\mathrm{d}}^{*})^{\mathrm{m}}\,\Gamma(\mathrm{i}\gamma^{*}/\alpha)\Gamma(-\mathrm{i}\gamma/\alpha)}{\Gamma(\mathrm{m}+\mathrm{i}\gamma^{*}/\alpha)\Gamma(\mathrm{n}-\mathrm{i}\gamma/\alpha)}\frac{{}_{0}\mathrm{F}_{2}\left(\mathrm{m}+\mathrm{i}\gamma^{*}/\alpha,\mathrm{n}-\mathrm{i}\gamma/\alpha,2\sigma^{2}|E_{\mathrm{d}}|^{2}\right)}{{}_{0}\mathrm{F}_{2}\left(\mathrm{i}\gamma^{*}/\alpha,-\mathrm{i}\gamma/\alpha,2\sigma^{2}|E_{\mathrm{d}}|^{2}\right)}, \tag{5}$$

where Γ is Gamma function and ${}_{0}\mathrm{F}_{2}$ is hypergeometric function.

Steady-state quasiprobability distribution (4) together with expression (5) provides complete information about the excitonic subsystem. However the exciton statistics is hardly accessible experimentally. The valuable information about the system state is usually obtained from the statistical properties of the emitted light. In particular, we are interested in the second order correlation function of the photonic field $g_{\mathrm{ph}}^{(2)}(\tau)=\dfrac{\left\langle\hat{\phi}^{+}(t)\hat{\phi}^{+}(t+\tau)\hat{\phi}(t)\hat{\phi}(t+\tau)\right\rangle}{\left\langle\hat{\phi}^{+}(t)\hat{\phi}(t)\right\rangle\left\langle\hat{\phi}^{+}(t+\tau)\hat{\phi}(t+\tau)\right\rangle}$. Early we claim that $\gamma_{\mathrm{ph}}\gg\gamma_{\mathrm{ex}}$. This condition guarantees that photonic field fluctuations are slaved by the excitonic mode [11]. Thus we express the value of the photonic field amplitude (in *c*-numbers) as

$$\phi=\frac{E_{\mathrm{d}}\left(\gamma_{\mathrm{ph}}-\mathrm{i}(\Delta-\Omega)\right)}{\left((\Delta-\Omega)^{2}+\gamma_{\mathrm{ph}}^{2}\right)}-\frac{\omega_{\mathrm{R}}\left(\mathrm{i}\gamma_{\mathrm{ph}}+(\Delta-\Omega)\right)}{\left((\Delta-\Omega)^{2}+\gamma_{\mathrm{ph}}^{2}\right)}\chi \tag{6}$$

and then obtain the following expression for the second order coherence function for zero delay $\tau=0$,

$$g_{\mathrm{ph}}^{(2)}(0)=\frac{E_{\mathrm{d}}^{4}-E_{\mathrm{d}}^{2}\omega_{\mathrm{R}}^{2}\left(G^{(02)}+G^{(20)}-4G^{(11)}\right)-2\mathrm{i}\omega_{\mathrm{R}}E_{\mathrm{d}}^{3}\left(G^{(01)}-G^{(10)}\right)+2\mathrm{i}\omega_{\mathrm{R}}^{3}E_{\mathrm{d}}\left(G^{(21)}-G^{(12)}\right)+\omega_{\mathrm{R}}^{4}G^{(22)}}{\left(E_{\mathrm{d}}^{2}+2\omega_{\mathrm{R}}E_{\mathrm{d}}\,\mathrm{Im}\left[G^{(01)}\right]+\omega_{R}^{2}G^{(11)}\right)^{2}}. \tag{7}$$

Behaviour of the second order correlation function of the fluctuating field in the presence of Kerr-like nonlinearity was investigated previously [2]. That results should be also valid for the considered two mode model provided that driving is tuned closely to the lower (or upper) polariton resonance thereby justifying single mode approximation, see [8,10]. The behaviour of $g_{\mathrm{ph}}^{(2)}(0)$ is determined by the stochastic dynamics of the field affected by the quantum noise. When the driving intensity is within the bistability cycle, quantum fluctuations initiate chaotic jumps between the two metastable states. The characteristic time scale of these jumps can be determined from the Liouvillian gap and crucially depends on the system parameters of the system [10]. The presence of such stochastic switching of the field intensity affects significantly the mean photon number and the statistical properties of the emitted light.

The unique quantum solution (4) predicts that the averaged value of the cavity mode population, $n_{\mathrm{ph}}=\left\langle\phi^{+}\phi\right\rangle$, lies between two metastable states predicted by the mean field theory – figure 1(b). The discrepancy between the quantum and the mean field solutions occurs for the driving intensities which are usually within the bistability loop. However for some specific parameters this variance appears also in the region where only single solution exists and any jumps between the metastable states are impossible. This very case is shown in figure 1(b). Such behaviour of the mean photon number indicates that the coherent state approximation underling the quasiclassical equations (3) is violated and the cavity mode is not in a coherent state due to the presence of quantum noise.

The nonclassical behaviour of $g_{\mathrm{ph}}^{(2)}(0)$ usually appears as a bunching peak, $g_{\mathrm{ph}}^{(2)}(0)>1$, under parameters where the quantum and quasiclassical values of the photon density differ, see the green dotted curve in figure 1(b).

The discussed results were obtained for the coupled exciton-photon system. However they are reproduced well with the single-mode polariton model provided that the driving field frequency is close to the lower dispersion branch. To demonstrate the capability of the developed two-mode approach we consider the region where the single mode approximation fails. In particular, we consider the domain around triple resonance of the cavity mode, the excitonic state and the driving frequency, $\Delta=0$ and $\Omega=0$. The map of the zero-delay second order coherence function $g_{\mathrm{ph}}^{(2)}(0)$ of the intracavity field on the parameter plane of Δ and Ω is shown in figure 2(a). The driving field is fixed, $E_{\mathrm{d}}=25$ ps^{-1}. The obtained dependence has several interesting peculiarities. Firstly, the giant bunching effect, $g_{\mathrm{ph}}^{(2)}(0)\gg1$, occurs for the parameters lying close to the line determined by the relation $\Delta=-\Omega$. Secondly, both bunching and antibunching effects are observed for the region where bistability is absent (above the white dotted line in figure 2(a)).

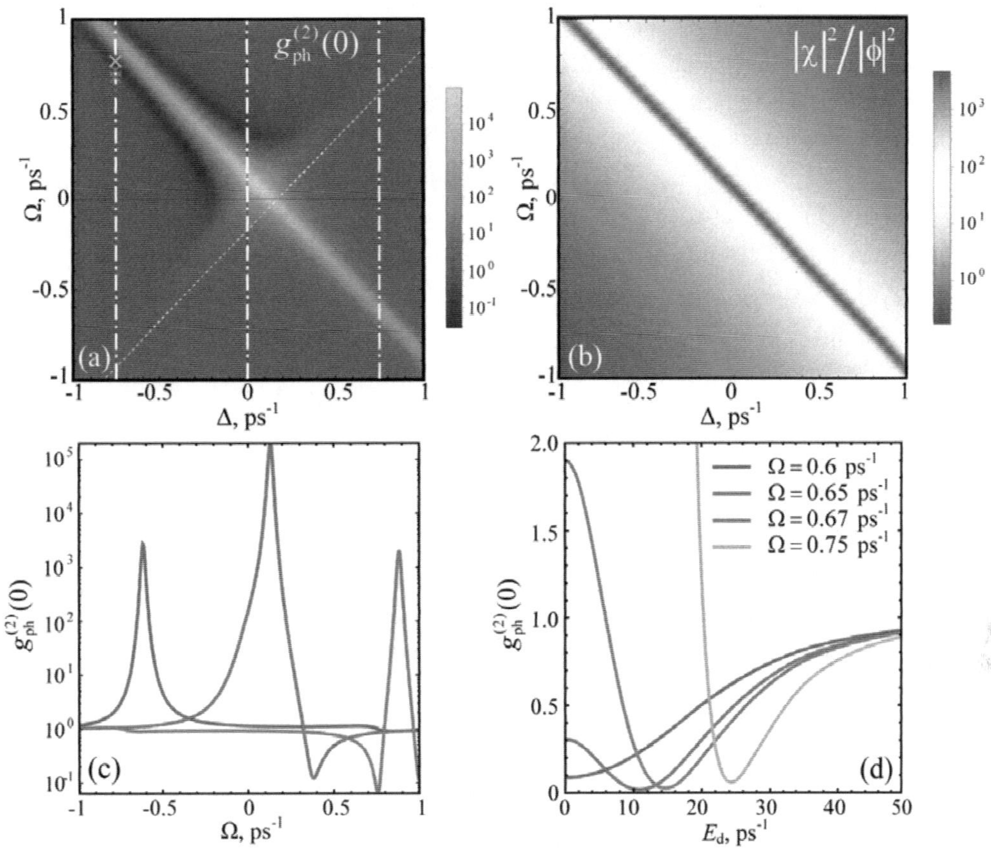

Figure 2. (a) The second order coherence function of the photon field for zero delay $g^{(2)}_{ph}(0)$ for various values of the driving detuning Ω and the exciton-photon detuning Δ. (b) The ratio between populations of the excitonic and the photonic modes $|\chi|^2/|\phi|^2$ on the same parameter plane as in the panel (a). (c) The dependence of $g^{(2)}_{ph}(0)$ on the driving detuning Ω for the fixed values of the exciton-photon detuning, shown by the vertical dot-dashed lines in the panel (a), $\Delta=-0.75$ ps^{-1} (green), $\Delta=0$ ps^{-1} (red) and $\Delta=0.75$ ps^{-1} (blue). (d) The dependence of $g^{(2)}_{ph}(0)$ on the driving amplitude E_d for $\Delta=-0.75$ ps^{-1} and the fixed values of the driving detuning (color crosses in the panel (a)), $\Omega\approx 0.6$ ps^{-1} (blue), $\Omega\approx 0.65$ ps^{-1} (red), $\Omega\approx 0.67$ ps^{-1} (green) and $\Omega\approx 0.75$ ps^{-1} (yellow).

The dependences of $g^{(2)}_{ph}(0)$ on the driving detuning Ω for the three different values of the exciton-photon detuning, $\Delta=-0.75$ ps^{-1}, $\Delta=0$ ps^{-1} and $\Delta=0.75$ ps^{-1}, are shown in figure 2(c). Formation of the giant bunching peaks can be attributed to a high difference between the populations of the photonic and the excitonic states. Our analysis reveals that the ratio $|\chi|^2/|\phi|^2$ maximizes under parameters corresponding to the giant bunching effect, see the panel (b) of figure 2. Since in this case the number of excitons greatly exceeds the population of the photonic mode, any fluctuation of the exciton state amplitude leads to the strong perturbation of the photon density because these subsystems are coherently coupled. This sudden inflation of the intracavity field manifests the presence of the photon bunching effect.

The antibunching behaviour, $g^{(2)}_{\mathrm{ph}}(0) < 1$, of the photon statistics appears in vicinity of the giant bunching peak in the region of single metastable solution, see the panel (a) of figure 2 and the green curve in the panel (c). The behaviour of the second order coherence function in this region strongly depends on the driving field. The dependences of $g^{(2)}_{\mathrm{ph}}(0)$ on the amplitude E_{d} for various driving detunings Ω and fixed $\Delta = -0.75$ ps^{-1} are demonstrated in figure 2(d). For the small values of the driving detuning (below $\Omega \approx 0.6$ ps^{-1} for the discussed parameters) the photon field demonstrates antibunching under weak driving, see the blue curve. The growth of the driving amplitude E_{d} brings the system into the coherent state, $g^{(2)}_{\mathrm{ph}}(0) \to 1$. However, as the driving detuning approaches the region of the high disbalance between the photon and the exciton populations (figure 2(b)) the effect of bunching increases (the green and the yellow curves).

4. Conclusion

Quantum fluctuations may significantly affect both the expectation values and the statistical properties of the light emitted from the coherently driving microcavity. The recent experimental studies elucidate an intriguing physics of dissipative phase transitions with the microcavity polaritons and its stochastic dynamics driven by the quantum noise. The approaches developed early were based on the single mode approximation and thus is relevant only in the case when the driving intensity is tuned in the vicinity of low polariton branch. In this paper we used the more general approach considering a coupled exciton-photon system. The rigorous quantum solution for the quasiprobability function demonstrates nontrivial statistical properties of the photon field. In particular both the antibunching and the giant bunching effects were predicted in the region close to the upper boundary of the mean field bistability domain.

Acknowledgment

This work was supported by the RFBR Grants No. 16-32-60102, 15-59-30406, 15-52-52001 and partially by the President of Russian Federation for state support of young Russian scientists, Grant MK-2988.2017.2. The work was also supported by Ministry of Education and Science of the Russian Federation, project no. 16.1123.2017/4.6.

References

[1] Drummond P D, Gardiner C W 1980 Generalized P-representations in quantum optics *Journal of Physics* **3** 2353–2368

[2] Drummond P D, McNeil K J, Walls D F 1981 Non-equilibrium transitions in sub/second harmonic generation. *Optica Acta: International Journal of Optics* **25**(2) 211–225

[3] Bass A, Karr J Ph, Giacobino E 2004 Optical bistability in semiconductor microcavities *Physical Review* A **69** 023809

[4] Abbaspour H, Sallen G, Trebaol S, Morier-Genoud F, Portella-Oberli M T and Deveaud B 2015 Effect of a noisy driving field on a bistable polariton system *Phys. Rev.* B **92** 165303

[5] Yulin A V, Egorov O A, Lederer F and Skryabin D V 2008 Dark polariton solitons in semiconductor microcavities *Phys. Rev.* A **78** 061801

[6] Werner A, Egorov O A and Lederer F 2014 Exciton-polariton patterns in coherently pumped semiconductor microcavities *Phys. Rev.* B **89** 245307

[7] Demirchyan S S, Khudaiberganov T A, Chestnov I Yu, Alodjants A P 2017 Quantum fluctuation in the system of exciton polaritons in semiconductor microcavities *J. Opt. Technol.* **84** 75-81

[8] Rodriguez S R K, Casteels W, Storme F, Zambon N C, Sagnes I, Le Gratiet L, Bloch J. 2017 Probing a dissipative phase transition via dynamical optical hysteresis *Phys. Rev. Lett.* **118**(24) 247402.

[9] Fink T, Schade A, Höfling S, Schneider C, İmamoğlu A 2017 Signatures of a dissipative phase

transition in photon correlation measurements *arXiv:*1707.01837
[10] Casteels W, Fazio R and Ciuti C 2017 Critical dynamical properties of a first-order dissipative phase transition *Phys. Rev.* A, **95**(1) 012128
[11] Gardiner C W 1986 Handbook of stochastic methods for physics, chemistry and the natural sciences *Applied Optics* **25** 3145

IOP Conf. Series: Journal of Physics: Conf. Series **951** (2017) 012033 doi:10.1088/1742-6596/951/1/012033

The method of surface plasmon-polariton pulses generation via cooperative effects in waveguide spaser

A. V. Shesterikov[1], **M. Yu. Gubin**[1], **M. G. Gladush**[2], **A. V. Prokhorov**[1]

[1]Vladimir State University named after Alexander and Nikolay Stoletovs, Vladimir, 600000 Russia
[2]Institute of Spectroscopy, Russian Academy of Sciences, Troitsk, Moscow, 108840 Russia

E-mail: avprokhorov33@mail.ru

Abstract. The problem of sub-picosecond plasmon-polariton pulse formation in metal/dielectric interface due to collective decay of excited quantum dots, placed in the dielectric layer near the metal surface, is considered. Theoretical approach to selection of semiconductor quantum dots and dielectric host medium to increase the energy transmission of quantum dot collective excitations into surface plasmon-polariton modes of waveguide spaser is developed.

1. Introduction

Collective processes in a system of quantum emitters for a long time remain a subject of intensive studying [1] with both theoretical and experimental points of view. New opportunities of the well-known cooperative effects in optics can be associated with the collective behavior of the plasmonic oscillators pumped by a near-field of excited chromophores (semiconductor quantum dots, dye molecules, etc. [2]). The kinematics of individual localized systems "quantum dot + metal nanoparticle" [3], the core-shell nanocrystals [4] is well described in the framework of spaser theory [2]. However, generated plasmons in such systems are strongly localized and their collective dynamic is restricted to the near-field area of the plasmonic nanoparticles [5]. Suitable interfaces for observing collective processes with surface plasmon-polariton (SPP) can be planar metal/dielectric waveguides [6] which were already implemented in practice [7]. One of approaches to solving the problem of fast damping of plasmons in such systems is connected with use of photonic crystals as a dielectric layer [8]. In this case, the long-range SPPs with a maximum energy of the field in the dielectric region are formed. On the other hand, compensation of damping of plasmons in metal can be realized in the model of the dissipative waveguide spaser with a near-field pumping from the chromophores placed in the dielectric layer near a metal surface.

In this work the approach to choosing specific chromophores and the dielectric host medium to increase the energy transmission efficiency of collective excitations of chromophores to SPPs modes in metal/dielectric waveguide is proposed. Considering that the refractive index of the dielectric host medium is a complex value, we have defined such conditions when the spontaneous emission rate of the chromophores near the metal-dielectric interface [9], as well as the collective optical processes with quantum emitters [10, 11] are almost completely suppressed

IOP Conf. Series: Journal of Physics: Conf. Series **951** (2017) 012033 doi:10.1088/1742-6596/951/1/012033

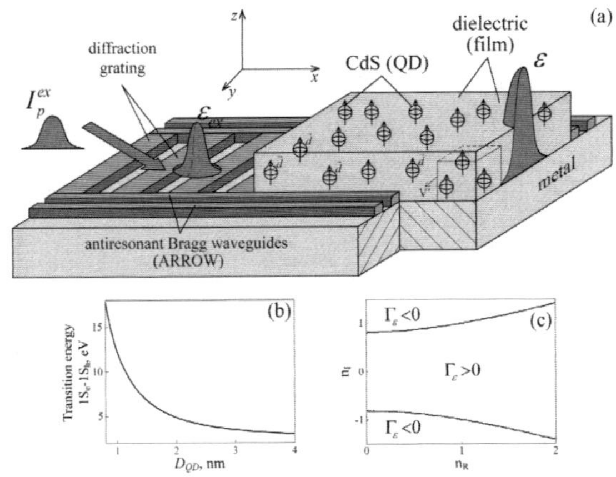

Figure 1. (a) Formation scheme of SPP pulses in a layered (planar) metal/dielectric waveguide pumped by CdS QDs; (b) dependence of the transition energy on the CdS QD size ($E_g = 2.42$ eV at 0 K for a bulk); (c) parametric plane of the complex refractive index $n = n_R + in_I$ of a dielectric medium with separatrices $\hat{}_{\varepsilon} = 0$ for the effective rate of radiative losses of QDs in this medium.

by the influence of the dielectric environment. Using model of the waveguide spaser the self-consistent system of the equations describing dynamics of excitons and propagated SPP pulses was obtained. It is shown that in the mean-field approximation the self-consistent problem can be reduced to a modified pendulum equation with an additional term of nonlinear losses. A separatrix solution of the nonlinear equation, which corresponds to the formation of the single SPP pulse in waveguide spaser model, was obtained. A model of the waveguide spaser with an ensemble of CdS quantum dots placed in the dielectric layer near the metal surface for the realization of the predicted effects was proposed.

2. Master equation for collective process of SPP generation in wavegide spaser

Consider the model of an interface in Fig. 1a in the form of a metal/dielectric waveguide [12] with two-level chromophores located inside a thin dielectric layer, the transition frequency between the two levels $\omega_a = 2\pi c/\lambda_a$ being resonant with the SPP frequency $\omega_{SPP} = 2\pi c/\lambda_{SPP}$ (1-ground and 2-upper levels, respectively). By selecting a dielectric medium with appropriate dispersion characteristics and providing the initial excitation (inversion) of a dense ensemble of chromophores in this model, it is possible to produce the collective decay of excitons. We assume that the characteristic size of the interaction region of the effective field of SPP and chromophores $h = L_x = L_y = L_z$ satisfies the inequality $h \ll \lambda_p$ and the inequality $L_z \ll l_d$ is also valid, where l_d is the SPP decay length along the z axis. The corresponding Rabi frequencies can be written in the form $\Omega = -\left(A\nabla\varphi\mu_{12}\varepsilon\right)/\hbar$, where $A = \sqrt{\hbar S/\left(\varepsilon_0\varepsilon_d\frac{\partial S}{\partial\omega}\right)}$, $\varepsilon = \sqrt{N_p}$ is the SPP amplitude, N_p is the number of SPP modes in the interaction region, μ_{12} is the transition dipole moment in a chromophore, and φ is the scalar potential of the plasmon field linearly decreasing with distance from the surface, $\hbar$ is the Planck's constant. In the case of excitation of a mode of the plasmon field at frequency ω, using the normalization $\int|\nabla\varphi|^2 dV = 1$ [13], the expression for the Rabi frequency can be approximated by the function

$$\Omega = \mu_{12}\sqrt{\frac{S_n}{\hbar\varepsilon_d\varepsilon_0 V\frac{\partial S_n}{\partial\omega}}}\varepsilon = g\varepsilon.$$

For a metal-dielectric boundary, the relation

$$\lambda_{SPP} = \sqrt{\frac{\mathrm{Re}(\varepsilon_m) + \mathrm{Re}(\varepsilon_d)}{\mathrm{Re}(\varepsilon_m)\mathrm{Re}(\varepsilon_d)}} \cdot \lambda_0$$

is valid, where the parameters ε_d and $\varepsilon_m(\bar{\omega}) = 1 - \omega_p^2/\left(\bar{\omega}^2 + i\gamma_s\bar{\omega}\right)$ are the dielectric permittivities of dielectric (with QD) and metal, respectively. Here, $\omega_p = \sqrt{4\pi n_m e^2/m_0}$ is the plasma frequency in a metal, m_0 and n_m are the electron mass and concentration, respectively, γ_s is the electron collision frequency in the metal, $\bar{\omega} = 2\pi c/\lambda_0$. We chose the wavelength $\lambda_0 = 387$ nm for our simulation. The spectral properties of the metal-dielectric interface can be described by use of the Bergman's parameter $S(\omega) = \mathrm{Re}\left(\varepsilon_d/(\varepsilon_d - \varepsilon_m(\omega))\right)$ [2].

We assume that the pumping volume V' is a dielectric containing QDs with the characteristic radius a and concentration $N \gg 10^{21}$ m^{-3}. Assuming that the refractive index $n = n_R + in_I$ of the dielectric environment of QDs is a complex quantity, where $n = \sqrt{\varepsilon_d}$ and ε_d is the complex permittivity, expressions for the radiative relaxation rate $\hat{\ }_a$, the Rabi frequency Ω, and the effective frequency detuning Δ_a can be written in the form [14]

$$\hat{\ }_\varepsilon = \hat{\ }_a \left(n_R l_R - n_I l_I + 2\frac{\delta_a}{\hat{\ }_a} l_I \right), \quad (1a)$$

$$\Omega_0 = \Omega \cdot \sqrt{l_R^2 + l_I^2}, \quad (1b)$$

$$\Delta_\varepsilon = \delta_a \left(l_R - \frac{\hat{\ }_a}{2\delta_a} \left(n_I l_R + n_R l_I \right) \right) + \Delta_a, \quad (1c)$$

where $l(n) = l_R + il_I$ is a complex function for which $l_R = \left(n_R^2 - n_I^2\right)/3$, $l_I = 2n_R n_I/3$; and δ_a is a small correction caused by the Lamb shift. It is assumed here that the function $l(n) = E_l/E_M$ coupling the Lorentz local E_l and Maxwell E_M fields will retain its structure in the case of the near field through which SPP are excited in the scheme in Fig. 1.

The parameter $\hat{\ }_a^* = 1/\tau_R + 1/\tau_F$ is the total rate of radiative (with the time $\tau_R = 1/\hat{\ }_a$) and nonradiative (with the time τ_F) losses for QDs in vacuum. In the semiclassical approximation, the system can be described similarly to the metal nanoparticle in a dielectric with chromophores spaser model [2] with the help of equations for elements of the density matrix ρ of a two-level chromophore:

$$\begin{aligned} \dot{\rho}_{12} &= -(i\Delta_\varepsilon + \frac{\hat{\ }_\varepsilon}{2})\rho_{12} \\ &+ (i\Omega_0^* + i\xi_0 u_R \rho_{12} + \xi_0 u_I \rho_{12})\, n_{21}, \quad (2a) \\ \dot{n}_{21} &= 2i\left(\Omega_0 \rho_{12} - \Omega_0^* \rho_{21}\right) \\ &- 4\xi_0 u_I \left|\rho_{12}\right|^2 - \hat{\ }_\varepsilon \left(1 + n_{21}\right), \quad (2b) \end{aligned}$$

where $\Delta_a = 2\pi c\left(1/\lambda_a - 1/\lambda_{SPP}\right)$, $n_{21} = \rho_{22} - \rho_{11}$. The Rabi frequency can be written as $\Omega_0 = g\varepsilon \cdot \sqrt{l_R^2 + l_I^2}$, where $g = \mu_{12}\sqrt{S_n/\left(\hbar\varepsilon_d\varepsilon_0 V\frac{\partial S_n}{\partial\omega}\right)}$ is the coupling constant and $\varepsilon = A_p\sqrt{\varepsilon_d\varepsilon_0 V\frac{\partial S_n}{\partial\omega}/\left(\hbar S_n\right)}$ is the normalized field with the amplitude A_p of the total field produced by the perturbed electron density in a metal and the electromagnetic field component in a dielectric.

The parameter $\xi_0 = N\mu_{12}^2/\left(3\hbar\varepsilon_0\right)$ in (2) determines the addition to the Rabi frequency appearing due to transition from the Maxwell E_M to the local field E_l [14] acting on a chromophore.

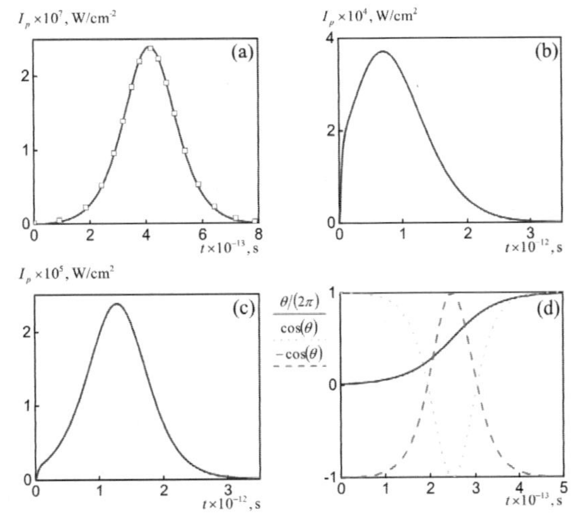

Figure 2. (a) Profiles of SPP pulse amplitude squared obtained by the numerical simulation of system (2)–(3) in the following regimes: (a) neglecting the dissipation parameters $\Gamma_\varepsilon = \Gamma_a = \gamma_p = 0$ s^{-1}; (b) the dissipative regime with $\Gamma_\varepsilon = \Gamma_a = 6.3 \times 10^{11}$ s^{-1}, $\gamma_p = 4.1 \times 10^{13}$ s^{-1}; (c) the regime of QD radiative decay suppression $\Gamma_\varepsilon = 0$ in the host medium with $n_R = 1.6$ and $n_I = 1.23$; (d) dynamics of the angle θ (solid curve) and coefficients $\cos(\theta)$ (dotted curve) and $-\cos(\theta)$ (dashed curve) for regime (a). The interaction parameters are $g = 8.1 \times 10^{11}$ s^{-1}, $\xi_0 = 6.3 \times 10^9$ s^{-1}. The initial polarization of the medium is $\rho_{12}(0) = i\theta_0 = i/\sqrt{N_a} = 0.14i$, the normalization parameter is $\Lambda = 5.7 \times 10^{12}$ s^{-1} for $N_a = 50$.

The dispersion and dissipative corrections

$$u_R = (l_R\varepsilon_R + l_I\varepsilon_I) / \left(\varepsilon_R^2 + \varepsilon_I^2\right),$$

$$u_I = (l_I\varepsilon_R - l_R\varepsilon_I) / \left(\varepsilon_R^2 + \varepsilon_I^2\right),$$

respectively, are expressed in terms of the real and imaginary parts of the permittivity of the host-medium [14] in which QDs are placed and have the physical meaning of the additional frequency modulation and the effects of absorption ($u_I < 0$) or amplification ($u_I > 0$) due to the local field (Fig. 2). The equation of motion for the Rabi frequency of SPP, which in the case of the exact plasmon resonance has the form

$$\dot{\Omega}_0 = -\frac{i}{t_R^2}\rho_{12} - \gamma_p\Omega_0, \tag{3}$$

where

$$t_R = \frac{1}{g\sqrt{N_a}} = \sqrt{\frac{\hbar\varepsilon_d\varepsilon_0\frac{\partial S_n}{\partial\omega}}{S_n\mu_{12}^2 N}}$$

determines the characteristic formation time for quantum correlations in Fig. 1a (compare with the optical problem [15] when emitters are located in the field formation region).

Note that the plasmon mode decay rate $\gamma_p = 1/\tau_J + 1/\tau_R$ is high and determined by the characteristic times τ_R and τ_J of radiative and joule losses, respectively. Under conditions $1/\tau_J \approx 30/\tau_R$ [16], radiative losses can be neglected, while joule losses are determined by the collision frequency in a metal, i.e., $\gamma_p \approx \gamma_s$, and in problem (3) in the absence of pump, the short-range SPP appear. The self-consistent problem (2)–(3) will be valid only under conditions when the characteristic establishment time t_R for correlations between QDs proves to be considerably shorter than τ_J. Because t_R is inversely proportional to the dipole moment of a chromophore, the relation $t_R < \tau_J$ can be valid for pumping a distributed waveguide spaser by QDs with their giant dipole transition moments.

We use the known dependence [17] of the $1S\,(e) \to 1S\,(h)$ transition energy on the QD diameter $D_{QD} = 2a$ (Fig. 1b) for regime of strong confinement

$$E_{1S(e)-1S(h)} = E_g + 2\frac{\hbar^2\pi^2}{D_{QD}^2}\left(\frac{1}{m_e} + \frac{1}{m_h}\right), \tag{4}$$

where e is the electron charge, m_e and m_h are the effective electron and hole masses, respectively, in the volume of the QD material with the permittivity ε_{qd} and band gap energy E_g [18, 19]. The corresponding parameters for CdS are $m_e = 0.19m_0$, $m_h = 0.8m_0$ and $\varepsilon_{qd} = 9$ [20], which gives $D_{QD} = 1.56$ nm. Bohr radius of exciton R_{ex} for CdS is 2.5 nm [21] therefore strong confinement regime [22] will be observed for the considered QDs, and energy sublevels of conductivity zone will be essentially separated. The dipole moment of the corresponding interband transition in QDs is assumed equal to $\mu = \mu_{12} = 5 \times 10^{-29}$ C $\cdot$ m [2].

For chosen model parameters and the QD concentration $N = 7 \times 10^{21}$ m^{-3}, the characteristic correlation time is $t_R = 124$ fs and the delay time $t_D = 414$ fs for the number of chromophores in the interaction region $N_a = 50$ (here we assume that $N_a = N_p$). The duration of a formed SPP monopulse is only about 200 fs for ideal conditions $\hat{}_{\varepsilon} = \hat{}_{a} = \gamma_p = 0$ s^{-1} in Fig. 2a. For dissipative regime in Fig. 2b we take into account the rate of decay $\hat{}_{\varepsilon} = 6.3 \times 10^{11}$ s^{-1} for QD near tye metal surface [23] and the rate of decay $\gamma_p = 4.1 \times 10^{13}$ s^{-1} for plasmon. In this case we also find the possibility for the formation of SPP pulses, since the characteristic time t_R shorter than the characteristic decay times in the system.

However, taking (1a) into account, the choice of the appropriate dielectric host-medium can partially or completely compensate the increase of $\hat{}_{a}$ (Fig. 1c), but it is also obvious that the properties of natural media are strongly restricted. Thus, for silica at the wavelength under study $\lambda_{SPP} = 192$ nm, we have $n_R = 1.6$, $n_I = 5 \times 10^{-7}$ [24] and $\hat{}_{\varepsilon} = 2.43\hat{}_{a}$. To completely compensate relaxation processes in (2a) ($\hat{}_{\varepsilon} \equiv 0$), the required combination of dispersion-dissipative parameters should satisfy the condition $n_R l_R - n_I l_I = 0$ (be neglecting a small Lamb shift), which is satisfied, for example, for the choice $n_R = 1.6$ and $n_I = 1.23$. Such conditions can be fulfilled for an artificial microstructured dielectric material with specified dispersion dissipative characteristics (the Cole-Cole diagram). They lead to the significant increase in the SPP pulse intensity, while energy transfer from chromophores to radiation proves to be suppressed (see Fig. 2c). In this case, the influence of the local field increases, the absolute values of its parameters increase (corrections $u_R = 0.37$ and $u_I = -0.158$ in (2)) and the formation dynamics of SPP pulses changes.

3. Collective dynamics of a waveguide spaser in the mean field approximation

To analyze the contribution of dissipative effects related to the imaginary part u_I of the local field correction, we can neglect the corresponding phase effects with u_R in (2) and decay in (2)–(3) and to pass in the mean field approximation to a simplified system of self-consistent equations for a medium

$$\dot{\rho}_{12} = \left(i\Omega_0^* + \xi_0 u_I \rho_{12}\right) n_{21}, \tag{5a}$$

$$\dot{n}_{21} = 2i\left(\Omega_0\rho_{12} - \Omega_0^*\rho_{21}\right) - 4\xi_0 u_I \left|\rho_{12}\right|^2 \tag{5b}$$

and the effective field

$$\dot{\Omega}_0 = -ig^2 N_a \rho_{12} \tag{6}$$

IOP Conf. Series: Journal of Physics: Conf. Series **951** (2017) 012033 doi:10.1088/1742-6596/951/1/012033

formed in it.

By passing to the representation for the Rabi frequency and polarization in the form

$$\Omega_0 = \frac{1}{2}\left(Ue^{-iK_0 t} + U^* e^{iK_0 t}\right),\ \rho_{12} = \frac{1}{2}R \cdot e^{-iK_0}$$

where $K_0 = \omega_{SPP}t - k_{SPP}z$, and assuming that $Z = n_{21}$, we can obtain the system of Maxwell-Bloch equations for a spaser taking into account the (dissipative) local response of the QD environment

$$\dot{Z} = -\frac{i}{2}\left(UR^* - U^*R\right) - \xi_0 u_I \left|R\right|^2, \tag{7a}$$

$$\dot{R} = i\left(U - i\xi_0 u_I R\right)Z, \tag{7b}$$

$$\dot{U} = -ig^2 N_a R. \tag{7c}$$

By passing to new dimensionless variables $\delta_0 = -i\frac{U}{\Lambda}$ and $\tau = t \cdot \Lambda$, where $\Lambda = g\sqrt{N_a}$ and setting $R^* = R$ and $\delta^* = \delta$, we represent system (7) in the form

$$\frac{\partial Z}{\partial \tau} = \delta_0 R - \frac{\xi_0 u_I}{\Lambda}\left|R\right|^2, \tag{8a}$$

$$\frac{\partial R}{\partial \tau} = -\delta_0 Z + \frac{\xi_0 u_I}{\Lambda} RZ, \tag{8b}$$

$$\frac{\partial \delta_0}{\partial \tau} = -R. \tag{8c}$$

The solution of system (8) can be written in the form $Z = \cos(\theta)$ and $R = \sin(\theta)$, where θ determine the angle of the so-called Bloch vector with coordinates Z and R and their substitution to (8) gives the equation for the angle

$$\dot{\theta} = -\delta_0 + \frac{\xi_0 u_I}{\Lambda}\sin(\theta). \tag{9}$$

By substituting the expression for δ_0 from (9) into (8c), we obtain a new variant of the pendulum equation with the nonlinear harmonic losses/decay term

$$\ddot{\theta} - K\cos(\theta) \cdot \dot{\theta} = \sin(\theta), \tag{10}$$

where the amplitude of the decay coefficient is defined as $K = \xi_0 u_I/\Lambda$. In the absence of the loss modulation, when $K\cos(\theta) \cdot \dot{\theta} = K \cdot \dot{\theta}$, Eq. (10) is reduced to the usual nonlinear pendulum equation with losses [25]. Taking the modulation into account under the same conditions $K < 0$ ($\varepsilon_I > 0$ and $u_I < 0$), the pendulum experiences the additional decay in intervals

$$\theta \in \left[0 + 2\pi m; \frac{\pi}{2} + 2\pi m\right],\ \theta \in \left[\frac{3\pi}{2} + 2\pi m; 2\pi + 2\pi m\right]$$

responsible for the formation of the leading and trailing edges of SPP pulse (see Fig. 2c), whereas in the interval

$$\theta \in \left[\frac{\pi}{2} + 2\pi m; \frac{3\pi}{2} + 2\pi m\right]$$

when the central part of SPP pulse is formed, the enhancement of pendulum oscillations is observed; $m = 0, 1, 2....$

In other words, the absorbing dielectric host-medium coherently preserves a part of the QD energy during the formation of the leading edge of the pulse and then coherently returns this energy to SPP pulse during formation of the pulse peak. As a result, taking into account the compensation of the spontaneous relaxation rate of QDs ($\hat{}_{\varepsilon} = 0$) and nonlinear terms with u_I in (5), the increase in the peak pulse intensity is observed with respect to the case when the response of the host-medium is neglected (see Figs. 2a,b).

The initial conditions in simulation of (10) are chosen equal to $\theta_0 = 1/\sqrt{N_a}$ for the initial oscillation angle and

$$v_\theta = \left.\frac{\partial\theta}{\partial t}\right|_{t=0} = \frac{2}{\cosh\left(\ln\frac{\theta_0}{4}\right)}$$

for the initial velocity of the pendulum.

Equation (10) is a particular case of the Lienard equation and its approximate analytic solution can be expressed in terms of elliptic integrals of the first kind. The numerical solution for the Rabi frequency of SPP pulse field obtained from (10) completely coincides with the results of the direct numerical simulation of system (5)–(6) under conditions of the suppression of spontaneous relaxation in QDs for the chosen values $n_R = 1.6$ and $n_I = 1.23$ ($K = -0.0147$) (see Fig. 2a).

4. Conclusions

We have proposed efficient method for the formation of short SPP pulses at the dielectric-metal interface containing QDs. The general conditions for selecting parameters of QDs and a dielectric host-medium are determined which provide the maximal collective energy transfer from a QD ensemble to SPP modes dominating over the radiative relaxation of individual chromophores. To tune the system parameters to the plasmon resonance more accurately, it is useful to employ experimental absorption and fluorescence spectra of of giant ensembles of emitters [26, 27] in different host medium. The models presented in the paper can be useful for practical implementation of multiqubits entanglements [28] and quantum computations in macroscopic and mesoscopic [29] systems. However, for the realization of the external control in such systems one additionally requires the use of multiwave schemes [30, 31] of nonlinear coherent interaction by analogy with optics [32, 33]. Further development of our research is related to the investigation of collective spin effects based on the photon echo [34] in plasmonic structures, as well as the possibilities of control such effects [35].

Acknowledgments

The work was supported by RFBR (17-42-330029) and the Ministry of Education and Science of the Russian Federation in the framework of the state task VlSU 2017 in the field of scientific research.

References

[1] Skribanowitz N, Hermann I P, MacGillivray J C and Feld M S 1973 *Phys. Rev. Lett.* **30** 309
[2] Stockman M 2010 *J. Opt.* **12** 024004
[3] Vinogradov A P, Andrianov E S, Pukhov A A, Dorofeenko A V and Lisyansky A A 2012 *Phys. Usp.* **55** 1046–1053
[4] Lambright S, Butaeva E, Razgoniaeva N, Hopkins T, Smith B, Perera D, Corbin J, Khon E, Thomas R, Moroz P, Mereshchenko A, Tarnovsky A and Zamkov M 2014 *ACS Nano* **8** 352–361
[5] Protsenko I and Uskov A V 2015 *Quantum Electron.* **45** 561–572
[6] Dzedolik I V 2014 *J. Opt.* **16** 125002
[7] Reinhardt C, Evlyukhin A B, Cheng W, Birr T, Markov A, Ung B, Skorobogatiy M and Chichkov B N 2013 *J. Opt. Soc. Am. B* **30** 2898–2905
[8] Konopsky V N and Alieva E V 2006 *Phys. Rev. Lett.* **97** 253904
[9] Klimov V V, Pavlov A A, Guzatov D V, Zabkov I V and Savinov V D 2016 *Phys. Rev. A* **93** 033831
[10] Golovanova A V, Gubin M Y, Gladush M G and Prokhorov A V 2016 *Bull. Russ. Acad. Sci. Phys.* **80** 808–813

[11] Kuraptsev A S and Sokolov I M 2016 *JETP* **123** 237–248
[12] MacDonald K F, Sámson Z L, Stockman M I and Zheludev N I 2009 *Nat. Photon.* **3** 55–58
[13] Stockman M I, Faleev S V and Bergman D J 2001 *Phys. Rev. Lett.* **87** 167401
[14] Gubin M Y, Leksin A Y, Gladush M G, Arakelian S M and Prokhorov A V 2015 *Opt. Spectrosc.* **119** 497–512
[15] Benedict M G and Trifonov E D 1988 *Phys. Rev. A* **38** 2854
[16] Dorofeenko A V, Zyablovsky A A, Vinogradov A P, Andrianov E S, Pukhov A A and Lisyansky A A 2013 *Opt. Express* **21** 14539–14547
[17] Mohamed N B H, Haouari M, Zaaboub Z, Nafoutti M, Hassen F, Maaref H and Ouada H B 2014 *J. Nanopart. Res.* **16** 2242
[18] Milekhin A G, Sveshnikova L L, Repinskii S M, Gutakovskii A K, Friedrich M and Zahn D R T 2002 *Phys. Solid State* **44** 1976–1980
[19] Guerrero S I C 2013 Doctoral Thesis (Micro Nanotechnol., Acoust., Telecommun.), Lille, France
[20] Pokutnii S I 2006 *Semiconductors* **40** 217–223
[21] Pokutnii S I 2010 *Semiconductors* **44** 488–493
[22] Fedorov A V and Baranov A V 2005 *Optics of quantum dots. In Proc. Optics of nanostructures.* (SPb.: Nedra)
[23] Larkin I A, Stockman M I, Achermann M and Klimov V I 2004 *Phys. Rev. B* **69** 121403(R)
[24] Kitamura R, Pilon L and Jonasz M 2007 *Appl. Opt.* **46** 8118–8133
[25] Benedict M G, Ermolaev A M and Malyshev V A 1996 *Superradiance. Multiatomic Coherent Emission* (Philadelphia, USA: IOP)
[26] Kamalieva A, Toropov N, Reznik I and Vartanyan T 2016 *Opt. Quant. Electron.* **48** 562
[27] Karimullin K R and Naumov A V 2014 *Journal of Luminescence* **152** 15–22
[28] Alodjants A P, Prokhorov A V and Arakelian S M 2003 *Laser Phys.* **13** 1264
[29] Alodjants A P, Leksin A Y, Prokhorov A V and Arakelian S M 2000 *Laser Phys.* **10** 603–613
[30] Prokhorov A V, Alodjants A P and Arakelian S M 2004 *JETP Letters* **80** 739–742
[31] Khokhlov N, Knyazev G, Glavin B, Shtykov Y, Romanov O and Belotelov V 2017 *Opt. Lett.* **42** 3558–3561
[32] Prokhorov A V, Gladush M G, Gubin M Y, Leksin A Y and Arakelian S M 2014 *Eur. Phys. J. D* **68** 158
[33] Prokhorov A V, Gubin M Y, Leksin A Y, Gladush M G, Alodzhants A P and Arakelyan S M 2012 *JETP* **115** 1–14
[34] Savochkin I V, Jäckl M, Belotelov V I, Akimov I A, Kozhaev M A, Sylgacheva D A, Chernov A I, Shaposhnikov A N, Prokopov A R, Berzhansky V N, Yakovlev D R, Zvezdin A K and Bayer M 2017 *Sci. Rep.* **7** 5668
[35] Melnikov A, Povolotskiy A and Bovensiepen U 2008 *Phys. Rev. Lett.* **100** 247401

IOP Conf. Series: Journal of Physics: Conf. Series **951** (2017) 012034 doi:10.1088/1742-6596/951/1/012034

Features of localized plasmons formation in four-particle spaser systems

M. Yu. Gubin, A. V. Shesterikov, S. N. Karpov, and A. V. Prokhorov

Vladimir State University named after Alexander and Nikolay Stoletovs, Vladimir, 600000 Russia

E-mail: avprokhorov33@mail.ru

Abstract. The article is focused on the management problem of quantum dynamics of localized plasmons in model of a four-particle spaser which consists of metal nanoparticles and semiconductor quantum dots. The conditions for observation stable stationary regime of plasmon formation in the presented model have been determined by the use of mean field approximation.

1. Introduction

The recent rapid progress in the field of nanotechnology has led to practical possibility for generation and control of N-photon states using single quantum emitters [1] and high-quality micro- [2] and nanoresonators [3]. The key position in the development of such devices is creation of a unique device – nanolaser [4]. Functional principles of the nanolaser require a reformulation of the known rules of laser generation for subwavelength scales [5]. This can be done for systems containing a localized spaser [6]. In the simple case this device consists of a semiconductor quantum dot (QD) and a metal nanoparticle (NP) coupled by nearfield interaction. QD is used here as an effective nearfield pump source, because the excitonic decay in the QD leads to perturbations of NP near field. Under the conditions of a plasmon-exciton resonance this perturbation leads to generation of the plasmons localized on the NP surface [7, 8]. Realistic model of the spaser can be based on a compound nanoobject consisting of a metal core and a semiconductor shell [9] or on a distributed system of nanoparticles with the complex geometry [10, 11, 12, 13]. Currently, both localized subwavelength [4] and waveguide [14] spaser-like systems realized in practice.

In this article we particularly consider the possibility dynamics of plasmons generated in a multiparticle spaser system. For this, the model suggested in [15] has been extended for a 2×2 spaser system, which consists of two closely located NPs and two QDs coupled by strong local fields. Importantly the considered spaser systems can be integrated in the individual plasmonic waveguide [16] and information plasmonic circuit [17, 18, 19].

2. The formation of localized plasmons in the system of two spasers coupled by strong dipole-dipole interactions

Let us consider the system of two spasers, consisting of 2 QDs and 2 NPs (so-called Spaser 2×2), see Fig. 1. First of all, the efficiency of interaction in the system will depend on the

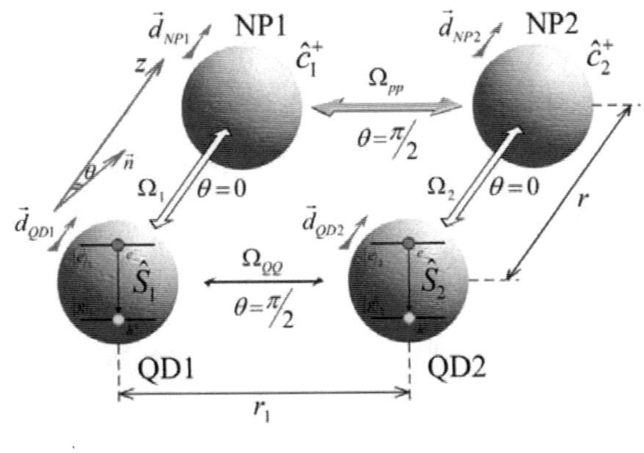

Figure 1. The model of four-particle 2×2 spaser, consisting of two NPs and two QDs.

geometry of the system, where characteristic lengths are r_{NP}, which is the distance between adjacent NPs, r_{QD}, which is the distance between adjacent QDs, and r_{QN}, which is the distance in a pair QD-NP. The vector $\vec{n} = \vec{r}/r$, making an angle θ with axis $\vec{z}$, determines the direction between the centers of any two interacting particles. We assume that all dipole moments $\hat{d}_{QD}$ of QD and $\hat{d}_{NP}$ of NP are collinear to each other and parallel to axis $\vec{z}$ [6].

The field $\hat{\mathrm{E}}_{NPi} = -\nabla \hat{A}_{NPi}$ on distance r from i-th NP of spherical shape with radius a_{NPi} can be expressed through the operator of vector potential $\hat{A}_{NPi} = \sum\limits_n \left(\frac{a_{NPi}}{r}\right)^{n+1} Y_{nm}(\theta,\varphi) E_{nm} \left(\hat{c}_i + \hat{c}_i^+\right) \vec{e}_{NPi}$ [7], where $\hat{c}_i(\hat{c}_i^+)$ are annihilation (creation) operators of plasmon mode in quasistatic approximation, $Y_{nm}(\theta,\varphi) = \sqrt{\frac{2n+1}{4\pi}\frac{(n-m)!}{(n+m)!}} P_n^m(\cos\theta)\, e^{im\varphi}$ are spherical functions expressed via Legendre polynomials, $\vec{e}_{NP}$ defines the orientation of NP dipole moment, $E_{nm} = \sqrt{\frac{\hbar\omega_{nm}}{2a_{NPi}(2n+1)\varepsilon_0}}$ is the dimensional factor, where n is the general quantum and m is the magnetic quantum numbers [8], a_{NPi} is the radius of i-th NP.

Then, near field of single QD is written as

$$\hat{\mathrm{E}}_{QDi} = \frac{1}{4\pi\varepsilon_0} \cdot \frac{3\vec{n}\left(\vec{n}\cdot\vec{e}_{QDi}\right) - \vec{e}_{QDi}}{r^3}\hat{d}_{QDi}, \quad (1)$$

where dipole moment operator $\hat{d}_{QDi} = \mu_{QDi}\left(\hat{S}_i + \hat{S}_i^+\right)\vec{e}_{QD}$ expressed via creation operator $\hat{S}_i^+ = |e\rangle_i \langle g|_i$ and annihilation operator $\hat{S}_i = |g\rangle_i \langle e|_i$ of excitons and matrix element μ_{QDi} corresponding interband transitions in QD, where $|e\rangle_i$ corresponds to exited and $|g\rangle_i$ ground states of the system. Presented operators satisfy commutation relationships $\left[\hat{S}_i^+, \hat{S}_i\right] = D_i$ and $\left[\hat{S}_i, D_i\right] = 2\hat{S}_i$, where $D_i = \hat{S}_i^+\hat{S}_i - \hat{S}_i\hat{S}_i^+$ is the population imbalance operator; $\vec{e}_{QDi}$ determines the QD dipole momentum orientation.

In conditions $\lambda_{1,2} \gg r > a_{NPi}, a_{QDi}$ where a_{QDi} is the QD radius, and $\lambda_{1,2}$ is the transition wavelength in QD, all pair interactions correspond to dipole-dipole energy exchange, in particular, interaction Hamiltonian between NP and QD can be written in the form $V_i^{QN} = -\hat{\mathrm{E}}_{NPi}^{\|}\hat{d}_{QDi}$. Presented in Fig. 1 geometry corresponds to the case $\theta = 0$ and therefore the Legendre polynomials take the form $P_1^0(\cos\theta) = 1$, $P_1^{m\neq 0}(\cos\theta) = 0$. Thus, the near-field of NP$_i$ in the position of QD$_i$ location has the form

$$\hat{\mathrm{E}}_{NPi}^{\|} = \sqrt{\frac{\hbar\omega_{pi}a_{NPi}^3}{2\pi\varepsilon_0}}\frac{1}{r^3}\left(\hat{c}_i + \hat{c}_i^+\right)\vec{e}_{NPi}, \quad (2)$$

where ω_{pi} determine plasmon frequencies of NP_i. Interaction Hamiltonian for adjacent NPs $V^{NN} = -\hat{\mathrm{E}}^{\perp}_{NP1}\hat{d}_{NP2}$ is determined by orientation $\theta = \pi/2$, for which $P_1^1(\cos\theta) = 1$ and $P_1^{m\neq 1}(\cos\theta) = 0$. In this case, the expression for the field takes the form

$$\hat{\mathrm{E}}^{\perp}_{NPi} = \sqrt{\frac{\hbar\omega_{pi}a^3_{NPi}}{4\pi\varepsilon_0}}\frac{1}{r^3}\left(\hat{c}_i + \hat{c}_i^+\right)\vec{e}_{NPi}, \tag{3}$$

and corresponding induced dipole moment of NP_i. This dipole moment can be obtained from comparison (3) and the expression for near field of NP_i in analogy with (1). This expression has the form $\hat{d}_{NPi} = \mu_{NPi}\left(\hat{c}_i + \hat{c}_i^+\right)\vec{e}_{NP}$, where $\mu_{NPi} = \sqrt{4\pi\varepsilon_0\hbar\omega_{pi}a^3_{NPi}}$. Finally, interaction Hamiltonian $V^{QQ} = -\hat{\mathrm{E}}_{QD1}\hat{d}_{QD2}$ between single QD in spaser structure is determined by the given geometry, for which $(\vec{n}\cdot\vec{e}_{QDi}) = 0$.

Based on the necessity of internal symmetry of layer arrangement for QD and NP in case of scaling spaser system to 2D massive of spasers we assume $r_{NP} = r_{QD} = r_1$, $r_{QN} = r$. The working regime of coupled spasers significantly depends on the ratio between frequencies $\omega_{1,2}$ of transitions in QDs and plasmon frequencies $\omega_{p1,p2}$. They are usually [6, 20] considered almost equal to each other so that there are mainly implemented linear plasmon interactions in the system. Then, in the case $\omega_i \approx \omega_{pi}$, corresponding Hamiltonian of interaction takes the form:

$$\begin{aligned} H =& \hbar\omega_{p1}\hat{c}_1^+\hat{c}_1 + \hbar\omega_{p2}\hat{c}_2^+\hat{c}_2 + \frac{\hbar\omega_1}{2}D_1 + \frac{\hbar\omega_2}{2}D_2 \\ &+ \hbar\Omega_1\left(\hat{c}_1\hat{S}_1^+ + \hat{c}_1^+\hat{S}_1\right) + \hbar\Omega_2\left(\hat{c}_2\hat{S}_2^+ + \hat{c}_2^+\hat{S}_2\right) \\ &+ \hbar\Omega_{QQ}\left(\hat{S}_1\hat{S}_2^+ + \hat{S}_1^+\hat{S}_2\right) + \hbar\Omega_{pp}\left(\hat{c}_1\hat{c}_2^+ + \hat{c}_1^+\hat{c}_2\right), \end{aligned} \tag{4}$$

where 5-th and 6-th terms with $\Omega_i = \sqrt{\frac{\omega_{pi}a^3_{NPi}}{2\pi\varepsilon_0\hbar}}\frac{\mu_{QD}}{r^3}$ correspond to V_i^{QN}, 7-th with $\Omega_{QQ} = \frac{\mu^2_{QD}}{4\pi\varepsilon_0\hbar r_1^3}$ appears from V^{QQ} and term with $\Omega_{pp} = \frac{\mu^2_{NP}}{4\pi\varepsilon_0\hbar r_1^3}$ is defined by Hamiltonian V^{NN}. The cross-interaction between NP and QD from neighboring spasers we neglect. In (4) we don't take into account the Ferster transfer of energy between QDs, although in a real situation it must be done.

As a model we choose a spaser consisting of metal NP and QD based on semiconductor CdS, for which the chosen states $|g\rangle_i$ and $|e\rangle_i$ correspond to hole level $1S(h)$ in valence band and electron level $1S(e)$ in conduction band. Estimates of QD sizes can be performed proceeding from the plasmon mode frequency $\omega_p = \omega_{p1} = \omega_{p2}$, which for spherical metal NP is determined by expression $\omega_p = \frac{\omega_{pl}}{\sqrt{3}}$. For satisfying exact resonance condition $\omega_p = \omega$ between NP and QD, the size of QD is determined by known dependence [21] of the interband transition energy $1S(e) \to 1S(h)$ on its diameter $D_{QD} = 2a_{QD}$:

$$\begin{aligned} E_{bb} &= \hbar\omega_p \\ &= E_g + 2\frac{\hbar^2\pi^2}{D^2_{QD}}\left(\frac{1}{m_e} + \frac{1}{m_h}\right) - \frac{3.56\cdot e^2}{\varepsilon\cdot D_{QD}}, \end{aligned} \tag{5}$$

where e is the electron charge, $\hbar$ is the Planck constant, m_e and m_h are the effective masses of electron and hole in bulk of QD material, respectively, with dielectric permittivity ε and bandgap E_g. For CdS corresponding parameters are $E_g/e = 2.42$ eV, $m_e = 0.19m_0$, $m_h = 0.8m_0$, $\varepsilon = 9$ and $D_{QD} = 1.88$ nm. Bohr radius of exciton R_{ex} for CdS is 2.5 nm [22] therefore strong confinement regime [23] will be observed for the considered QDs, and energy sublevels of conductivity zone will be essentially separated. Therefore, it can be assumed that the two-level model will be valid for QDs. However, the size of QDs is sufficient for multiple generation of electron-hole pairs under the influence of an external pump. The subsequent decay of excitons leads to the formation of multi-particle plasmon states, which are localized on the NPs. However, assuming that the electron-hole pairs generated in the QDs have similar values of the energy, the summation symbol in (4) can be eliminated by use of a single-mode Hamiltonian.

The value of dipole moment for corresponding interband transition [24] can be approximately determined as $|\mu_{QD}|^2 = \frac{e^2}{6m_0\omega_p^2}\left(\frac{m_0}{m_e}-1\right)E_{bb}$ and in selected conditions equals to $\mu_{QD} = 0.16\cdot 10^{-28}$ C $\cdot$ m. The dipole moment of NP with radius, which exactly coincide with radius of QD, equals to $\mu_{NP} = 1.58\cdot 10^{-28}$ C $\cdot$ m. We suppose that spaser 2×2 is a square with characteristic sizes $r_1 = r = 3$ nm and $a_{NP} = a_{QD}$, corresponding Rabi frequencies in (4) take a value $\Omega_1 = \Omega_2 = \Omega = 2\cdot 10^{13}$ s^{-1}, $\Omega_{QQ} = 8.46\cdot 10^{11}$ s^{-1}, $\Omega_{pp} = 1.54\cdot 10^{14}$ s^{-1}. It is shown that the efficiency of dipole-dipole interaction between QDs in the presented geometry is significantly lower for the similar efficiency both between the adjacent NPs, and in a pair of NP-QD. Thus, for further consideration of the limit of a single-particle states, the term with Ω_{QQ} can be neglected, and then we can proceed to the consideration of the following Heisenberg-Langevin system of equations obtained from (4):

$$\dot{\hat{c}}_1 = i\left(\Delta_1 + \frac{i}{\tau_{C1}}\right)\hat{c}_1 - i\Omega_1\hat{S}_1 - i\Omega_{pp}\hat{c}_2 + \hat{F}_{c1}, \quad (6a)$$

$$\dot{\hat{c}}_2 = i\left(\Delta_2 + \frac{i}{\tau_{C2}}\right)\hat{c}_2 - i\Omega_2\hat{S}_2 - i\Omega_{pp}\hat{c}_1 + \hat{F}_{c2}, \quad (6b)$$

$$\dot{\hat{S}}_1 = i\left(\delta_1 + \frac{i}{\tau_{S1}}\right)\hat{S}_1 + i\Omega_1 D_1\hat{c}_1 + \hat{F}_{S1}, \quad (6c)$$

$$\dot{\hat{S}}_2 = i\left(\delta_2 + \frac{i}{\tau_{S2}}\right)\hat{S}_2 + i\Omega_2 D_2\hat{c}_2 + \hat{F}_{S2}, \quad (6d)$$

$$\dot{D}_1 = -2i\Omega_1\left(\hat{S}_1^+\hat{c}_1 - \hat{S}_1\hat{c}_1^+\right) - \frac{D_1 - D_{01}}{\tau_{D1}} + \hat{F}_{D1}, \quad (6e)$$

$$\dot{D}_2 = -2i\Omega_2\left(\hat{S}_2^+\hat{c}_2 - \hat{S}_2\hat{c}_2^+\right) - \frac{D_2 - D_{02}}{\tau_{D2}} + \hat{F}_{D2}, \quad (6f)$$

where $\Delta_1 = \bar{\omega} - \omega_{p1}$, $\Delta_2 = \bar{\omega} - \omega_{p2}$, $\delta_1 = \bar{\omega} - \omega_1$, $\delta_2 = \bar{\omega} - \omega_2$, and parameters $\bar{\omega}$ and $D_{01(02)}$ correspond to the frequency and value of spaser pump, respectively. In the deriving system (6) we used rotating-wave approximation $\hat{c} = \hat{c}\cdot\exp(-i\bar{\omega}t)$ and $\hat{S} = \hat{S}\cdot\exp(-i\bar{\omega}t)$ upon passage to the new slow-varying operators $\hat{c}\,(\hat{c}^+)$ and $\hat{S}\left(\hat{S}^+\right)$. In equations (6) the characteristic parameters of decay rate for plasmons $\frac{1}{\tau_{c1(c2)}}$ in NP, the decay rate of excitons $\frac{1}{\tau_{S1(S2)}}$ in excited QDs, and also the operators of Langevin noises $\hat{F}_{c1(c2)}$ ($\hat{F}_{S1(S2)}$, $\hat{F}_{D1(D2)}$) are introduced phenomenologically [25], proceeding from a condition of system interaction with the reservoir. The pump operator and appropriate time, generally, have the forms $D_{01(02)} = \frac{2\tau_{S1(S2)} - \tau_{p1(p2)}}{2\tau_{S1(S2)} + \tau_{p1(p2)}}\hat{I}$ and $\tau_{D1(D2)} = \left(\frac{1}{2\tau_{S1(S2)}} + \frac{1}{\tau_{p1(p2)}}\right)^{-1}$ [26], where parameters $\tau_{p1(p2)}$ are presented by characteristic pumping times. Further, we assume that both NPs (and also both QDs) in spaser 2×2 are identical to each other, consequently, it is possible to suppose $\Omega_1 = \Omega_2$, $\tau_{c1} = \tau_{c2} = \tau_c$, $\tau_{S1} = \tau_{S2} = \tau_S$, $\Delta_1 = \Delta_2 = \Delta$ ($\omega_{p1} = \omega_{p2} = \omega_p$), $\delta_1 = \delta_2 = \delta$ ($\omega_1 = \omega_2 = \omega$). Besides, we assume that populations in both QDs change synchronously, i.e. $D_1 = D_2 = D$.

Going over to the estimates of relaxation parameters of the problem, it should be noted, that decay rate of plasmon mode $\gamma_p = \frac{1}{\tau_c} = \frac{1}{\tau_J} + \frac{1}{\tau_R}$ is determined by characteristic time of radiation

IOP Conf. Series: Journal of Physics: Conf. Series **951** (2017) 012034 doi:10.1088/1742-6596/951/1/012034

τ_R and Joule τ_J losses. However, under the condition $\frac{1}{\tau_J} \approx 30\frac{1}{\tau_R}$ [27] we do not have to take radiation losses into account and Joule losses are normally determined by electron collisional frequency in metal γ_s, i.e. $\gamma_p \approx \gamma_s$, where $\gamma_s = 4 \cdot 10^{13}$ s^{-1}. A more correct calculation of the decay rate of plasmons localized on the NPs requires consideration of its geometrical features, as well as a more precise chemical composition [28].

The parameter $\frac{1}{\tau_S} = \frac{1}{\tau_R} + \frac{1}{\tau_F}$ presents the total rate of radiative (with time τ_R) and nonradiative (with time τ_F) losses in QD. At the same time, parameter $\frac{1}{\tau_F}$ gives the main contribution, since the processes of nonradiative recombination of excitons (with creating phonon modes) occur on short times τ_F when annealing technology [29] is used, this parameter can be essentially increased [30]. Therefore, following the work [6] we accept the estimate $\frac{1}{\tau_S} = 4 \cdot 10^{10}$ s^{-1}. A more correct approach requires taking into account the binding energy of excitons and the accurate derivation of radiative damping rate dependencies by analogy with quantum well model [31].

3. The stationary solutions for four-particle spaser

We determine characteristic spaser generation frequency (spasing frequency) $\bar{\omega}$, spasing threshold D_{th}, and also find the possible stationary regimes of its time evolution. For this purpose, at the initial stage we write the system of equations for average values, analogical to operator system (6), assuming that now $c = \langle \hat{c} \rangle$, $c^* = \langle \hat{c}^+ \rangle$ and consider that Langevin noise operators are equal to zero. Then, assuming $\dot{c}_{1(2)} = \dot{D} = \dot{S}_{1(2)} = 0$, we express c_2 from equation on average values (6a) and substitute obtained expression in equations on average values (6b) and (6d). After it, we express parameter S_2 from (6d) and substitute it in (6b) and (6c). Obtained new system of algebraic equations on the average values will be written in the form:

$$\Omega D c_1 + B S_1 = 0, \tag{7a}$$

$$\left(\frac{A^2}{\Omega_{pp}} + \frac{A\Omega^2 D}{B\Omega_{pp}} - \Omega_{pp}\right) c_1 + \left(-\frac{\Omega A}{\Omega_{pp}} - \frac{\Omega^3 D}{\Omega_{pp} B}\right) S_1 = 0, \tag{7b}$$

where new definitions $A = A_R + iA_I = \Delta + \frac{i}{\tau_c}$ and $B = B_R + iB_I = \delta + \frac{i}{\tau_S}$ are introduced. The system (7) can have non-trivial solutions in the case, when matrix determinant for the left side of system of equations is equal to zero. Thus, we obtain the system of two self-consistent equations for the real and imaginary parts of such determinant

$$\Omega^4 D^2 + 2\left(A_R B_R - A_I B_I\right)\Omega^2 D + \left(B_I^2 - B_R^2\right)\left(A_I^2 - A_R^2 + \Omega_{pp}^2\right) - 4A_I A_R B_I B_R = 0, \tag{8a}$$

$$2\left(A_R B_I + A_I B_R\right)\left(A_R B_R - A_I B_I + \Omega^2 D\right) - 2\Omega_{pp}^2 B_I B_R = 0. \tag{8b}$$

Solving these equations (8a) and (8b) together, it is possible to obtain expressions for spasing frequency $\bar{\omega}$ and for threshold value D_{th}. In this case, the system of equations has two roots in the form

$$\bar{\omega}_{\pm} = \frac{\tau_S \omega + \tau_c\left(\omega_p \pm \Omega_{pp}\right)}{\tau_c + \tau_S}, \tag{9a}$$

$$D_{th,\pm} = \frac{1 + \left(\frac{\tau_c \tau_S}{\tau_c + \tau_S}\right)^2 \left(\omega - \omega_p \pm \Omega_{pp}\right)^2}{\tau_c \tau_S \Omega^2} \tag{9b}$$

one of which $(\bar{\omega}_-, D_{th,-})$ will not give stable solution for (6), and another we will assume that $D_{th} = D_{th,+}$.

It is should be noted, that in the limit case $\Omega_{pp} = 0$ solutions (9) coincide with a known model of spaser 1×1 which consists of one QD and one NP [7]. However, the presence of near-field interaction between two NPs significantly increases the spasing threshold. The obtained in (9) solutions correspond only to the situation for the regime of linear energy exchange between NP and QD determined by Hamiltonian (4), which can be implemented under conditions of small frequency detunings Δ and δ. Under selected parameters of interaction, the values of thresholds are $D_{\rm th}^{1\times1} = D_{th}(\Omega_{\rm pp} = 0) = 0.004$ and $D_{\rm th}^{2\times2} = D_{th}\left(\Omega_{\rm pp} = 1.54 \cdot 10^{14}\ {\rm s}^{-1}\right) = 0.0632$, and in further simulation pumps are supposed to equal each other for both types of spasers and are chosen according to the conditions $D_0 > \max(D_{\rm th}^{1\times1}, D_{\rm th}^{2\times2})$ and $D_0 = 0.1$.

Assuming $c_1 = c_2 = c$ $(S_1 = S_2 = S)$, we express c from equation (7b) and substitute obtained expression in the result of summing equations on average values (6e) and (6f), which are obtained from condition $\dot{D} = 0$. As a result, steady-state solutions for the amplitudes of generating on NPs plasmons and forming in QD excitons take forms which are determined up to the phase ϕ:

$$\bar{c} = e^{i\phi_i}\sqrt{\frac{\tau_c}{4\tau_D}\left(D_0 - D_{th}\right)}, \tag{10a}$$

$$\bar{S} = \frac{i + \frac{\tau_c\tau_S}{\tau_c+\tau_S}\left(\omega - \omega_p - \Omega_{pp}\right)}{\tau_c\Omega}\bar{c}. \tag{10b}$$

The test on stability of obtained solutions was carried out by analysis of eigenvalues λ_i of the linearized system of equations on average values (6) near fixed points (10), as well as by using a direct numerical simulation of this system. At the same time, for obtained solutions (10) one of the roots of characteristic equation for linearized system (6) on average values is always equal to zero, while other eigenvalues are satisfied to inequality $\mathrm{Re}\left(\lambda_i\right) < 0$. Thus, the system moves to the boundary of aperiodic stability in the absence of the external synchronization. From the mathematical point of view, in such case further analysis full nonlinear system (6) should be carried out. Nevertheless, numerical analysis of the system (6) demonstrates the stability of obtained solutions (10).

The parametric plane formed by the combination of the parameters pump value D_0 and the characteristic time of excitons decay τ_S in the system with the depicted stability area for the solutions (10) verified by numerical simulation of system (6) is presented in Fig. 2. Obviously, in the case of using chemical inhomogeneous (not pure in composition) QDs jointly leading to decreasing value τ_S, it is necessary to increase pump value D_0 for supporting stationary conditions for spasing. At the same time, increasing D_0 leads to the linear growth of the number of plasmons $|\bar{c}|^2$ which are generated by spaser provided that the chosen characteristic pumping time is $\tau_D = 2.7 \cdot 10^{-15}$ s, the inset in Fig. 2.

The validity test of the obtained solutions (10) with taking into account correction Ω_{pp} has been carried out by their comparing with direct numerical simulation of the system (6) for average values. In particular, in Fig. 3, performed for parameter D, the consent of analytical and numerical results is clearly observed, when choosing the parameters of interaction for point A from presented area of stability in Fig. 2. At the same time, the dynamics of the transition process to the stationary values for D and $|c|^2$ also has a significant dependence on Ω_{pp}. In particular, fast synchronization regime of energy exchange processes between QD and NP under condition $\Omega_{pp} = 0$ is replaced by more "protracted" in time stabilization process taking into account the contribution of Ω_{pp}, from Fig. 3, 4. Besides, when accounting Ω_{pp} more slowly and intensive energy exchange is established in the system, which can be observed by using temporal dynamics of changing amount $\sin\left(\Delta\phi\left(t\right)\right)$, where parameter $\Delta\phi\left(t\right) = Arg\left(\frac{c(t)}{S(t)}\right)$ determines a

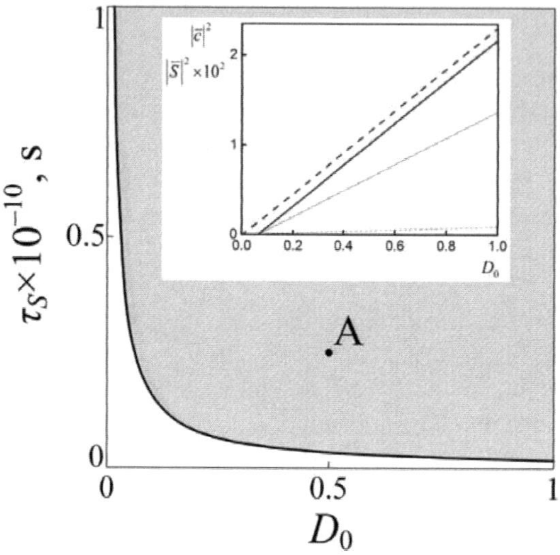

Figure 2. The parametric plane (pump value D_0, decay time of excitons in QD τ_S) with depicted stability area of 2×2 spaser and a point A $(0.5,\ 0.25 \cdot 10^{-10}$ s). In the inset: the dependences of average numbers of plasmons $|\bar{c}|^2$ (blue thick lines) and excitons $|\bar{S}|^2$ (red thin lines) on the value of pump for spaser 2×2 with accounting Ω_{pp} (solid lines) and for spaser 1×1 without accounting Ω_{pp} (dashed lines). The parameters of interaction: $\omega_p = \omega = 7.9 \cdot 10^{15}$ s^{-1}, $\Omega = 2 \cdot 10^{13}$ s^{-1}, $\Omega_{pp} = 1.54 \cdot 10^{14}$ s^{-1}, $\tau_c = 0.25 \cdot 10^{-13}$ s, $\tau_D = 2.7 \cdot 10^{-15}$ s.

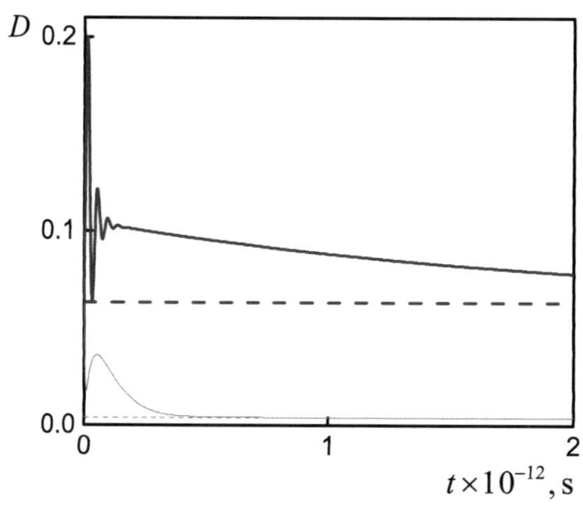

Figure 3. The time dependences of population imbalance parameter D for spasers 1×1 (red thin lines) and 2×2 (blue thick lines), calculated by using formulas (9) (dashed lines) and by using direct numerical simulation (solid lines) of system (6). The initial values: $c(0) = \bar{c} \cdot 1.05$, $S(0) = \bar{S} \cdot 0.95$, $D(0) = 0.1$, where steady-state solutions $\bar{c} = 0.7110 + 0.7110i$, $\bar{S} = -6.8917 - 4.0475i$ for spaser 1×1 and $\bar{c} = 0.7577 + 0.7577i$, $\bar{S} = -1.5154 + 1.5154i$ for spaser 2×2 when choosing phase $\phi(0) = \pi/4$. The simulation parameters correspond to point A from Fig. 2.

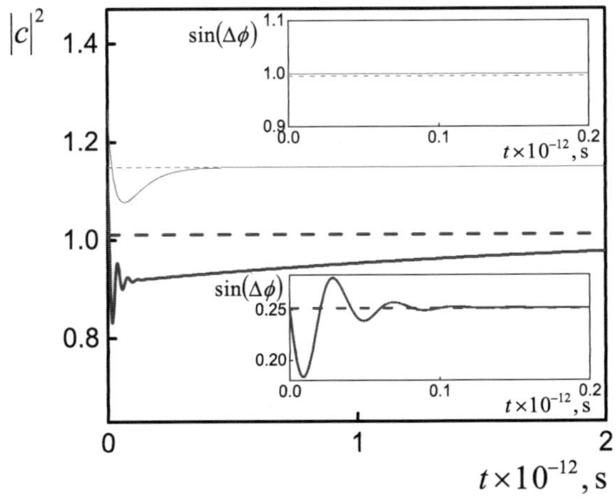

Figure 4. The time dependences of average number of plasmons $|c|^2$ for spaser 1×1 (red thin lines) and for spaser 2×2 (blue thick lines), calculated by using formulas (10a) (dashed lines) and by using direct numerical simulation (solid line) of system (6). In the insets: dependences of the relative phase between plasmon and exciton modes for spaser 2×2 (bottom inset) and for spaser 1×1 (top inset). The initial values and parameters of simulation correspond to Fig. 3.

relative phase between plasmons and excitons, the inset in Fig. 4. The formed constant phase shift $\Delta\phi_{\mathrm{st}}$, as a result of the dynamics when the system reaches the steady-state value, is shifted to new value in relation to a case $\Omega_{pp} = 0$, for which it is always valid $\Delta\phi_{\mathrm{st}} = \pi/2$ [7]. At the same time, the amount $\Delta\phi_{\mathrm{st}}$ doesn't depend on initial values $c(0)$ and $S(0)$.

4. Conclusions

In this paper we describe the scheme of double spaser and observe the stationary regime formation of localized plasmons in this model. The scheme consists of two NPs and two QDs located in the vertices of a square and coupled with each other by means of a nearfield. It is shown that the plasmon-exciton kinematics in the scheme significantly depends on contribution of the energy of dipole-dipole interactions between the pair of NPs, while the contribution of dipole-dipole interactions between the pair of QDs can be neglected. Based on the realistic parameters of metal NPs and semiconductor QDs of CdS, the optimal conditions for observation of steady-states solutions for average numbers of plasmons and excitons in proposed scheme of spaser have been determined. Further development of this work is aimed at a complex simulation of quantum dynamics and a nonclassical states control of localized plasmons by analogy with optics [32].

Acknowledgments

We would like to thank A.B. Evlyukhin for helpful discussions. This work was supported by Ministry of Education and Science of the Russian Federation in the framework of state task VlSU 2017 in the field of scientific research and the programme Michail-Lomonosov-Programme - Linie A, 2017 (ID 57320203) of German Academic Exchange Service and Ministry of Education and Science of the Russian Federation (No. 2.9953.2017/DAAD).

IOP Conf. Series: Journal of Physics: Conf. Series **951** (2017) 012034 doi:10.1088/1742-6596/951/1/012034

References

[1] Bimberg D, Kirstaedter N, Ledentsov N N, Alferov Z I, Kop'ev P S and Ustinov V M 1997 *IEEE Journal of Selected Topics in Quantum Electronics* **3** 196–205
[2] Gorodetsky M L 2011 *Optical Microcavities with Giant Q-Factor* (FIZMATLIT, Moscow)
[3] Vesseur E J R, de Waele R, Lezec H J, Atwater H A, García de Abajo F J and Polman A 2008 *Appl. Phys. Lett.* **92** 083110
[4] Noginov M A, Zhu G, Belgrave A M, Bakker R, Shalaev V M, Narimanov E E, Stout S, Herz E, Suteewong T and Wiesner U 2009 *Nature* **460** 1110–1112
[5] Stockman M I 2008 *Nature Photonics* **2** 327–329
[6] Stockman M I 2010 *J. Opt.* **12** 024004
[7] Vinogradov A P, Andrianov E S, Pukhov A A, Dorofeenko A V and Lisyansky A A 2012 *Phys. Usp.* **55** 1046–1053
[8] Klimov V V 2014 *Nanoplasmonics* (Pan Stanford; 1 edition)
[9] Lambright S, Butaeva E, Razgoniaeva N, Hopkins T, Smith B, Perera D, Corbin J, Khon E, Thomas R, Moroz P, Mereshchenko A, Tarnovsky A and Zamkov M 2014 *ACS Nano* **8** 352–361
[10] Pavlov A A, Klimov V V, Vladimorova Y V and Zadkov V N 2013 *Quantum Electronics* **43** 496–501
[11] Evlyukhin A B, Reinhardt C, Zywietz U and Chichkov B N 2012 *Phys. Rev. B* **85** 245411
[12] Shesterikov A V, Gubin M Y, Gladush M G and Prokhorov A V 2017 *JETP* **124** 18–31
[13] Martín-Cano D, Martín-Moreno L, JGarcía-Vidal F and Moreno E 2010 *Nano Lett.* **10(8)** 3129–3134
[14] Bermúdez-Ureña E, Gonzalez-Ballestero C, Geiselmann M, Marty R, Radko I P, Holmgaard T, Alaverdyan Y, Moreno E, García-Vidal F J and Bozhevolnyi S I 2015 *Nature Commun.* **6** 7883
[15] Andrianov E S, Pukhov A A, Dorofeenko A V, Vinogradov A P and Lisyansky A A 2015 *Quantum Electronics* **45** 245–249
[16] Reinhardt C, Evlyukhin A B, Cheng W, Birr T, Markov A, Ung B, Skorobogatiy M and Chichkov B N 2013 *J. Opt. Soc. Am. B* **30** 2898–2905
[17] Fedyanin D Y, Yakubovsky D I, Kirtaev R V and Volkov V S 2016 *Nano Lett.* **16** 362–366
[18] Lemke C, Schneider C, Leißner T, Bayer D, Radke J W, Fischer A, Melchior P, Evlyukhin A B, Chichkov B N, Reinhardt C, Bauer M and Aeschlimann M 2013 *Nano Lett.* **13** 1053–1058
[19] Lemke C, Leißner T, Evlyukhin A, Radke J W, Klick A, Fiutowski J, Kjelstrup-Hansen J, Rubahn H G, Chichkov B N, Reinhardt C and Bauer M 2014 *Nano Lett.* **14** 2431–2435
[20] Bergman D J and Stockman M I 2003 *Phys. Rev. Lett.* **90** 027402
[21] Bel Haj Mohamed N, Haouari M, Zaaboub Z, Nafoutti M, Hassen F, Maaref H and Ben Ouada H 2014 *J. Nanopart. Res.* **16** 2242
[22] Pokutnii S I 2010 *Semiconductors* **44** 488–493
[23] Fedorov A V and Baranov A V 2005 *Optics of quantum dots* (SPb: Nedra)
[24] Melikyan A O and Minasyan G R 2000 *Semiconductors* **34** 386–388
[25] Rosenthal A S and Ghannam T 2009 *Phys. Rev. A* **79** 043824
[26] Andrianov E S, Pukhov A A, Dorofeenko A V, Vinogradov A P and Lisyansky A A 2012 *Phys. Rev. B* **85** 035405
[27] Dorofeenko A V, Zyablovsky A A, Vinogradov A P, Andrianov E S, Pukhov A A and Lisyansky A A 2013 *Optics Express* **21** 14539–14547
[28] Bosbach J, Hendrich C, Stietz F, Vartanyan T and Träger F 2002 *Phys. Rev. Lett.* **89** 257404
[29] Ik Kim J, Kim J, Lee J, Jung D R, Kim H, Choi H, Lee S, Byun S, Kang S and Park B 2012 *Nanoscale Research Letters* **7** 482
[30] Sadhu S and Patra A 2008 *Chem. Phys. Chem.* **9** 2052–2058
[31] Khramtsov E S, Belov P A, Grigoryev P S, Ignatiev I V, Verbin S Y, Efimov Y P, Eliseev S A, Lovtcius V A, Petrov V V and Yakovlev S L 2016 *Journal of Applied Physics* **119** 184301
[32] Losev A S, Tikhonov K S, Golubeva T Y and Golubev Y M 2016 *J. Phys. B: At. Mol. Opt. Phys.* **49** 195501